NEUROMETHODS

Series Editor
Wolfgang Walz
University of Saskatchewan
Saskatoon, SK, Canada

For further volumes:
http://www.springer.com/series/7657

Transcranial Magnetic Stimulation

Edited by

Alexander Rotenberg

Neuromodulation Program, Department of Neurology, Boston Children's Hospital, Harvard Medical School, Boston, MA, USA

Jared Cooney Horvath

Psychological Sciences, University of Melbourne, Melbourne, VIC, Australia

Alvaro Pascual-Leone

Berenson Allen Center for Noninvasive Brain Stimulation, Department of Neurology, Beth Israel Deaconess Medical Center, Harvard Medical School, Boston, MA, USA

Humana Press

Editors
Alexander Rotenberg
Neuromodulation Program
Department of Neurology
Boston Children's Hospital
Harvard Medical School
Boston, MA, USA

Jared Cooney Horvath
Psychological Sciences
University of Melbourne
Melbourne, VIC, Australia

Alvaro Pascual-Leone
Berenson Allen Center for Noninvasive Brain Stimulation
Department of Neurology
Beth Israel Deaconess Medical Center
Harvard Medical School
Boston, MA, USA

ISSN 0893-2336 ISSN 1940-6045 (electronic)
ISBN 978-1-4939-0878-3 ISBN 978-1-4939-0879-0 (eBook)
DOI 10.1007/978-1-4939-0879-0
Springer New York Heidelberg Dordrecht London

Library of Congress Control Number: 2014939513

Printed on acid-free paper

Humana Press is a brand of Springer
Springer is part of Springer Science+Business Media (www.springer.com)

Series Preface

Experimental life sciences have two basic foundations: concepts and tools. The *Neuromethods* series focuses on the tools and techniques unique to the investigation of the nervous system and excitable cells. It will not, however, shortchange the concept side of things as care has been taken to integrate these tools within the context of the concepts and questions under investigation. In this way, the series is unique in that it not only collects protocols but also includes theoretical background information and critiques which led to the methods and their development. Thus it gives the reader a better understanding of the origin of the techniques and their potential future development. The *Neuromethods* publishing program strikes a balance between recent and exciting developments like those concerning new animal models of disease, imaging, in vivo methods, and more established techniques, including, for example, immunocytochemistry and electrophysiological technologies. New trainees in neurosciences still need a sound footing in these older methods in order to apply a critical approach to their results.

Under the guidance of its founders, Alan Boulton and Glen Baker, the *Neuromethods* series has been a success since its first volume published through Humana Press in 1985. The series continues to flourish through many changes over the years. It is now published under the umbrella of Springer Protocols. While methods involving brain research have changed a lot since the series started, the publishing environment and technology have changed even more radically. Neuromethods has the distinct layout and style of the Springer Protocols program, designed specifically for readability and ease of reference in a laboratory setting.

The careful application of methods is potentially the most important step in the process of scientific inquiry. In the past, new methodologies led the way in developing new disciplines in the biological and medical sciences. For example, physiology emerged out of anatomy in the nineteenth century by harnessing new methods based on the newly discovered phenomenon of electricity. Nowadays, the relationships between disciplines and methods are more complex. Methods are now widely shared between disciplines and research areas. New developments in electronic publishing make it possible for scientists that encounter new methods to quickly find sources of information electronically. The design of individual volumes and chapters in this series takes this new access technology into account. Springer Protocols makes it possible to download single protocols separately. In addition, Springer makes its print-on-demand technology available globally. A print copy can therefore be acquired quickly and for a competitive price anywhere in the world.

Wolfgang Walz
Saskatoon, SK, Canada

Preface

Transcranial magnetic stimulation (TMS) is no longer a novel experimental method. TMS is an established therapeutic and diagnostic technique in clinical practice. Hundreds of clinical patients a year undergo TMS to treat their medication-resistant depression or to establish detailed cortical motor and language maps prior to surgical or other therapeutic interventions. In addition, TMS is a valuable neuroscientific tool, and many patients and healthy volunteers enroll each year into research studies that utilize TMS to characterize cortical reactivity and plasticity, evaluate corticospinal and cortico-cortical connectivity, explore causal relations between brain activity and behavior, assess the impact of pharmacologic and other interventions, etc. According to PubMed, more TMS studies have been published in the last 5 years than in the previous 20 years, and 2013, at the writing of this preface, was on track to break the 1,000 papers in a year mark. Clinical trials are currently underway around the globe exploring the effects of TMS in diverse disease states including autism, epilepsy, migraine, tinnitus, stroke recovery, schizophrenia, Parkinson's, and Alzheimer's disease.

As with any tool, the rapidly growing use of TMS is a mixed blessing. On the one hand, an expanded TMS practitioner base allows for more, better, and deeper exploration of the technological, scientific, diagnostic, and therapeutic possibilities. On the other hand, as the number of TMS users grows, it becomes more and more difficult to maintain a keen grasp of foundational and emerging methodologies. Without care, the TMS research field can easily divide into a number of "camps" with each utilizing and purporting the benefits of their own devices, stimulation protocols, and methodologies. Fractionation based on informed practice and therapeutic evolution is not necessarily a bad thing; however, fractionation due to non-standardized or incoherent education and communication is potentially dangerous for the future of TMS.

This book aims to enable new and existing practitioners to learn and follow established TMS protocols. We describe many tried and true techniques: from single to multiple pulse TMS paradigms; from clinical to academic pursuits; from electromyographic to neuroimaging measurements. We hope that this work will serve not only as a good methodological introduction to those new to the TMS field, but also as a source of continual reference for experienced practitioners.

Boston, MA, USA — *Alexander Rotenberg*
Melbourne, VIC, Australia — *Jared Cooney Horvath*
Boston, MA, USA — *Alvaro Pascual-Leone*

Contents

Contributors

DAVID G. BROCK • *Neuronetics Inc, Malvern, PA, USA*
JOAN A. CAMPRODON • *Massachusetts General Hospital, Boston, MA, USA*
ZAIRA CATTANEO • *University of Pavia, Pavia, Italy*
ROCH COMEAU • *Rogue Research Inc., Montreal, QC, Canada*
MARK A. DEMITRACK • *Neuronetics Inc., Malvern, PA, USA*
MARK ELDAIEF • *Berenson-Allen Center for Noninvasive Brain Stimulation, Beth Israel Deaconess Medical Center, Boston, MA, USA*
FARANAK FARZAN • *Centre for Addiction and Mental Health, Toronto, ON, Canada*
SHIRLEY FECTEAU • *Laval University, Quebec City, QC, Canada*
MARK A. HALKO • *Berenson-Allen Center for Noninvasive Brain Stimulation, Beth Israel Deaconess Medical Center, Boston, MA, USA*
HENRI HANNULA • *Nexstim Ltd., Helsinki, Finland*
JARED COONEY HORVATH • *Psychological Sciences, University of Melbourne, Melbourne, VIC, Australia*
JARI KARHU • *Nexstim Ltd., Helsinki, Finland*
JARMO LAINE • *Nexstim Ltd., Helsinki, Finland*
UMER NAJIB • *Ruby Memorial Hospital, Morgantown, WV, USA*
LINDSAY OBERMAN • *E.P. Bradley Hospital, East Providence, RI, USA*
ALVARO PASCUAL-LEONE • *Berenson Allen Center for Noninvasive Brain Stimulation, Department of Neurology, Beth Israel Deaconess Medical Center, Harvard Medical School, Boston, MA, USA*
DANIEL PRESS • *Berenson-Allen Center for Noninvasive Brain Stimulation, Beth Israel Deaconess Medical Center, Boston, MA, USA*
ALEXANDER ROTENBERG • *Neuromodulation Program, Department of Neurology, Boston Children's Hospital, Harvard Medical School, Boston, MA, USA*
YIFTACH ROTH • *Ben-Gurion University of the Negev, Be'er Sheva, Israel*
JARMO RUOHONEN • *Nexstim Ltd., Helsinki, Finland*
JUHA SILVANTO • *O.V. Lounasmaa Laboratory, Aalto University, Aalto, Finland*
PHIROZ E. TARAPORE • *University of California San Francisco, San Francisco, CA, USA*
GREGOR THUT • *University of Glasgow, Glasgow, UK*
ANDREW VAHABZADEH-HAGH • *David Geffen School of Medicine, UCLA, Los Angeles, CA, USA*
JOSEP VALLS-SOLE • *Hospital Clinic, Barcelona, Spain*
MARINE VERNET • *University Pierre et Marie Curie, Paris, France*
ABRAHAM ZANGEN • *Ben-Gurion University of the Negev, Be'er Sheva, Israel*

Part I

Transcranial Magnetic Stimulation Fundamentals

Chapter 1

The Transcranial Magnetic Stimulation (TMS) Device and Foundational Techniques

Alexander Rotenberg, Jared Cooney Horvath, and Alvaro Pascual-Leone

Abstract

Transcranial magnetic stimulation (TMS) is a technique that is constantly evolving. Today, not only are there a number of technical options to consider, but also a number of methodological and experimental options. In this chapter, we supply a comprehensive overview of these many considerations. We first examine the physical and hardware foundations of TMS (including electromagnetic induction, stimulator characteristics, and coil variations). Following this, we briefly outline the most utilized and efficacious stimulation paradigms (including varied single and repetitive pulse patterns). Finally, we offer several practical procedural techniques universal to all devices and protocols.

Key words Transcranial magnetic stimulation, Coils, Hardware, Techniques

1 Introduction

Transcranial magnetic stimulation (TMS) is a neurophysiologic technique that allows for noninvasive stimulation of the human brain. Since its introduction close to 30 years ago [1], TMS, often in conjunction with other neuroscientific methods, has been used to study intracortical, cortico-cortical, and cortico-subcortical interactions (for review: [2–4]), assess causal relations between brain activity and behavior, and investigate the neurophysiologic substrate of the symptoms and pathophysiology of various neurological and psychiatric disorders (for review: [2–4]). In addition, repetitive transcranial magnetic stimulation (rTMS) has the capacity to modulate brain activity beyond the duration of application and holds therapeutic promise in a range of neuropsychiatric conditions such as major depression, chronic pain, and epilepsy (for review: [5–8]). In the past 5 years, two TMS devices and protocols have received United States Food and Drug Administration (FDA) for the treatment of medication-refractory depression (FDA approval

Alexander Rotenberg et al. (eds.), *Transcranial Magnetic Stimulation*, Neuromethods, vol. 89,
DOI 10.1007/978-1-4939-0879-0_1, © Springer Science+Business Media New York 2014

K061053; FDA approval K122288) and one TMS device has been approved for presurgical motor and speech mapping (FDA approval K112881). In Europe several devices have been awarded European CE Mark approval and are increasingly used for diagnostic and therapeutic indications in clinical practice.

2 Electromagnetic Induction

TMS induces electrical currents in the brain via Faraday's principle of electromagnetic induction [9]. Put simply, Faraday discovered that a pulse of electric current sent through a wire coil generates a magnetic field. The rate of change of this magnetic field determines the induction of a secondary current in a nearby conductor. With regard to TMS, an electric pulse, which grows to peak strength and diminishes back to zero in a short period of time (<1 ms), is sent through the conductive wiring within the TMS coil. The rapid fluctuation of this current produces a magnetic field perpendicular to the plane of the coil that similarly rises (up to about 2.5 T) and falls rapidly in time. This rapidly fluctuating magnetic field passes unimpeded through the subject's scalp and skull and induces a current in the brain that flows in a plane parallel to that of the coil but in the opposite direction of the original current [5, 10]. Thus, TMS might be best conceptualized as "electrodeless electric stimulation of the brain via electromagnetic induction."

Electromagnetic induction adheres to the inverse cube law: namely, the power of the magnetic field decreases exponentially as the distance from the original current increases. Thus, the induced current in the brain also decreases rapidly with distance from the coil. Because of this, the majority of TMS stimulation is restricted to superficial layers on the convexity of the brain (1.5–2 cm deep from the scalp: see [11]). Although techniques do exist that allow for deeper brain stimulation, with current TMS devices and coils, superficial areas of the brain closer to the plane of the coil will always be exposed to greater induced currents than deeper brain regions [12–14].

3 Inside the Brain

Currents induced in the brain by TMS primarily flow parallel to the brain's cortical surface (when the coil is held tangentially to the scalp). Exactly which neural elements are activated by TMS, however, remains unclear and might be variable across different brain regions and different subjects ([14, 15]; for review: [16]). Ultimately, the effect of TMS might be conceptualized as an interaction between the induced current and the affected brain tissue (both in terms of its specific structure as well as in terms of its state of activation). Therefore it is important not only to consider the anatomy of the neural structures, but also the state of activity in the neural elements affected by the TMS.

4 TMS Hardware

The design and components of TMS devices are relatively straightforward and universal. Each machine consists of a main unit and a stimulating coil. The main unit is composed of several components (*see* Fig. 1):

- *Charging system*—The charging system generates the current used to generate the magnetic field essential to TMS. A typical charging system can generate 8,000 A within several 100 ms.
- *One or more energy storage capacitors*—Capacitors allow for multiple energetic pulses to be generated, stored, and discharged in quick succession (typical voltage rating of 7.5 kV). Multiple storage capacitors are required for repetitive TMS protocols.
- *Energy recovery circuitry*—Energy recovery units allow for the main unit to recharge following discharge.
- *Thyristor*—Thyristors are electrical devices capable of switching large currents over a short period of time. In this case, the Thyristor acts as the bridge between the capacitor and coil, transferring 500 J between the two in less than 100 ms.
- *Pulse-shape circuitry*—Specialized circuitry can be used to generate either monophasic or biphasic pulses (for discussion: [17]).

The stimulating coil consists of one or more well-insulated coils of copper wire (frequently housed in a molded plastic cover). As current passes through these coils, varied patterns of magnetic fields are generated which, in turn, generate a current in the opposing direction in any nearby conductor. Coils can be arrayed in a variety of shapes and sizes (Fig. 2). The specific geometry of each

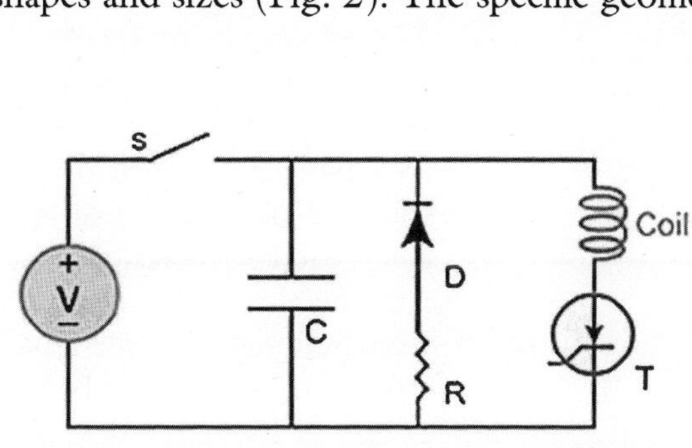

Fig. 1 Simplified circuit diagram of a single-pulse magnetic stimulator (*V* voltage source, *S* switch, *C* capacitor, *D* diode, *R* resistor, *T* thyristor. Modified from [16])

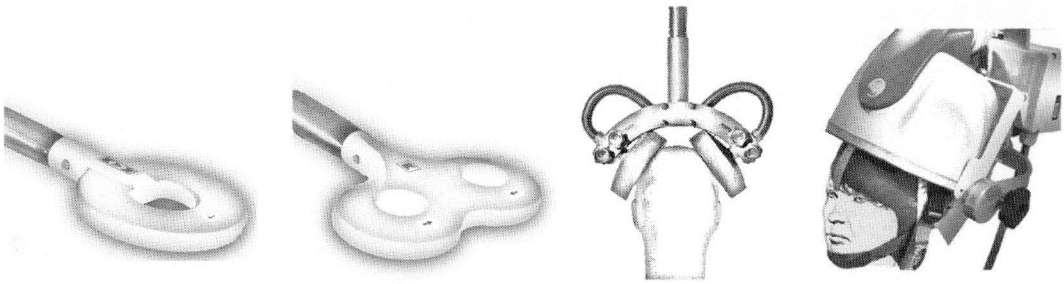

Fig. 2 Schematic drawings of different types of TMS coils. (From *left* to *right*) Round Coil, Figure-of-Eight Coil, Double Coil, H-Coil

coil determines the shape, strength, and overall focality of the resultant induced electric field, and thus of the brain stimulation.

- *Circular or Round Coil*—The circular coil is the oldest and simplest TMS coil design. A single, centrally located coil generates a spherical magnetic field perpendicular to the coil itself (as such, a magnetic sink will occur in the middle). Although not very focal, this type of coil is useful for protocols requiring single pulses and peripheral stimulation.
- *Figure-of-8 Coil* (also referred to as Butterfly Coil)—Easily the most recognizable and utilized of coil designs, the figure-of-8-coil is formed by abutting two single, circular coils against one another. Although the pattern from each individual coil may be relatively non-focal, the combined magnetic field (summed at the contact point between coils) is stronger than in surrounding regions and is relatively easy to determine spatially. This type of coil is often preferred for most clinical and academic uses of TMS—including repetitive and chronometric measures. Although not fully experimentally confirmed, mathematical modeling suggests a small figure-of-8 coil (each wing measuring 4 cm in diameter) can achieve a spatial resolution of approximately 5 mm^3 of the brain volume.
- *H-Coil* (*see* Chap. 3)—Recently obtaining FDA approval for the treatment of medication-resistant depression, the H-Coil aims to stimulate deeper, non-superficial cortical layers. This is achieved by having a more complex coil design with several planes such that the decay function of the generated magnetic field is less steep and the current reaches deeper into the brain (although the superficial cortical layers still are exposed to the strongest field). Research suggests the H-Coil may be able to stimulate neural structures up to 6 cm below the cortical surface [13].

Coils can be made to specification. As such, a number of experimental coil designs have been created, including the horizontal racetrack, the vertical racetrack, and small-animal sized coils.

When used for a repetitive TMS paradigm, the conductive material of the stimulation coils heat up potentially limiting the duration of rTMS trains. Therefore, it is essential to keep coils cool during stimulation. To this end, several coils have been developed with an internal liquid or air cooling systems, as well as other methods to offer heat sinks and reduce coil heating.

5 Pulse Waveforms

As noted above, specialized circuitry within the TMS main unit can generate varied TMS pulse shapes (*see* Fig. 3).

- *Monophasic*—Monophasic pulses generate only unidirectional voltage. As the initial course of voltage (positive) through a coil would induce an opposing (negative) oscillation, in order to generate a monophasic pulse a shunting diode and power resistor must be used to dampen this natural cycle [18]. Due to this pulse-shaping, monophasic pulses can only be delivered singularly (unless multiple energy sources are utilized).
- *Biphasic (Polyphasic)*—Biphasic pulses generate full positive/negative voltage oscillations [19]. This oscillation, in turn, causes a rapid directional shift of the initial and induced currents. This type of pulse can be terminated after a single cycle (biphasic) or after several oscillatory cycles (polyphasic pulses: [20]).

6 Pulse Strength

The amount of voltage passed through the stimulating coil can be adjusted up or down as needed. The strength of this initial current contributes to the strength of the induced current (however, the rate of change of the magnetic field amplitude is still tantamount in stimulation). Devices often express output current as percent of

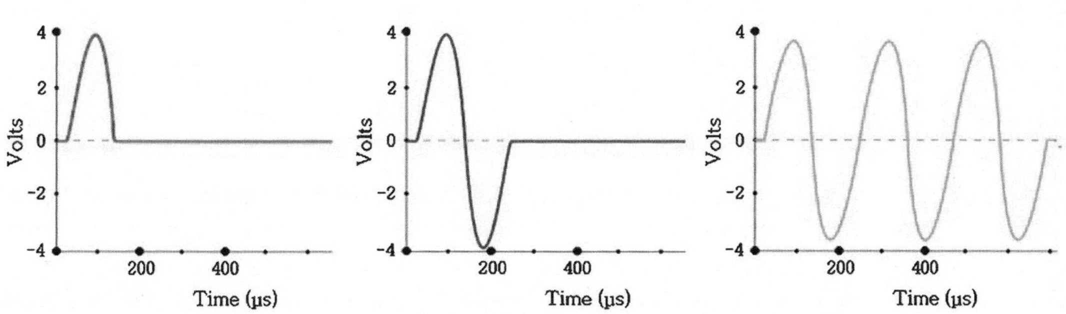

Fig. 3 Pulse-shaped graphs. (From *left* to *right*) Monophasic, biphasic, and polyphasic

maximal output, rather than as absolute current values as these are different depending on the coils used. Frequently the applied stimulation strength will be referenced in terms of a subject's motor threshold at a specified baseline.

7 Stimulation Paradigms

An important feature of TMS is the variability of its functional parameters. Through different pulse patterns and durations, clinicians and researchers can examine a vast number of interesting questions.

– *Single-Pulse* (*see* Chap. 5): Single-pulse TMS paradigms utilize *isolated*, uniquely modulated pulses applied to a specific cortical location. Single-pulse TMS paradigms are useful for the diagnostic (for review: [21]) and exploratory measurement of cortical reaction to each pulse [22–24]. Important parameters to consider when utilizing single-pulse stimulation include cortical location, pulse intensity, and response measurement.

When applied to the primary motor cortex, single-pulse stimulation can induce contralateral muscle activity which can be recorded by electromyography (EMG) as motor-evoked potentials (MEPs). MEPs serve as a quantification of the TMS effect and can be utilized both in mapping and intercessory protocols [25, 26]. When applied to the primary visual cortex, single-pulse stimulation can produce phosphenes (subjective sensations of flashes of light) in the visual field. Similar to MEPs, phosphenes can be used to determine the threshold for cortical activation by TMS (for review: [27]). The effects of single-pulse stimulation can also be recorded using electroencephalography (EEG) as TMS-evoked potentials (TEPs) which provide a metric of cortical reactivity.

– *Paired-Pulse* (*see* Chap. 6): Paired-pulse TMS paradigms utilize two isolated pulses delivered in close succession. Each pulse can be applied to the same cortical region or to separate regions and used to assess their functionally connectivity [28]. In paired-pulse paradigms, the cortical effects of the first (or *conditioning*) pulse can be measured via variations in the effect of the second (or *test*) pulse. The effects of both pulses will depend on their unique intensities and the duration of the inter-pulse interval [29].

Paired-pulse paradigms are useful for examining cortical excitation/inhibition ratio in healthy subjects and in patients. Important parameters to consider when utilizing paired-pulse stimulation include cortical location, conditioning and test pulse intensities, and the inter-pulse interval duration [29].

- *Repetitive TMS* (*see* Chap. 7): Repetitive TMS (rTMS) paradigms utilize trains of pulses to induce cortical effects that outlast the stimulation duration [6]. As with paired-pulse, the stimulation intensity, the frequency of stimulation, the overall duration of the trains and their pattern (continuous or intermittent burst interrupted by pauses) determine the effect of each particular rTMS protocol.
- Low-frequency rTMS (typically 1 Hz) applied for several minutes will typically lead to a suppression of cortical activity at the stimulation location for a period lasting for about half the duration of the stimulation train [30]. High-frequency rTMS (typically >5 Hz) is typically applied in bursts interrupted by pauses in order to comply with current safety guidelines and prevent complications (particularly the induction of a seizure). Such high-frequency or fast rTMS trains typically lead to an increase in cortical activity at the stimulation location for a period lasting for about half the duration of the stimulation train [31].

rTMS allows clinicians to effect long lasting changes in cortical reactivity and plasticity. Additionally, rTMS allows researchers to modify cortical function during task performance, which can help reveal causal relations between brain activity and behavior, and might impair or enhance behavior and cognition [16].

In a specialized form of rTMS, theta-burst stimulation (TBS) protocols deliver a novel pattern of stimulation that mimics neural oscillatory patterns thought to correspond with effective cognitive processing [32]. As other forms of rTMS, TBS protocols can be used to induce cortical plasticity across both healthy subjects and clinical pathologies and to modify brain activity for scientific and clinical applications [33].

- *Virtual Lesion* (*see* Chap. 8): Virtual lesion paradigms aim to temporarily disrupt cortical processing within specific cortical regions via the introduction of extraneous noise. These “lesions” can be generated in any number of ways: single pulses generated at the proper time, low-frequency rTMS generated for an effective duration, continuous TBS for 600 pulses, etc. Virtual lesions allow researchers to examine both functional connectivity and unique cognitive processing [34].
- *Online and Off-Line Protocols*: Online TMS protocols refer to cognitive studies undertaken whilst TMS is being administered. For instance, generating pulses during a language task to disrupt speech production would be considered an “online” paradigm [35].

Off-line TMS protocols refer to cognitive studies undertaken after TMS has been administered. For instance, testing language skills after 15 min of 1 Hz. TMS would be considered an "off-line" paradigm [35].

- *Chronometric Protocols*: Chronometric paradigms seek to explore *when* during a given task specific neural regions become critical. Exclusively online and often single or paired-pulse, chronometric designs contain arguably the greatest potential for determining the time course of neural processing [34].
- *Triple-Pulse Stimulation*: Triple-pulse stimulation combines TMS with electrical stimulators to examine corticospinal tract integrity. By correctly timing a TMS pulse to M1 with two peripheral stimulatory pulses (typically at Erb's point and the ulnar nerve), clinicians can accurately measure the percentage of corticospinal fibers excited by the TMS pulse and thus calculate the integrity of the corticospinal tract. The lower the percentage, the more corticospinal tract compromise can be assumed [36].
- *Quadripulse Stimulation*: Quadripulse stimulation is a unique pattern of rTMS that can be utilized to explore and induce neural plasticity [37]. During quadripulse stimulation, 300–400 trains (separated by ~5 s) of four monophasic pulses are delivered to the neural region of interest. If the interstimulus interval of each of the four pulses is short (<15 ms), facilitation is typically engendered for a period lasting 45–75 min. If the interstimulus interval of each of the four pulses is long (25–100 ms), depression is typically engendered for a period lasting 45–75 min.

8 Basic TMS Techniques

Although TMS devices and design necessarily differ between manufacturers, there are several procedural techniques universal to practitioners irrespective of which device is utilized.

- *Activating the Pulse*: The TMS pulse can typically be activated in one of the three ways: via an activation button on the coil handle, through an activation pedal controlled by the foot, or through an activation button on the TMS device itself. Some devices, in order to increase safety and prevent accidental pulse discharges, actually require several of these steps to be done together. Although many consider the button on the coil handle to be the simplest, this configuration is not always available (as some devices do not include an on handle button and others must be disengaged when connecting coil trackers for neuronavigation). When utilizing the foot pedal, it is a good idea to place the pedal in a location that is comfortable for your foot (some practitioners prefer to use the ball of their foot, others the heel) and allows room for coil movement and repositioning. Finally, as can be predicted, when utilizing the onboard activation button, it is advisable to have two operators: one to hold the coil steady, the other to manipulate the buttons on the device (this ensures no coil movement during stimulation).

- *Localization* (*see* Chap. 3): Stimulation localization can be achieved in any number of ways. The first involves measuring and marking the head (typically using a washable grease pencil) according to the common 10–20 international system of EEG electrode placement. Once established, the 10–20 landmarks can be utilized to roughly determine cerebral regions and develop small pulse grids to find specific neural locations. Another localization technique involves utilizing a tight-fitting swimming cap. With this, any number of scalp references can be noted (such as vertex, inion, 10–20 points) using a dark and easily observed marker. In addition, the swim cap allows for the creation of small point-grids (again, using a marker) around neural regions of interest to obtain specific localization. Finally, several neuronavigation systems have recently been developed. With these systems, anatomical landmarks are co-registered with a participant's structural MRI or PET scan and the head and coil are both tracked in time and space utilizing either infrared or ultra-frequency pulses. In addition to real-time navigation, several of these systems allow for the tracking of each pulse and the modeling of the stimulated region within the brain itself (as determined via the anatomical scan).
- *Long-Duration Stimulation Paradigms*: When administering long-duration stimulation protocols, it is common for the arm and/or shoulder supporting the coil to tire. To combat this, practitioners have tried a number of tricks—from hooking the arm into a neck sling to allow for relaxation to locking the coil into a T-Stand pressed against the participant's scalp. As can be imagined, nothing yet tried has proven ideal and, unfortunately, arm fatigue may lead to subtle coil shifting over the course of stimulation. Although there are no fast answers to this issue, it is important to keep this in mind during long-duration stimulation and try to combat fatigue and coil shift as best as possible.

9 Conclusion

TMS has proven useful and efficacious in numerous medical and research settings (as will be explored and expanded upon throughout this book). As such, it appears to have staying power and will soon likely complement the set of more established imaging and brain mapping tools utilized in most hospitals and academic laboratories. TMS is deceivingly simple, but appropriate utilization requires a keen and up-to-date understanding of TMS devices as they evolve, methodological techniques as they develop, and stimulation paradigms as they emerge. Hopefully this and subsequent chapters will facilitate these goals.

References

1. Barker AT, Jalinous R, Freeston IL (1985) Non-invasive magnetic stimulation of human motor cortex. Lancet 325:1106–1107
2. Pascual-Leone A, Davey M, Wassermann EM, Rothwell J, Puri B (eds) (2002) Handbook of transcranial magnetic stimulation. Edward Arnold, London
3. Rotenberg A (2010) Prospects for clinical applications of transcranial magnetic stimulation and real-time EEG in epilepsy. Brain Topogr 22(4):257–266
4. Horvath JC, Perez JM, Forrow L, Fregni F, Pascual-Leone A (2011) Transcranial magnetic stimulation: a historical evaluation and future prognosis of therapeutically relevant ethical concerns. J Med Ethics 37(3):137–143
5. Kobayashi M, Pascual-Leone A (2003) Transcranial magnetic stimulation in neurology. Lancet Neurol 2(3):145–156
6. Fitzgerald PB, Fountain S, Daskalakis ZJ (2006) A comprehensive review of the effects of rTMS on motor cortical excitability and inhibition. Clin Neurophysiol 117(12):2584–2596
7. Valero-Cabré A, Pascual-Leone A, Coubard OA (2011) Transcranial magnetic stimulation (TMS) in basic and clinical neuroscience research. Neurol Rev 167(4):291–316
8. Najib U, Bashir S, Edwards D, Rotenberg A, Pascual-Leone A (2011) Transcranial brain stimulation: clinical applications and future directions. Neurosurg Clin N Am 22(2):233
9. Faraday M (1839) Experimental researches in electricity, vol 1. Bernard Quaritch, London
10. Hallett M (2000) Transcranial magnetic stimulation and the human brain. Nature 406(6792):147–150
11. Deng ZD, Lisanby SH, Peterchev AV (2012) Electric field depth–focality tradeoff in transcranial magnetic stimulation: simulation comparison of 50 coil designs. Brain Stimul 6(1):1–13
12. Wagner TA, Zahn M, Grodzinsky AJ, Pascual-Leone A (2004) Three-dimensional head model simulation of transcranial magnetic stimulation. Biomed Eng IEEE Trans 51(9): 1586–1598
13. Roth Y, Amir A, Levkovitz Y, Zangen A (2007) Three-dimensional distribution of the electric field induced in the brain by transcranial magnetic stimulation using figure-8 and deep H-coils. J Clin Neurophysiol 24(1):31–38
14. Amassian VE, Eberle L, Maccabee PJ, Cracco RQ (1992) Modelling magnetic coil excitation of human cerebral cortex with a peripheral nerve immersed in a brain-shaped volume conductor: the significance of fiber bending in excitation. Electroencephalogr Clin Neurophysiol 85(5):291–301
15. Maccabee PJ, Amassian VE, Eberle LP, Cracco RQ (1993) Magnetic coil stimulation of straight and bent amphibian and mammalian peripheral nerve in vitro: locus of excitation. J Physiol 460(1):201–219
16. Wagner T, Valero-Cabre A, Pascual-Leone A (2007) Noninvasive human brain stimulation. Annu Rev Biomed Eng 9:527–565
17. Peterchev AV, Jalinous R, Lisanby SH (2008) A transcranial magnetic stimulator inducing near-rectangular pulses with controllable pulse width (cTMS). Biomed Eng IEEE Trans 55(1):257–266
18. Barker AT, Freeston IL, Jalinous R, Jarratt JA (1987) Magnetic stimulation of the human brain and peripheral nervous system: an introduction and the results of an initial clinical evaluation. Neurosurgery 20:100–109
19. Dantec (2003) Instruction manual. Dantec Medical A/S, Skovlunde, DK
20. Sommer M, Alfaro A, Rummel M, Speck S, Lang N, Tings T, Paulus W (2006) Half sine, monophasic and biphasic transcranial magnetic stimulation of the human motor cortex. Clin Neurophysiol 117(4):838–844
21. Chen R, Cros D, Curra A, Di Lazzaro V, Lefaucheur JP, Magistris MR, Mills K, Rösler KM, Triggs WJ, Ugawa Y, Ziemann U (2008) The clinical diagnostic utility of transcranial magnetic stimulation: report of an IFCN com mittee. Clin Neurophysiol 119(3):504–532
22. Muri RM, Vermersch AI, Rivaud S, Gaymard B, Pierrot-Deseilligny C (1996) Effects of single-pulse transcranial magnetic stimulation over the prefrontal and posterior parietal cortices during memory-guided saccades in humans. J Neurophysiol 76(3):2102–2106

23. Amassian VE, Cracco RQ, Maccabee PJ, Cracco JB, Rudell AP, Eberle L (1998) Transcranial magnetic stimulation in study of the visual pathway. J Clin Neurophysiol 15(4):288–304
24. Thut G, Northoff G, Ives JR, Kamitani Y, Pfennig A, Kampmann F, Schomer DL, Pascual-Leone A (2003) Effects of single-pulse transcranial magnetic stimulation (TMS) on functional brain activity: a combined event-related TMS and evoked potential study. Clin Neurophysiol 114(11):2071–2080
25. Rothwell JC (1997) Techniques and mechanisms of action of transcranial stimulation of the human motor cortex. J Neurosci Methods 74(2):113–122
26. Chen R (2000) Studies of human motor physiology with transcranial magnetic stimulation. Muscle Nerve Suppl 9:S26–S32
27. Merabet LB, Theoret H, Pascual-Leone A (2003) Transcranial magnetic stimulation as an investigative tool in the study of visual function. Optom Vis Sci 80(5):356–368
28. Rothwell JC (1999) Paired-pulse investigations of short-latency intracortical facilitation using TMS in humans. Electroencephalogr Clin Neurophysiol Suppl 51:113–119
29. Currà A, Modugno N, Inghilleri M, Manfredi M, Hallett M, Berardelli A (2002) Transcranial magnetic stimulation techniques in clinical investigation. Neurology 59(12):1851–1859
30. Chen R, Classen J, Gerloff C, Celnik P, Wassermann EM, Hallett M, Cohen LG (1997) Depression of motor cortex excitability by low-frequency transcranial magnetic stimulation. Neurology 48(5):1398–1403
31. Guse B, Falkai P, Wobrock T (2010) Cognitive effects of high-frequency repetitive transcranial magnetic stimulation: a systematic review. J Neural Transm 117(1):105–122
32. Huang YZ, Edwards MJ, Rounis E, Bhatia KP, Rothwell JC (2005) Theta burst stimulation of the human motor cortex. Neuron 45(2):201–206
33. Chen R, Udupa K (2009) Measurement and modulation of plasticity of the motor system in humans using transcranial magnetic stimulation. Mot Control 13(4):442–453
34. Pascual-Leone A, Walsh V, Rothwell J (2000) Transcranial magnetic stimulation in cognitive neuroscience–virtual lesion, chronometry, and functional connectivity. Curr Opin Neurobiol 10(2):232–237
35. Terao Y, Ugawa Y (2006) Studying higher cerebral functions by transcranial magnetic stimulation. Suppl Clin Neurophysiol 59:9–17
36. Ni Z, Müller-Dahlhaus F, Chen R, Ziemann U (2011) Triple-pulse TMS to study interactions between neural circuits in human cortex. Brain Stimul 4(4):281–293
37. Hamada M, Terao Y, Hanajima R, Shirota Y, Nakatani-Enomoto S, Furubayashi T, Matsumoto H, Ugawa Y (2008) Bidirectional long-term motor cortical plasticity and metaplasticity induced by quadripulse transcranial magnetic stimulation. J Physiol 586(16):3927–3947

Chapter 2

Transcranial Magnetic Stimulation (TMS) Safety Considerations and Recommendations

Umer Najib and Jared Cooney Horvath

Abstract

Ensuring patient and participant safety during transcranial magnetic stimulation (TMS) is of paramount importance. In this chapter, we begin by exploring a number of general safety concerns and the prevalence of reported side-effects in the TMS literature. Next, we outline contraindications and the recommended safety parameters for each of the major stimulation paradigms (including single and repetitive pulse patterns). Finally, we offer several practical tips to ensure TMS is delivered in the safest and most ethical manner.

Key words Transcranial magnetic stimulation, Safety, Ethics, Side-effects, Contraindications

1 Introduction

Transcranial magnetic stimulation (TMS) is a noninvasive neuromodulatory and neurostimulatory technique increasingly utilized in clinics and research laboratories around the world. Exploiting the properties of electromagnetic induction, TMS can transiently or lastingly modulate cortical excitability via the application of localized, time-varying magnetic field pulses.

TMS has been used in a growing number of laboratories and clinics worldwide since 1984. Since then, a number of adverse events have been reported and thoroughly reviewed. In 1996 and 2008, consensus conferences were held to establish safe-use recommendations for both clinical and academic TMS. The resulting publications were, and remain today, the mainstay safety guidelines for TMS therapeutics and research [1, 2]. In this chapter, we will outline the major issues and recommendations set forth by the TMS Safety Consensus Group and we will also briefly examine TMS-related ethical concerns.

Alexander Rotenberg et al. (eds.), *Transcranial Magnetic Stimulation*, Neuromethods, vol. 89,
DOI 10.1007/978-1-4939-0879-0_2, © Springer Science+Business Media New York 2014

2 Safety Concerns

Heating—Brain tissue heating caused by single-pulse TMS has been shown to be less than 0.1 °C [3]. To put this into context, estimated heating in the tissue surrounding deep brain stimulatory electrodes can reach as high as 0.8 °C [4]. When TMS is combined with electrode recording/imaging devices, induced eddy currents can cause heating of the electrode surface. The temperature increase depends on the shape, size, orientation, conductivity, and surrounding tissue properties of the electrode or implant as well as the TMS coil type, position, and stimulation parameters. As can be intuited, excessive electrode heating can lead to skin burns. To avoid this possible risk, it is recommended low-conductivity plastic electrodes be used whenever possible. Radial notching of electrodes and skull plates can also reduce heating by interrupting the eddy current path [2].

In addition to surface heating, implanted metallic devices—such as aneurysm clips or titanium skull plates—can theoretically generate temperature increases; although, recent evidence suggests these increases are negligible [5, 6]. Even so, when confronted with this scenario, it is recommended ex vivo heating be examined before commencing TMS.

Magnetization—As the magnetic field generated by TMS exerts an attractive/repellant force on all point charges (due to the Lorentz force: [7]), any implanted medical or therapeutic device sensitive to these fields may shift during treatment. Implanted devices, including aneurysm clips, implanted electrodes, and cochlear implants, could potentially suffer movement or demagnetization during stimulation. Again, whenever possible, it is recommended ex vivo effects be measured prior to TMS and that watches, jewelry, glasses, and other potentially conductive or magnetic objects worn on or close to the head be removed to prevent interactions with the magnetic field [2].

Induced Voltages—Any wires or electronic devices near the discharging TMS coil may suffer deleterious induced voltages. To avoid this, it is recommended all proximal wires be kept free of loops and bound in a twisted fashion [8]. In addition, induced voltages may occur in any implanted device containing circuitry, such as cochlear implants, deep brain stimulation (DBS) systems, and epidural electrode arrays for cortical stimulation. TMS can induce voltages in the electrode wires whether the implant is turned ON or OFF, and this can result in unintended brain stimulation at the electrode site. TMS pulses can also damage the internal circuitry of electronic implants near the coil, causing them to malfunction or permanently break down [9, 10]. The same recommendations outlined above apply here.

Implanted Electrodes—Based on several ex- and in vivo studies, TMS appears to be safe for patients with implanted stimulators as long as the "internal pulse generator" systems are not in close proximity to the TMS coil [8–11]. However, exact parameters for "close proximity" have not yet been determined (see [10, 12]). As such, TMS should *only* be applied in patients with implanted stimulators if there are sound medical justifications, with appropriate oversight by the Institutional Review Board or Ethic Committee. With regard to peripheral devices (vagal nerve stimulators, pacemakers, spinal cord stimulators, etc.), TMS is considered safe so long as coil discharge is not initiated near said device components [2, 8].

3 Side Effects (Table 1)

3.1 Common

Headache/Neck Pain—Headache and/or neck pain have been reported in an estimated 20–40 % of subjects undergoing TMS [13, 14]. This is the most commonly reported side effect of TMS (Table 1). The intensity of pain experienced varies from subject to subject, depending on individual susceptibility, coil design, stimulation location, intensity, and frequency. Reported head/neck pains are largely believed to occur due to muscle tension, generated either by the stimulation itself or the posture assumed during longer protocols [13]. A single dose of acetaminophen or aspirin may be recommended if pain persists beyond stimulation duration. No migraine attacks have been described following TMS, either in healthy controls or in migraine patients who underwent rTMS applications as treatment [15]. It has also been reported that the local painfulness of prefrontal rTMS declines over the first few days of daily treatment [16]. To take advantage of this finding, ramping algorithms (where investigators intentionally start below target dose and gradually increase over the first week of treatment) are used by some practitioners.

Acoustic Trauma—During discharge, the TMS coil produces a deceptively loud clicking noise (on the order of 120–140 dB: [17]). This exceeds the recommended safety levels for the auditory system (OSHA). Although seemingly innocuous, repeated exposure to this intense sound can lead to acoustic trauma. To date, several reports of a shift in auditory threshold following stimulation have occurred [18, 19] and one report of a permanent threshold shift following H-Coil stimulation without the use of ear protection has been published [20]. In order to prevent these potential adverse effects, it is recommended subjects and operators wear earplugs during the full duration of treatment. Furthermore, prompt referral for auditory assessment of individuals who complain of hearing loss, tinnitus, or aural fullness following completion of

Table 1
Possible side effects from TMS according to protocol

Side effect	Single-pulse	Paired-pulse	Low frequency rTMS	High frequency rTMS	Theta-burst
Transient headache/ neck pain	Possible/rare	Not reported	20–40 %	20–40 %	Possible
Transient hearing change	Possible	Not reported	Possible	Possible	Not reported
Seizure induction	Rare	Not reported	Rare	<1 % in healthy subjects	1 reported
Syncope	Possible as epiphenomenon	Possible as epiphenomenon	Possible as epiphenomenon	Possible as epiphenomenon	Not reported
Transient hypomania	No	No	Rare	Possible (following left PFC stimulation)	Not reported
Transient cognitive changes	Not reported	Not reported	Rare/negligible	Rare/negligible	Transient WM impairment reported
Induced current in electrical circuits	Theoretically possible	Theoretically possible	Theoretically possible	Theoretically possible	Theoretically possible

TMS is advised. Patients with known preexisting noise-induced hearing loss or concurrent treatment with ototoxic medications (Aminoglycosides, Cisplatin) should only receive TMS in cases of a favorable risk/benefit ratio, as when rTMS is used for the treatment of tinnitus [2].

3.2 Rare

Seizure—The induction of seizure, although exceedingly rare, is of major concern when utilizing TMS. Seizures can be induced by rTMS when pulses are applied with relatively high-frequencies and short interval periods between trains of stimulation. rTMS can theoretically induce seizures during two different periods of stimulation: (a) during or immediately after rTMS trains and (b) post-stimulation due to the modulation of cortical excitability (i.e., kindling effect: [21]). Although the first has been seen, there is no evidence that the latter has ever occurred. From the several thousands of TMS studies reported to date, a total of 16 seizures had been reported through 2008 [2]. Since then, four reports of seizures have been reported [22–25]. Based on this data, the reported risk of seizure stands conservatively at 1 in 1,000 applications. It is important a plan be established prior to treatment to deal with any induced seizure. In the case of a seizure, treatment should be ceased immediately and the subject should be treated as any other patient with a witnessed seizure.

Syncope/Fainting—Syncope during TMS can occur for several reasons in addition to the stimulation: anxiety, physical discomfort, psychological discomfort, etc. Syncope has been reported less frequently than seizure; however, the true number of occurrences is not known. It is recommended all subjects be monitored closely for any signs of syncope (dizziness, light-headedness, faint feelings). In the event of syncope, stimulation should be ceased immediately, assistance offered, and a full neurological evaluation undertaken.

Mood—Acute mania emergence during rTMS over the left prefrontal cortex in patients with uni- and bipolar depression has been reported. However, the prevalence of this occurrence (13 reports in 53 RCTs) appears to be below natural switch rates of bipolar patients taking mood stabilizers (0.84 % with rTMS vs. 2.3–3.45 % with mood stabilizers: [26]). In addition, transient psychotic symptoms, anxiety, insomnia, suicidal ideation, and extreme agitation have all been reported following rTMS in psychiatric patient populations, but it remains to be established whether these symptoms occur at a higher rate than during the natural course of each disease state [27, 28]. Although psychotic symptoms and suicidal ideation have never been reported in healthy subjects, it is important to inform all potential subjects of these possible acute side effects. Should acute mood shifts occur, cease stimulation immediately, monitor the subject closely until normal function returns, and administer a full neurological exam.

Dental Pain—Although rare, several instances of induced dental pain during TMS have been reported [2]. If this occurs, it will happen during stimulation and may be the sign of a tooth cavity or loose cap/filling. If dental pain is reported, cease stimulation immediately and encourage the subject to seek a dental evaluation.

4 Contraindications

Decisions regarding TMS treatment should be made on a case-by-case basis. Contraindication guidelines should be followed closely and deviations made only under extremely compelling medical conditions.

- *Implanted Cranial Electrodes*—As discussed above, TMS may cause heating or induced voltages within ferro-magnetic electrodes or medical devices implanted in the cranium. The presence of such metallic hardware in close contact to the discharging coil is therefore an absolute contraindication to TMS/rTMS. Accordingly, it is recommended TMS treatment be avoided in these cases.
- *Cochlear Implants*—Again, as TMS can cause heating and induced voltages within the electronics of cochlear implants, it is recommended TMS treatment be avoided under this circumstance.
- *Personal History of Syncope or Seizure*—Subjects with a history of syncope/seizure may be at higher risk for seizure induction. Currently, TMS is under investigation for seizure treatment. Accordingly, under circumstances of treatment, seizure history may not be considered a contraindication. However, under "non-treatment" circumstances, it is recommended treatment be avoided.
- *Patients with epilepsy*—As seizure induction is a real, if rare, possible side effect of TMS, it is typically not recommended to utilize TMS on subjects with a history of epilepsy. With this said, several studies have suggested beneficial effects of low-frequency TMS on intractable epilepsy (for review: [29]). Again, possible benefit must be weighed against the likely risk before treatment is recommended.
- *Cerebral Lesion*—Due to issues of induced current shunting, it is recommended any patient suffering from a vascular, traumatic, tumoral, infectious, or metabolic lesion—even without a history of seizure—avoid TMS unless medically compelling reasons exist.
- *Drug/Medication Interactions*—Because active or recent intake of several drug classes may lower a subject's seizure threshold, it is recommended a full and detailed medication history be obtained (Table 2). Treatment decisions should

Table 2
Drugs with potential Hazards for rTMS

Strong potential hazard	Relative hazard
Alcohol	Ampicillin
Amitriptyline	Anticholinergics
Amphetamines	Antihistamines
Chlorpromazine	Aripiprazole
Clozapine	BCNU
Cocaine	Bupropion
Doxepine	Cephalosporins
Ecstasy	Chlorambucil
Foscarnet	Chloroquine
Gamma-hydroxybutyrate (GHB)	Citalopram
Ganciclovir	Cyclosporine
Imipramine	Cytosine arabinoside
Ketamine	Duloxetine
Maprotiline	Fluoxetine
MDMA	Fluphenazine
Nortriptyline	Fluvoxamine
Phencyclidine (PCP)	Haloperidol
Ritonavir	Imipenem
Theophylline	Isoniazid
	Levofloxacin
	Lithium
	Mefloquine
	Methotrexate
	Metronidazole
	Mianserin
	Mirtazapine
	Olanzapine
	Paroxetine
	Penicillin
	Pimozide

(continued)

Table 2 (continued)

Strong potential hazard	Relative hazard
	Quetiapine
	Reboxetine
	Risperidone
	Sertraline
	Sympathomimetics
	Venlafaxine
	Vincristine
	Ziprasidone

only be made following a careful and medically responsible evaluation.

- *Recent Drug Withdrawal*—Recent withdrawal from alcohol, barbiturates, benzodiazepines, meprobamate, and/or chloral hydrate may significantly reduce a subject's seizure threshold. Accordingly, treatment is not recommended until a suitable time following drug cessation.
- *Pregnancy*—As the magnetic field generated by TMS decays rapidly with distance, any fetal exposure to TMS effects is very unlikely. However, under this circumstance it is arguably best to take a conservative stance and avoid treatment for women of term. In addition, any TMS operators should maintain a 0.8 m distance (conservative) from the discharging coil [2].
- *Children*—To date, there have been nearly 100 studies reporting TMS application to pediatric populations (for review: [30]). Although no serious adverse effects have been reported, special consideration should be taken when considering TMS in children [31]. It has been *cautiously* concluded that single- and paired-pulse TMS is safe in children above the age of two [2]. For children younger than 2 years, data about risk for acoustic injury are not available, and therefore specialized hearing protection may be required. In absence of an appreciable volume of data on the potential for adverse effects with rTMS, children should not be used as subjects for rTMS without compelling clinical reasons, such as the treatment of refractory epilepsy or particular psychiatric conditions.
- *Illness*—Since the effects of TMS are dependent upon the baseline activation state of the targeted cortical region, it is important to monitor and consider the effects of both physical

Table 3
Example of TMS screening questionnaire

	Yes	No	Notes
Have you ever had TMS before?			
Have you ever had an adverse reaction to TMS?			
Have you ever had a seizure?			
Is there any family history of epilepsy?			
Have you ever had an unexplained loss of consciousness?			
Do you suffer from chronic or severe headaches?			
Have you ever had a stroke?			
Do you have any brain-related neurological illness?			
Have you ever had any serious head injury or concussion?			
Have you ever had any surgery to your head?			
Have you ever had any illness that may cause brain damage?			
Do you have any metal in your head outside of your mouth?			
Do you have any implanted medical devices?			
Are you taking any medications (including OTCs)?			
Are you pregnant or potentially pregnant?			

and mental illness on this basal activity. For instance, Fitzgerald and colleagues [32] found that schizophrenic patients show a decreased response to low-frequency TMS while Oberman and colleagues [33] found that autistic patients show a prolonged response to theta-burst stimulation. Accordingly, it is important to monitor subject condition and consider the neural effects prior to treatment (Table 3).

5 Stimulation Parameters

The low number of severe side effects reported with the use of TMS is due, in large part, to the exacting stimulation guidelines first laid out by Wassermann et al. in 1998 [1] and refined by Rossi et al. in 2009 [2]. Strict adherence to the guidelines will help ensure subject safety and maintain the strong track record of TMS as a safe form of noninvasive treatment.

When considering issues of parameter safety, there are four important quantities to consider: intensity, frequency, train duration, and inter-train interval (Tables 4 and 5). Although these

Table 4
Parameter safety issues: maximum recommended stimulation duration of single TMS trains (in seconds)

Freq (Hz)	90 % MT	100 % MT	110 % MT	120 % MT	130 % MT	140 % MT	150 % MT	160 % MT	170 % MT	180 % MT	190 % MT	200 % MT
1	>1,800	>1,800	>1,800	360	>50	>50	>50	>50	27	11	11	8
5	>10	>10	>10	>10	>10	7.6	5.2	3.6	2.6	2.4	1.6	1.4
10	>5	>5	>5	4.2	2.9	1.3	0.8	0.9	0.8	0.5	0.6	0.4
20	2.05	2.05	1.6	1.0	0.55	0.36	0.25	0.25	0.15	0.2	0.25	0.2
25	1.28	1.28	0.84	0.4	0.24	0.2	0.24	0.2	0.12	0.08	0.12	0.12

Table 5
Parameter safety issues: commonly employed stimulation parameters

rTMS Frequency	No. of studies	Average train duration	Average inter-train interval	Average no. of trials
4–9 Hz	>10	Variable	Variable	Variable
10 Hz	>50	5–6 pulse-trains for 400–500 ms	3.2 s	250
20–25 Hz	>20	10 pulse-trains for 400–500 ms	17.1 s	80

values will necessarily differ across varied circumstances, it is important to remain conservative when constructing a paradigm so as to maintain maximum subject safety.

rTMS for Cognitive Research—rTMS applied shortly preceding or during a cognitive task has been shown to modulate subject performance [34]. Although low-frequency TMS typically inhibits neural activity and high frequency excites neural activity, investigations of cognitive nature see a wide variation in neural response across subjects. Furthermore, in several studies, certain cognitive tasks have been demonstrated to be enhanced by inhibitory rTMS, revealing the potential of TMS-induced paradoxical functional facilitation (for review: [35, 36]). As such, when determining TMS parameters for a cognitive study, it is important that intensity, train duration, and inter-train interval be established before study commencement and not amended simply to evoke a desired effect in non-responding subjects.

rTMS for Therapeutics—The cumulative effects of repeated rTMS sessions can be at once beneficial and detrimental. Whereas many studies have shown an ameliorative effect of TMS on numerous neurological symptoms, several side-effects—including fatigue, difficulty concentrating, and neck pain—have been reported. It is important that determined treatment parameters remain within recommended safety boundaries and patient status be assessed both before and following each treatment.

rTMS of the Motor Cortex—It is recommended rTMS of the motor cortex not exceed 130 % resting motor threshold [2]. If a subject's motor threshold cannot be determined, it is recommended an intensity corresponding to the lower 95 % confidence interval of the average MT of the other subjects be used.

Theta-Burst Stimulation—The use of TBS protocols in both therapeutics and research is increasing rapidly. Although there have been no formal safety guidelines issued from the TMS Safety Consensus Group, it is strongly recommended TBS intensity be derived from a subject's *active* motor threshold (rather than the resting motor threshold). This lower number will increase overall safety during the rapid stimulation paradigm. Also, until further research is conducted exploring safe inter-TBS session durations, it is recommended subjects not undergo TBS more than once during a 7-day period [37].

6 Physical and Neuropsychological Monitoring

It is strongly recommended that practitioners administer both pre- and post-stimulation physical and neuropsychological evaluations. These evaluations should be short and easy to administer yet sensitive enough to detect subtle deficits possibly brought on due to TMS. Possible evaluations include (but are not limited to) the Mini-Mental State Examination [38], the Montreal Cognitive Assessment [39], the Beck Depression Inventory [40], the Autism Diagnostic Interview—Revised [41], and any standardized IQ test. Although points of interest will vary according to utilized paradigms, it is important that issues regarding both physical and cognitive status be examined (Table 6).

7 The TMS Lab/Clinic

Space—For diagnostic and therapeutic applications of TMS, a medical setting with attending physicians is required [2]. However, for studies with normal subjects utilizing the prescribed parameters, a medical setting may not be necessary. Each institution's IRB

Table 6
Example physical/mental status questionnaire

	Severity (1–5)	Relationship	Notes
Headache			
Neck pain			
Scalp pain			
Seizure			
Scalp burn (if EEG utilized)			
Hearing impairment			
Impaired cognition			
Trouble concentrating			
Acute mood change			
Other			

should be the final decider regarding this issue. It is suggested that, regardless of setting, appropriate life-support equipment be available onsite at each TMS clinic/lab.

Practitioners—The TMS Consensus Group is currently working on recommendations for practitioner training and certification. Although work is ongoing, several suggestions regarding the practitioner team are relevant. First, any clinical application of TMS should be overseen by a trained and certified neurologist. On the other hand medical assistants, including nurses and nurse practitioners, are also highly recommended during any clinical utilization of TMS [2].

All technicians administering stimulation should be BLS certified and well trained in both stimulation techniques and patient assessment. Although there are no official certification classes, it is recommended each center adheres to a strict "internal" certification system for new technicians, which includes ample observation and supervised treatments. In addition, a well-defined plan of action in the event of a seizure or syncope should be developed and well learned by each practitioner.

8 General Ethical Concerns

Informed Consent—Subjects must be provided with all information regarding procedure, risks, and/or any possible discomfort associated with treatment in order to supply the practitioner with informed consent. This information must be presented in easy-to-understand language without equivocation [42].

Risk-to-Benefit Ratio—When considering treatment options, informed consent does not constitute sufficient reason to forge ahead. Instead, a careful analysis of the possible benefits of therapy must be undertaken and shown to clearly outweigh the possible risks. The same risk-to-benefit ratio assessment stands in matters regarding research and data collection [42].

9 Conclusion

The use of TMS has grown dramatically in the past decade. New protocols of TMS have been developed and changes in the devices have been implemented. Furthermore, TMS is being increasingly combined with other brain imaging and neurophysiologic techniques including fMRI and EEG, and a growing number of subjects and patients are being studied with expanding numbers of longer stimulation sessions. A further increase in the widespread use of TMS in medical therapeutic applications and research is expected. This makes the need for clear and updated safety guidelines and recommendations of proper practice of application critical.

Over the years, safety and ethical considerations have been generally guided by the consensus statements [1, 2]. This chapter reflects not only on safety guidelines, including the appropriate training of TMS personnel, but also many other ethical issues raised in both clinical and research applications of TMS. As in any evolving field, the most essential are the questions we're still trying to answer.

References

1. Wassermann EM, Grafman J, Berry C, Hollnagel C, Wild K, Clark K et al (1996) Use and safety of a new repetitive transcranial magnetic stimulator. Electroencephalogr Clin Neurophysiol 101:412–417
2. Rossi S, Hallett M, Rossini PM, Pascual-Leone A (2009) Safety, ethical considerations, and application guidelines for the use of transcranial magnetic stimulation in clinical practice and research. Clin Neurophysiol 120(12):2008–2039
3. Ruohonen J, Ilmoniemi RJ (2002) Physical principles for transcranial magnetic stimulation. In: Pascual-Leone A, Davey NJ, Rothwell J, Wassermann EM, Puri BK (eds) Handbook of transcranial magnetic stimulation. Oxford University Press, New York
4. Elwassif MM, Kong Q, Vazquez M, Bikson M (2006) Bio-heat transfer model of deep brain stimulation-induced temperature changes. J Neural Eng 3:306–315
5. Rotenberg A, Harrington MG, Birnbaum DS, Madsen JR, Glass IE, Jensen FE, Pascual-Leone A (2007) Minimal heating of titanium skull plates during 1Hz repetitive transcranial magnetic stimulation. Clin Neurophysiol 118(11):2536–2538
6. Hsieh TH, Dhamne SC, Chen JJJ, Carpenter LL, Anastasio EM, Pascual-Leone A, Rotenberg A (2012) Minimal heating of aneurysm clips during repetitive transcranial magnetic stimulation. Clin Neurophysiol 123(7):1471
7. Darrigol O (2000) Electrodynamics: from Ampère to Einstein. Oxford University Press, Oxford
8. Schrader LM, Stern JM, Fields TA, Nuwer MR, Wilson CL (2005) A lack of effect from transcranial magnetic stimulation (TMS) on the vagus nerve stimulator (VNS). Clin Neurophysiol 116:2501–2504
9. Kumar R, Chen R, Ashby P (1999) Safety of transcranial magnetic stimulation in patients with implanted deep brain stimulators. Mov Disord 14:157–158
10. Kühn AA, Trottenberg T, Kupsch A, Meyer BU (2002) Pseudo-bilateral hand motor responses evoked by transcranial magnetic stimulation in patients with deep brain stimulators. Clin Neurophysiol 113:341–345
11. Kofler M, Leis AA, Sherwood AM, Delapasse JS, Halter JA (1991) Safety of transcranial magnetic stimulation in patients with abdominally implanted electronic devices. Lancet 338:1275–1276
12. Hidding U, Bäumer T, Siebner HR, Demiralay C, Buhmann C, Weyh T et al (2006) MEP latency shift after implantation of deep brain stimulation systems in the subthalamic nucleus in patients with advanced Parkinson's disease. Mov Disord 21:1471–1476
13. Machii K, Cohen D, Ramos-Estebanez C, Pascual-Leone A (2006) Safety of rTMS to nonmotor cortical areas in healthy participants and patients. Clin Neurophysiol 117:455–471
14. Loo CK, McFarquhar TF, Mitchell PB (2008) A review of the safety of repetitive transcranial magnetic stimulation as a clinical treatment for depression. Int J Neuropsychopharmacol 11:131–147
15. Brighina F, Piazza A, Vitello G, Aloisio A, Palermo A, Daniele O et al (2004) rTMS of the prefrontal cortex in the treatment of chronic migraine: a pilot study. J Neurol Sci 227:67–71
16. Anderson BS, Kavanagh K, Borckardt JJ, Nahas ZH, Kose S, Lisanby SH et al (2009) Decreasing procedural pain over time of left prefrontal rTMS for depression: initial results from the open-label phase of a multi-site trial (OPT-TMS). Brain Stimul 2:88–92
17. Counter SA, Borg E (1992) Analysis of the coil generated impulse noise in extracranial magnetic stimulation. Electroencephalogr Clin Neurophysiol 85:280–288
18. Pascual-Leone A, Houser CM, Reese K, Shotland LI, Grafman J, Sato S, Valls-Sol J, Brasil-Neto JP, Wassermannn EM, Cohen LG (1993) Safety of rapid-rate transcranial magnetic stimulation in normal volunteers. Electroencephalogr Clin Neurophysiol 89:120–130

19. Loo C, Sachdev P, Elsayed H, McDarmont B, Mitchell P, Wilkinson M et al (2001) Effects of a 2- to 4-week course of repetitive transcranial magnetic stimulation (rTMS) on neuropsychologic functioning, electroencephalogram, and auditory threshold in depressed patients. Biol Psychiatry 49:615–623
20. Zangen A, Roth Y, Voller B, Hallett M (2005) Transcranial magnetic stimulation of deep brain regions: evidence for efficacy of the H-coil. Clin Neurophysiol 116:775–779
21. Wassermannn EM (1998) Risk and safety of repetitive transcranial magnetic stimulation: report and suggested guidelines from the International Workshop on the Safety of Repetitive Transcranial Magnetic Stimulation, June 5–7, 1996. Electroencephalogr Clin Neurophysiol 108:1–16
22. Hu SH, Wang SS, Zhang MM, Wang JW, Hu JB, Huang ML, Wei N, Zhou WH, Qi HL, Xu WJ, Xu Y (2011) Repetitive transcranial magnetic stimulation-induced seizure of a patient with adolescent-onset depression: a case report and literature review. J Int Med Res 39(5): 2039–2044
23. Kratz O, Studer P, Barth W, Wangler S, Hoegl T, Heinrich H, Moll GH (2011) Seizure in a nonpredisposed individual induced by single-pulse transcranial magnetic stimulation. J ECT 27(1):48–50
24. Gómez L, Morales L, Trápaga O, Zaldívar M, Sánchez A, Padilla E, Grass D, Díaz A, Morales H (2011) Seizure induced by sub-threshold 10-Hz rTMS in a patient with multiple risk factors. Clin Neurophysiol 122(5): 1057–1058
25. Oberman LM, Pascual-Leone A (2009) Report of seizure induced by continuous theta burst stimulation. Brain Stimul 2(4): 246–247
26. Xia G, Gajwani P, Muzina DJ, Kemp DE, Gao K, Ganocy SJ et al (2008) Treatment-emergent mania in unipolar and bipolar depression: focus on repetitive transcranial magnetic stimulation. Int J Neuropsychopharmacol 11:119–130
27. Zwanzger P, Ella R, Keck ME, Rupprecht R, Padberg F (2002) Occurrence of delusions during repetitive transcranial magnetic stimulation (rTMS) in major depression. Biol Psychiatry 51:602–603
28. Janicak PG, O'Reardon JP, Sampson SM, Husain MM, Lisanby SH, Rado JT et al (2008) Transcranial magnetic stimulation in the treatment of major depressive disorder: a comprehensive summary of safety experience from acute exposure, extended exposure, and during reintroduction treatment. J Clin Psychiatry 69:222–232
29. Hsu WY, Cheng CH, Lin MW, Shih YH, Liao KK, Lin YY (2011) Antiepileptic effects of low frequency repetitive transcranial magnetic stimulation: a meta-analysis. Epilepsy Res 96(3):231–240
30. Frye RE, Rotenberg A, Ousley M, Pascual-Leone A (2008) Transcranial magnetic stimulation in child neurology: current and future directions. J Child Neurol 23:79–96
31. Gilbert DL, Garvey MA, Bansal AS, Lipps T, Zhang J, Wassermannn EM (2004) Should transcranial magnetic stimulation research in children be considered minimal risk? Clin Neurophysiol 115:1730–1739
32. Fitzgerald PB, Brown TL, Marston NA, Oxley T, De Castella A, Daskalakis ZJ, Kulkarni J (2004) Reduced plastic brain responses in schizophrenia: a transcranial magnetic stimulation study. Schizophr Res 71(1):17–26
33. Oberman LM, Horvath JC, Pascual-Leone A (2010) TMS: using the theta-burst protocol to explore mechanism of plasticity in individuals with Fragile X syndrome and autism. J Vis Exp 28(46):2272
34. Walsh V, Pascual-Leone A (2003) Transcranial magnetic stimulation: a neurochronometrics of mind. Cambridge Press, Cambridge, MA
35. Fecteau S, Pascual-Leone A, Théoret H (2006) Paradoxical facilitation of attention in healthy humans. Behav Neurol 17(3–4):159–162
36. Théoret H, Kobayashi M, Valero-Cabré A, Pascual-Leone A (2003) Exploring paradoxical functional facilitation with TMS. Suppl Clin Neurophysiol 56:211–219
37. Oberman L, Edwards D, Eldaief M, Pascual-Leone A (2011) Safety of theta burst transcranial magnetic stimulation: a systematic review of the literature. J Clin Neurophysiol 28(1): 67–74
38. Folstein MF, Folstein SE, McHugh PR (1975) Mini-mental state: a practical method for grading the cognitive state of patients for the clinician. J Psychiatry Res 12(3): 189–198
39. Nasreddine ZS, Phillips NA, Bedirian V, Charbonneau S, Whitehead V, Collin I, Cummings JL, Chertkow H (2005) The Montreal cognitive assessment, MoCA: a brief screening tool for mild cognitive impairment. J Am Geriatr Soc 53(4): 695–699

40. Beck AT, Ward CH, Mendelson M, Mock J, Erbaugh J (1961) An inventory for measuring depression. Arch Gen Psychiatry 4(6): 561–571
41. Lord C, Rutter M, Le Couteur A (1994) Autism diagnostic interview-revised: a revised version of a diagnostic interview for caregivers of individuals with possible pervasive developmental disorders. J Autism Dev Disord 24(5): 659–685
42. Horvath JC, Perez JM, Forrow L, Fregni F, Pascual-Leone A (2011) Transcranial magnetic stimulation: a historical evaluation and future prognosis of therapeutically relevant ethical concerns. J Med Ethics 37(3): 137–143

Chapter 3

Neuronavigation for Transcranial Magnetic Stimulation

Roch Comeau

Abstract

In order to ensure successful application of transcranial magnetic stimulation (TMS), practitioners must be certain that they are targeting the correct cortical location. To aid in this, a number of clinicians and practitioners have begun utilizing various neuronavigation systems to track coil and participant-head position in space for the duration of a stimulation session. In this chapter, I explore the history of neuronavigation and the developments that made combining this technology with TMS possible. Following this, I discuss the practical aspects of properly utilizing a neuronavigation system: including MRI acquisition, 3D-reconstruction, head and coil co-registration, cortical targeting, motion tracking, and electric field modeling. I conclude with a brief discussion of incorporating robotic assistance in coil positioning and tracking.

Key words Transcranial magnetic stimulation, Neuronavigation, Modeling, Targeting

1 Introduction

The application of transcranial magnetic stimulation (TMS) to the subject head is rarely done in a vacuum. More likely, it is done as part of a carefully choreographed experiment or clinical paradigm. Successful use of TMS depends on applying the correct stimulus parameters at the correct location within the brain, sometimes at the correct time in relation to an external stimulus, or synchronization clock. The most straightforward method of placing the coil over the correct area of the brain can be as simple as moving the coil around on the head, firing it at different locations and observing a simple response, for example, the twitch of a finger. Once the desired response has been obtained, the "real" paradigm would begin. In many cases however, there is no obvious or immediate response one can rely on to confirm correct coil placement. This chapter will describe how a neuronavigator can be used to define a target using a variety of criteria, correctly place the coil over a target, and provide the added benefit of aiding the operator in ensuring the coil is kept over the correct location for the duration of the session.

Alexander Rotenberg et al. (eds.), *Transcranial Magnetic Stimulation*, Neuromethods, vol. 89,
DOI 10.1007/978-1-4939-0879-0_3,

Before going further, a disclaimer is in order. The author is the owner of a company that developed the first commercially available neuronavigation system specifically for use in TMS. At the time of this writing, at least 5 neuronavigation systems were available commercially for use specifically for TMS. Each one will have features or methods that differ from one another. This chapter will attempt to present practical methods that are applicable to all neuronavigation systems, and when relevant, highlight the differences between them without judgment on which one may be better. Judgement will be left to the reader.

Since the focus of this book is methods, rather than providing an exhaustive reference of publications referring to TMS and neuronavigation, the focus will be placed on the practical aspects of using neuronavigation for TMS.

1.1 History of Neuronavigation

Neuronavigation is a modern name for a method of tool placement often referred to as frameless stereotaxy or image-guided *something*, the something being TMS here, but includes neurosurgery, therapy etc..... Frameless stereotaxy has its roots in the neurosurgical world, where for a long time, frame-based stereotaxy was the state of the art. While many of us know this lineage, to the new, young TMS users, referring to neuronavigation as frameless stereotaxy is akin to referring to the automobile as the horseless carriage. Other terms often used for image-guided TMS is Navigated Brain Stimulation (NBS)

Stereotaxy, be it frame or frameless, usually refers to the act of creating a common coordinate grid that can be applied to a set of anatomical images of a subject, and, using a special tool and/or method, to the physical head of that subject. In the surgical context, one would be able to determine the coordinate of some pathology in the brain as observed on medical images, and using the tool guide, be able to reach the homologous location in the actual brain, either to extract a sample for biopsy, inject a medicinal product, insert a deep brain stimulator or remove a lesion in its entirety. Stereotaxy is also used for, among other things, guidance of placement of radioactive seeds in Brachytherapy or ionizing X-ray beams in radiosurgery.

1.1.1 From Frame to Frameless Stereotaxy for Neurosurgery

The stereotactic frame was used to define the coordinate system. The frame was fixed to the subject's head using pins that went through the scalp to make direct contact with the skull. The critical assumption was that the frame was rigidly and "permanently" fixed to the skull and thus the brain. That meant that the basic principle of the stereotactic coordinates of the images and the head being homologous was valid as long as the frame was fixed. Imaging-specific fiducial markers were placed on the frame at known stereotactic coordinates and would be visible on any images acquired of the patient with the frame. The markers were identified on the images allowing the coordinates of any point in the images to be calculated in the frame's reference space. Rulers were embedded in the frame's rails to act as tool guides to allow tools to be held and guided into the head (after a suitable hole was drilled into the skull) using the same coordinates.

1.2 Basic Principles of Frameless Stereotaxy

While the frame revolutionized neurosurgery and is still used today when high accuracy is required ($\approx$1 mm), it does suffer drawbacks, including discomfort for the subject and limited access to the head due to the bulk of the frame. These limitations inspired the development of frameless stereotaxy. The basic principles are similar, except that instead of a rigid frame being the link between the image and real-world coordinate systems (and lasting only as long as the frame was fixed to the head), a coordinate system based on the images is defined, and transient mappings from the real-world space to the image space is established when needed (e.g., at surgery). The mapping in surgery is established using a position sensor, which can monitor and report the location of one or more tools in 3D space, and a registration method that calculates the mapping from that sensor space to the image space. The sensor space can be defined by the location of the sensor itself, or one of the trackers can be defined to be the reference (e.g., after fixing to the head), and the location of the other tools can be reported in the reference space. The mapping then calculates the homologous image coordinates of the tool that the display software uses to display a representation of the tool on the images.

These same principles hold true for TMS. Figure 1 shows a typical image-guided TMS setup. The red arrows illustrate the various coordinate systems that are defined and used.

Not surprisingly, once the availability of neuronavigation became more widespread for neurosurgery, other areas of application began to emerge. One of the first reports of neuronavigation specifically for TMS (where some systematic evaluation of utility was reported) was by Gugino et al. [1], who performed a detailed comparison of guided vs. blind coil positioning. Paus [2] reported on using a modified neurosurgical neuronavigator to place a TMS coil on the subject's head during PET scanning. The result showed cortical activation beneath the TMS coil and synchronous activation elsewhere in the brain, demonstrating the utility of combining image-guided TMS and neuroimaging to map out connected regions of the brain. Once the utility became apparent in the TMS field, neuronavigation systems specific for TMS were subsequently developed [3–5].

Since these first applications of image-guided TMS, hundreds of TMS laboratories around the world have made neuronavigation a routine part of TMS protocols. Many comparisons of traditional positioning techniques with neuronavigation have been performed [6–8] (to name a few).

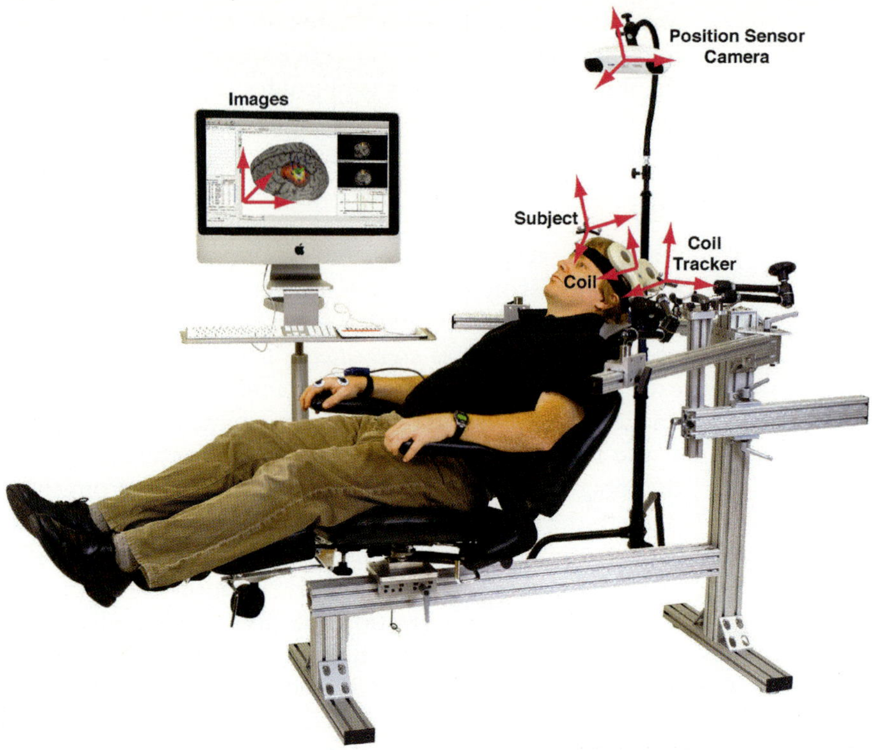

Fig. 1 Typical setup from one of the commercially available neuronavigation systems for TMS

1.3 Basic Mathematical Principles

In order to display a representation of a tool in the image space, the position of the tool must be determined in the image space. The tool will have a point of reference (the point we are interested in) and an orientation. A series of transformations from one coordinate system to the other is performed until the transform from the tool's coordinate system to the images is determined. For example, the TMS coil has a reference point and direction, usually defined by the "hot spot" of the coil, with a vector projecting away from the coil face (Fig. 2).

The tracker fixed to the coil defines a second coordinate space. As long as the tracker is rigidly fixed to the coil, the transformation from the coil to tracker coordinate space can be defined by a rigid body transform and represented by a 4×4 matrix (${}^{\text{coil}}_{\text{tr}}T$). The position sensor can report the position and orientation of the tracker in the position sensor's reference space (the tracker fixed to the subject's head) and represented as another 4×4 matrix (${}^{\text{tr}}_{\text{ref}}T$). The registration procedure (various methods of determining the registration will be presented in a subsequent section) yields the transformation from the reference space to the image space (${}^{\text{ref}}_{\text{image}}T$). So the transform from the tool's coordinate space to the image space can be calculated by multiplying the matrices through

$${}^{\text{coil}}_{\text{tr}}T \circ {}^{\text{tr}}_{\text{ref}}T \circ {}^{\text{ref}}_{\text{image}}T = {}^{\text{coil}}_{\text{image}}T$$

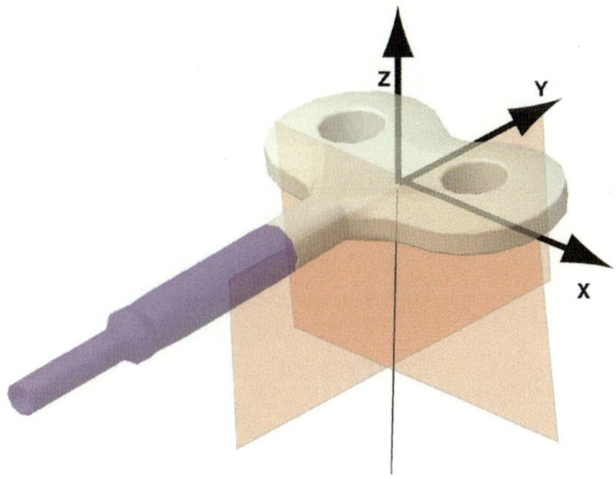

Fig. 2 Illustration of the coil and the concept of defining a local coordinate system to the face of the coil

Once ${}^{coil}_{image}T$ is known, a representation of the coil can be drawn in a 3D view using common 3D graphics techniques. In addition, 2D image planes can be defined as each 2D vector pair of the coil's coordinate system and the image voxels that line in each plane can be determined and displayed as a 2D image.

"Under the hood," the position sensor usually defines its own coordinate space that is physically linked to the sensor itself. For example, a camera-based sensor will calculate the coordinates of the trackers w.r.t. the camera. If the coil tracker to camera transform can be defined as ${}^{camera}_{tr}T$, and the reference to camera defined as ${}^{camera}_{ref}T$, the position of the coil tracker can be calculated by inverting the reference to camera transform and multiplying the two:

$${}^{tr}_{camera}T \circ {}^{camera}_{ref}T = {}^{tr}_{ref}T$$

2 Practical Use of Neuronavigation in TMS

This section will outline the typical steps in preparing for, and conducting an image-guided TMS procedure. Most of the work is done prior to the subject arriving, and much of it is performed once for a subject, as the data can be stored in a file for use in multiple sessions. The navigation system will also require some non-subject-specific preparations. These are usually coil-specific (e.g., fixing a tracker to the coil and defining a reference point) and will be described in a subsequent session. The steps can be roughly defined as

- Prepare neuronavigation equipment
- Prepare subject-specific pretrial data
- Perform TMS session
- Review results

2.1 Prepare Neuronavigation System (NOT Subject-Specific)

In many TMS labs, multiple TMS experiments are performed throughout a given day. Many of these experiments make use for different equipment in addition to the TMS. Care should be taken to ensure that the operator can reach the relevant equipment quickly during the TMS session. In the context of the neuronavigator, the position sensor should be placed in a location that allows unobstructed views of the subject tracker and the TMS coil tracker. Take the expected orientation of the coil and the operator into account.

When organizing the layout, take a moment to decide on the location of the tracker on the coil (if this can be manipulated). Make sure that the tracker will be facing the camera when in the expected orientation on the subject's head.

Some navigation systems use the coil's magnetic field output for display or as input for induced current modeling. Refer to the manufacturer's instructions in getting the coil characterized, which may require sending the coil to them.

2.2 Prepare Pretrial Data (Subject-Specific)

There are essentially three main steps in preparing the data for a TMS study. The actual steps to be performed within each main step will depend on your protocol. For completeness, the most common ones are all listed here. Your protocol will likely only require a subset of these. Also, your specific neuronavigator may offer additional functionality not described here; however the basic organization of the steps should be essentially the same.

- Gather and process subject-specific data
  - Select anatomical images
  - Select overlay images (e.g., fMRI, PET)
  - Segment images (e.g., define ROI)
  - Generate 3D reconstructions
  - Co-register to standard space (e.g., MNI or Talairach)
- Define Target(s) (points, vectors, grids)
  - From anatomy
  - From overlay
  - From atlas or standard space coordinates
- Perform neuronavigator-specific preparations
  - Identify and record registration landmarks (e.g., anatomical landmarks)

This workflow assumes the target can be defined based on anatomical or functional data acquired prior to the first TMS session, or based on an atlas coordinate (e.g., MNI coordinate). It is also possible to skip many of these steps and perform an initial pilot TMS session without a preset target, and use some physiological response that can be recorded during the pilot session to define a target for subsequent sessions.

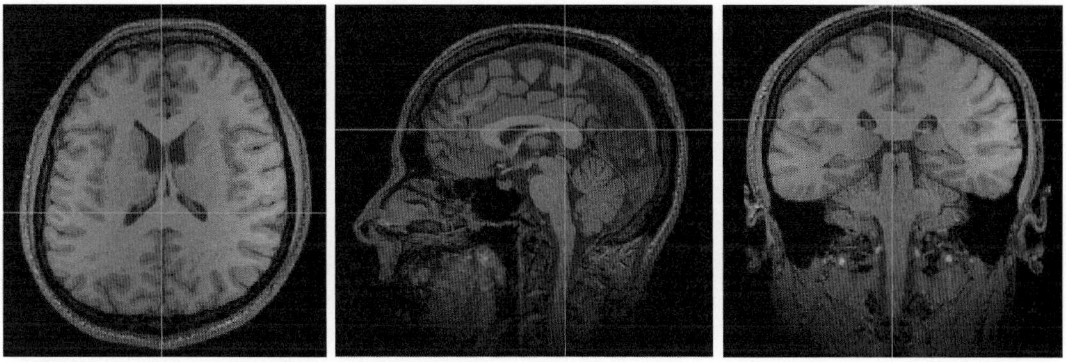

Fig. 3 Typical tri-planar view of an anatomical MRI

2.2.1 Selecting Anatomical Images

The structural images form the backbone of the navigation system as the individual subject's native coordinate system is usually based on this. All other information is mapped to the anatomical images for display and interpretation. Anatomical images are usually T1 weighted MR images of the subject. In cases where the subject-specific images are not available, a data set of a model head may be used (e.g., MNI average brain). The advantage of having the individual's images is that you will be able to use the anatomy, and any information that is mapped to the anatomy, as a means of selecting a target. Using a model head does not provide the benefit of the subject-specific anatomy for targetting, but preserves the reproducability of coil placement between sessions. Presumably, a pilot session would be performed to determine the correct location for the coil, and the navigation system would be used to return to that location (or a location at a fixed distance from that initial location) without having to repeat the hunting. It may be possible that the navigator can perform a registration between the subject's head and the standard space so that the standard images can be used; however care should be taken to understand the accuracy limitations introduced by using a model head rather than the subject-specific images.

Anatomical Images

The anatomical images usually consist of a T1 weighted, volumetric acquisition (Fig. 3). The actual protocol used varies for different scanner models. A 3D CT scan may also be used, but it would be difficult to justify the radiation exposure from a high resolution CT scan for research purposes. For typical MR images, the following points should be considered:

- The voxel size should be small enough to visualize the brain's sulci as well as the anatomical landmarks for co-registration. The most common voxel size is 1 mm isotropic. Thicker slices may be used, but slices thicker than 1.5 mm will start to create a banding effect for oblique images. Smaller voxels usually don't yield more useful images (in this context) and, in addition to increasing acquisition time, will yield slower performance of the navigation system due to the higher number of voxels to process for display.

- The field of view of the images should be large enough to encompass the outer anatomical landmarks that will be used to co-register the subject to the images (e.g., nasion, ears).
- It is also helpful to place an external fiducial marker on one side of the head (and be consistent with which side), particularly if you are going to convert the anatomical images to other file formats. This will allow you to verify, at a glance, that the images were not inadvertently swapped for left and right.

Model Head

In some cases though images of the subject may not be available, a model head is used. In this instance, the main goal of the use of the navigation system should be to record the location of the coil in one session, and accurately and quickly reproduce it in a future session.

- *Use a model MRI volume* (e.g., *ICBM average brain* [9]). This will allow you to co-register the subject to "a head"; however the internal anatomy will not be a reliable map for a particular subject. This is due to the lack of reliable, validated mapping of internal brain structures to the external landmarks or overall shape of the head. In addition, after the subject has been registered to the model head, there will likely be visual discrepancies between the location of the tool on the subject's head, and the appearance of the tool displayed on the model head. This is because the actual subject's head will have a different shape than the model head. This visual discrepancy will not affect the accuracy of the coil placement because the discrepancy is a visual artefact. The "error" will be the same for each session (within the inter-trial accuracy), so the mapping from the subject space to the model image space will be consistent.
- *Use of a generic head surface.* In some cases, the software would provide tools to measure the shape of the individual's head to tailor the model head to more closely match the subject's head shape. While this may provide a more reassuring image on the screen, it has no impact on the overall placement accuracy.

2.2.2 Load Functional Data

There are many applications where fMRI can be used to identify a potential target for TMS stimulation (see [10–13] for examples). If your target is based on an activation derived from fMRI data, it can be useful to load the data for display when placing the coil. Typically, the fMRI data would have been pre-processed and co-registered to the anatomical data using your favorite fMRI tools. Take care to ensure that you use the same anatomical images as were used with the fMRI processing/analysis software.

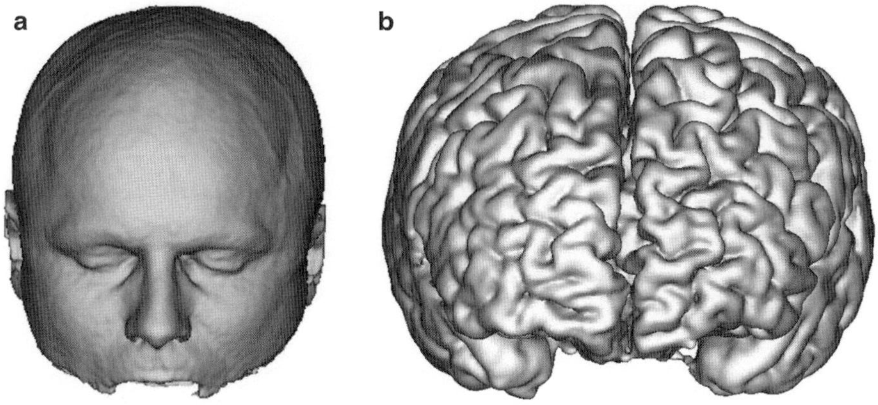

Fig. 4 Typical 3D surfaces represented by a mesh. (**a**) The skin. (**b**) The brain surface

In some cases, the neuronavigation software also functions as the fMRI processing tool. In this case, follow the instructions that refer to fMRI image processing.

2.2.3 Perform 3D Reconstructions

Since manipulating the coil is inherently a 3D operation, it is helpful to be able to visualize the brain and important structures derived from the MR images in 3D. As with many image display software packages, the neuronavigation system usually includes the ability to create 3D versions of the brain and subject skin. The skin is helpful in identifying anatomical landmarks that are often used to co-register the subject to the images, or as a source of curvature data for curve-based registration.

2.2.4 Surface Mesh

Surface meshes are used to create 3D representations of the skin and brain (Fig. 4). These may be created automatically, or using segmentation tools provided by the software.

2.2.5 Curvilinear Reconstruction

3D surface meshes are used by many visualization applications to provide a 3D representation of the cortical surface. In TMS, we are interested in more than the cortical surface, but rather we are interested in visualizing the 10–20 mm of cortex, often referred to as the cortical ribbon (Fig. 5).

Curvilinear reconstruction was originally developed to provide such a display for Epilepsy research, specifically to aid in the detection of subtle lesions such as focal cortical dysplasia [14]. It has also found use in Image-guided TMS due to its ability to display the cortical ribbon. Many of the currently available Image-guided TMS systems provide curvilinear reconstructions (Fig. 6).

The curvilinear reconstruction can be explained using a peeler metaphor. Imagine staring with an image of an extracted brain (e.g., the dura), and being able to mathematically peel a constant thickness of voxels off the surface (like peeling a potato) to reveal the structures below (a potato is rather plain on the inside, but

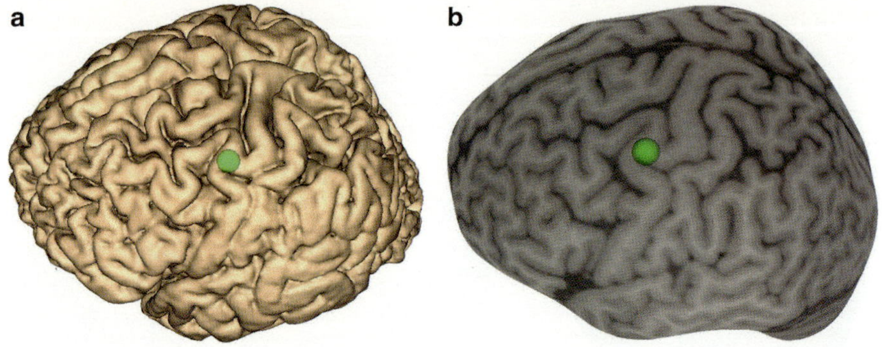

Fig. 5 Two types of 3D representations of the brain. (**a**) Surface mesh. (**b**) Curvilinear reconstruction

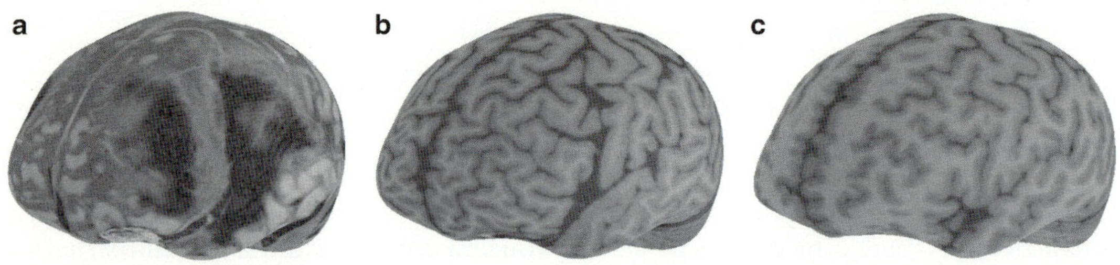

Fig. 6 Curvilinear reconstruction of the brain. (**a**) Reference, or "0" surface. (**b**) Surface cut 6 mm from the reference surface. (**c**) 12 mm from the reference surface

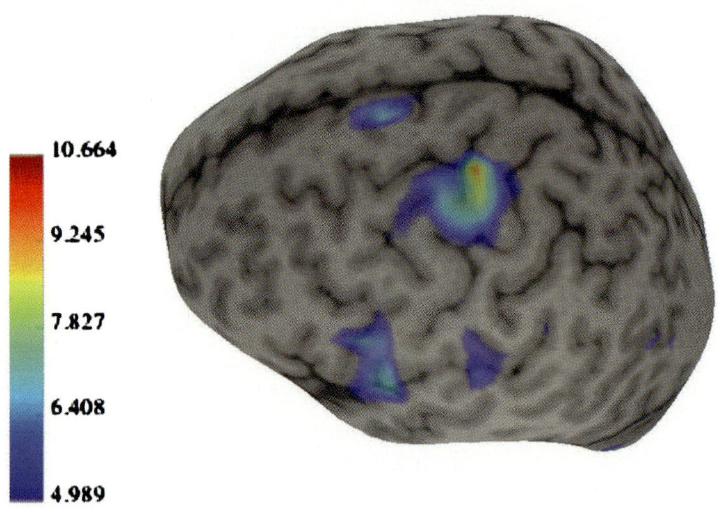

Fig. 7 Typical fMRI from a finger-related motor task co-registered to the anatomical images

hopefully you get the idea). This allows for better visualization of the relative position of the various cortical structures below the surface. This is also useful when visualizing functional overlays, as it allows you to peel off the surfaces to reveal the activations within the cortex, where they were measured (Fig. 7).

2.2.6 Co-register to Atlas Space

In many areas of research and clinical applications, it is desirable to be able to compare homologous anatomical and functional areas of an individual subject to that of a more general population. This led to the development of anatomically based reference grids, or coordinate spaces [9, 15] (e.g., Talairach or MNI) that enable one to assign a 3D coordinate to an individual structure within that reference brain. By co-registering the individual brain (as represented by the image data) to fit onto the reference grid, the standardized coordinates can be translated to the individual, and anatomical and functional observations made within the individual's brain can be mapped back to the standard space. Atlases based on anatomical and/or functional observations can be created and used as templates to be overlaid on the reference brain.

In the context of TMS, the standard space can be used as both a source of potential targets for stimulation and as a means of comparing results from a population of subjects in a common space. For example, one might use a common atlas, the Brodman atlas, to pick an anatomical region known for speech generation, and use it as a TMS target to disrupt speech. Conversely, one can try stimulating various anatomical regions of a subject for a particular effect. Once the effect has been observed, the coil's orientation can be projected into the cortex to derive a coordinate within the cortex directly beneath the coil and that coordinate can be mapped back to a standard space for interpretation using an atlas, or for comparison with a pool of subjects (whose stimulation coordinates were mapped to the same standard space).

One of the main differences between how the standardized space is used for TMS and in other applications is the direction of the transformation. In fMRI research it is common to transform the individual MR images of the subject into the standardized space soon after acquisition, and to perform much of the analysis and interpretation with the warped images. Part of this methodology is historical in nature as it dates back to the days before fMRI when positron emission tomography (PET) was the main tool for functional neuroimaging. PET often lacked the signal to noise (SNR) to obtain significant results on single subjects, so pooling of subjects into standard space provided the SNR required to obtain reliable results. In neuronavigation, where the images are to be co-registered to the actual subject's head, it is preferable to keep the images undistorted, but to record the mathematical transformation between the native images and the standardized space. This allows one to pick a native image coordinate and display the homologous standard space coordinate. The inverse transform (from standard to individual space) can be used to warp an atlas from standard space to the individual's space.

Currently the most common standardized spaces are Talairach and MNI spaces. They are both very similar (the MNI space is essentially a modernized version of the Talairach space). Most neuronavigation systems provide a means to co-register the subject's images to the MNI or Talairach space, or import the registration matrix obtained from commonly used software designed for that purpose (e.g., SPM [16]). There are different methods of characterizing the transformation from native to standard space. The simplest and most common method is a linear transformation. This defines the transformation using translation, rotation, and scaling (usually three scaling in the three orthogonal directions). This is the simplest, but the least accurate as it does not account for skew, or asymmetry of the subject brain vs. the standard brain.

2.3 Define and Record Target(s)

2.3.1 Scalp or Cortical Target?

Before defining the target, it is important to decide on the type of target. Historically, targets for TMS stimulation have been defined at the scalp. This was for practical reasons. First, one could wear a swimming cap and draw a dot on the cap to define the target and the coil could then be placed on the cap for stimulation. Conversely, one can record the location of the coil at any time for future reference by drawing a dot during a pilot session. Second, an EEG-based anatomical reference system (e.g., 10–20 grid) could be used to define the target by simply wearing an EEG electrode cap (without the electrodes) and use the holes in the cap as target markers. It is important to consider than a scalp-based target is not a "real" target in that one does not perform TMS to stimulate the scalp (although it does). Rather, there is an inherent assumption that the coil is oriented in such a way that the cortex immediately below the scalp is stimulated. This assumption depends on the skill of the operator to be valid.

One of the most significant benefits to a neuronavigation system is that targets can now be set *in the cortex*. The target can be reached from multiple scalp locations, although one location will be more optimal than the others. This means that cortex-based targets will be more tolerant of subtle errors in scalp locations because the navigator will always display the optimal tilt for the coil to ensure that the target is reached (Fig. 8).

2.3.2 Target Type: Simple Marker or Marker and Trajectory?

Ultimately, the target is set of neurons that we wish to stimulate. For simplicity, this is usually defined by a discrete location within the cortex, presumably in the center of the pool of neurons to stimulate. The target definition can be augmented by supplying an approach angle and coil twist to ensure that an optimal, unambiguous approach to the target is defined (Fig. 9). Whether to include the approach angle and twist depends on the operator. Many of the navigators provide real-time display of the coil's orientation, both on the 3D brain and with 2D oblique slices centered on the coil. In many cases, it is sufficient to define a single point within the cortex and to simply find the location on the scalp where the coil sits well (given the local scalp shape) and the coil points to the target. In many cases however, the person defining the target and the person

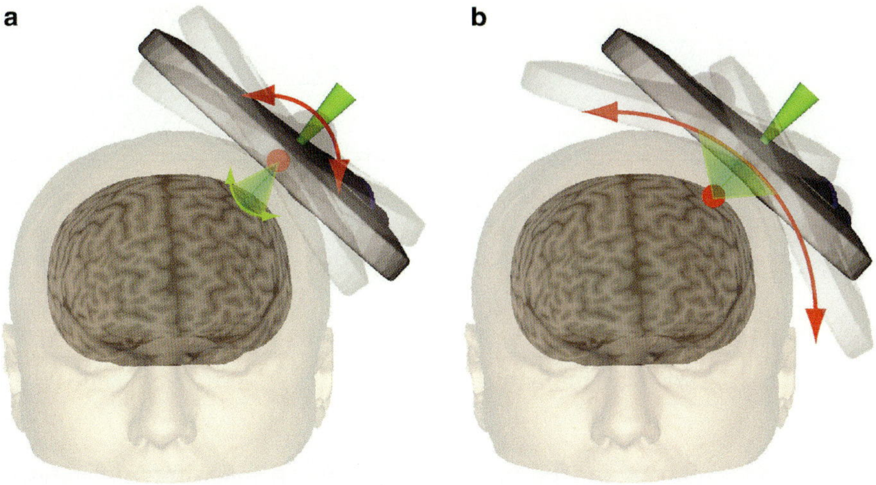

Fig. 8 The difference between scalp-based targets (**a**) and cortex-based targets (**b**). (**a**) Using a target in the scalp is an indirect way to define the true target and assume that the coil is tilted properly to ultimately hit the undefined, yet true target, the cortex. Small errors in tilt can significantly change the actual cortex that is stimulated because the pivot point is on the scalp and thus far from the cortex. (**b**) Defining the true target in the cortex means that regardless of where the coil is on the scalp, the navigation system will guide the user to the proper coil tilt to point to the correct part of the cortex. Relatively large variations in the scalp location yield small changes in the actual stimulation because the coil pivot is in the cortex and the correct tilt is inherently shown

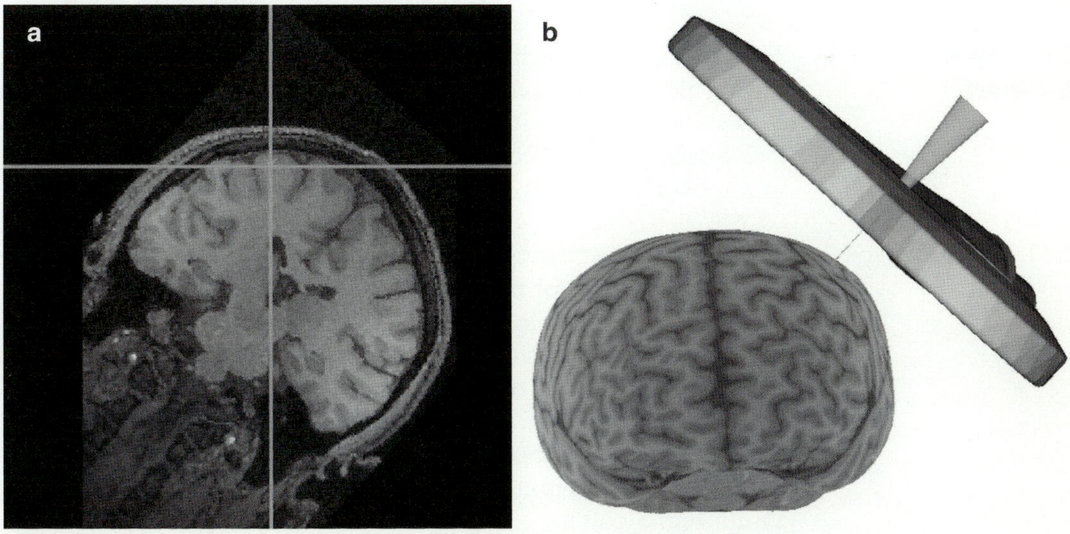

Fig. 9 (**a**) Oblique image showing the coil's approach angle into the brain. (**b**) 3D version of the coil sitting over the brain

ultimately manipulating the coil are different people (the latter may actually be a robot), so to remove inter-operator variability, defining the approach angle as well may be advantageous.

All navigation systems allow multiple targets to be defined. In some applications, multiple targets can be defined as a grid.

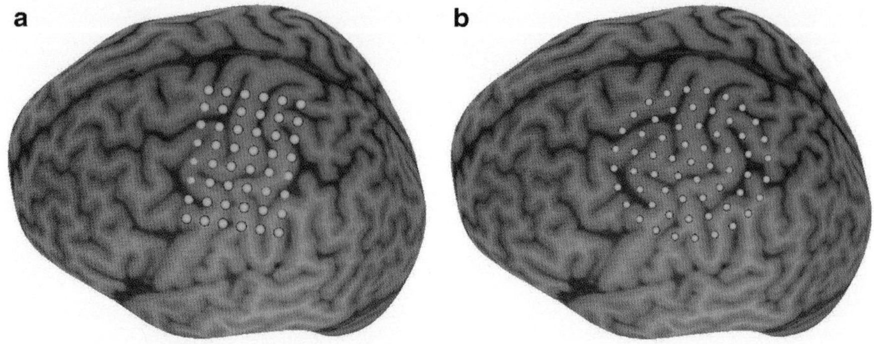

Fig. 10 Rectangular and circular grid target arrays

Grids can be particularly useful in mapping applications and can have two benefits. First, it allows you to define a series or uniformly spaced targets that simplify the task of stimulating a series of targets to cover a region of the brain. Second, if carefully thought out, it allows you to refer to these targets in a 2D sense using the grid node location within the grid coordinate system (e.g., row, column for rectangular grids, or ring and node on a circular grid). This is essentially a marriage of old and new. The grid can be defined on the cortex providing the advantages of the cortex-based target definitions with the simplicity of cartographic (2D) naming convention (Fig. 10).

2.4 How to Define the Target

The criteria for recording the target depend on your protocol, and this would have been considered during the data preparation phase discussed in the previous sections. For example, if you are using fMRI data as your selection criteria, then you should have arranged to import the relevant functional images already.

2.4.1 Anatomical Targets

This is the most straightforward target. Using the visualization tools, the location of the structure is visualized, and clicking on it identifies the target location. Use the navigation tools to record the target and, if desired, the approach angle to the target. The figure below shows a target selection screen from a common neuronavigator (Fig. 11). Most navigators will have a similar screen. Note the display of both 2D and 3D images and the inclusion of slider tools to manipulate the proposed coil orientation (shown as a mini Fig. 8 coil for clarity).

2.4.2 Functional Data

Many TMS experiments involve stimulating a functionally specific part of the cortex. In some cases, the anatomical region may be poorly defined by obvious landmarks, or may have a high degree of variability (w.r.t. visible landmarks) between subjects, particularly in cases involving pathology that may have altered the functional representation (Fig. 12). In these cases, it may be desirable to devise a task that can be performed during fMRI to localize the region defining the target [10–13, 17, 18].

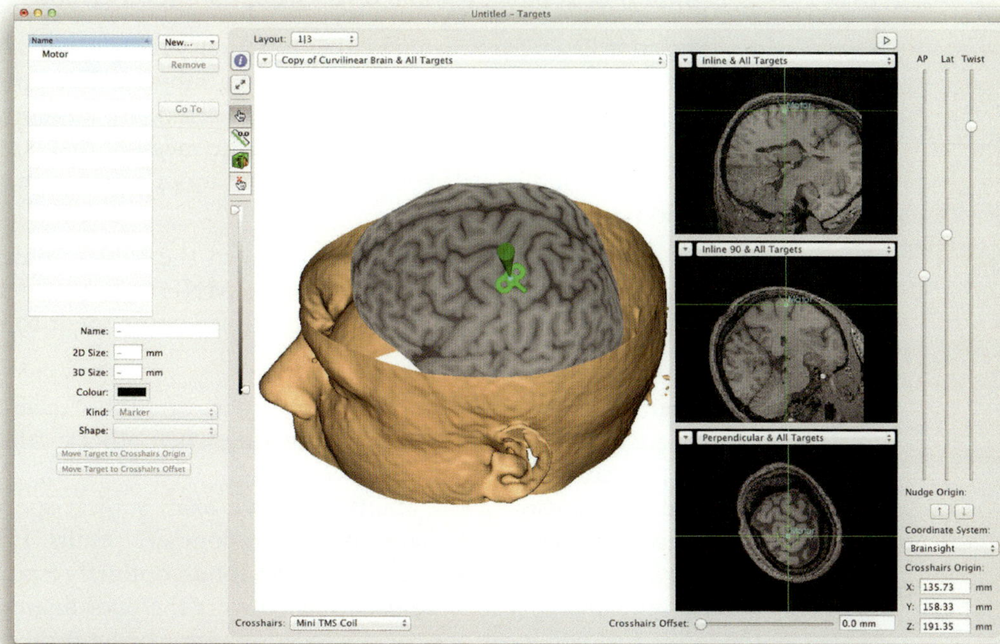

Fig. 11 Anatomical display to select and record targets for stimulation

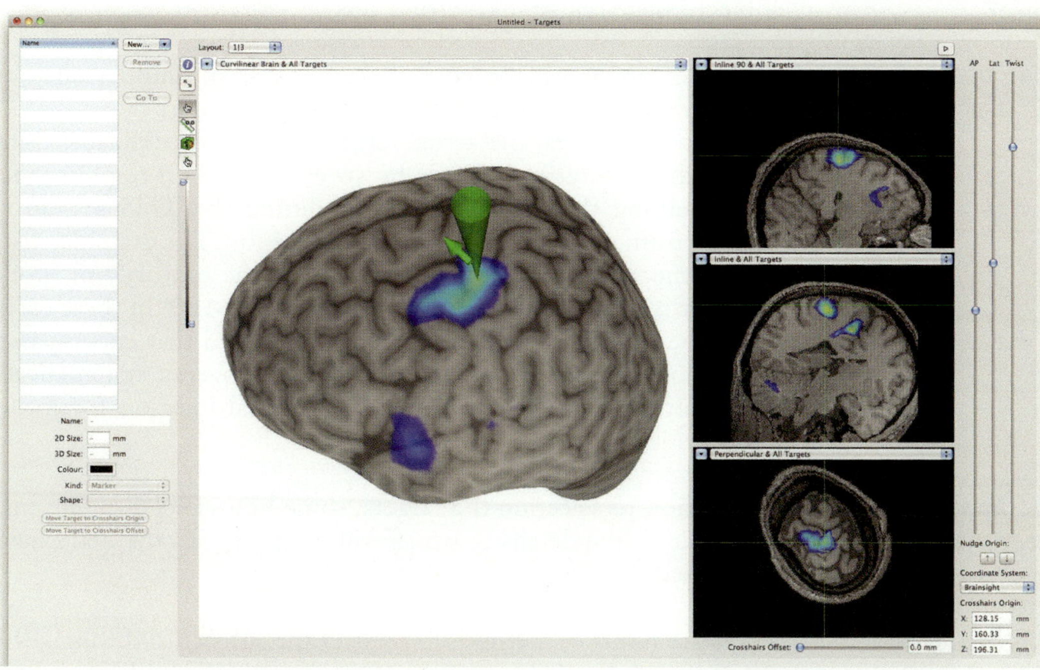

Fig. 12 fMRI overlay for target selection and recording

2.4.3 Atlas Coordinates

This should be used with some reservation for a few reasons [19]:

- The coordinate from an atlas either represents a single individual (e.g., Talairach) or an average of multiple subjects. The potential individual variability should be considered. By targeting a "representative" location rather than some measure that is specific to the individual, you may observe more variance in the response.
- The accuracy of the registration between the subject-specific MRI and the normalized brain (e.g., MNI) will be on the order of a few mm. This adds an additional component to the coil positioning error.

Given those caveats, any functional atlas, or MNI/Talairach coordinate can be used as an aid for targeting when other methods are not practical.

One of the uses of the ability to move from subject-specific MRI coordinates to normalized space is to compare results. If for example, a search is performed using TMS for an optimal response among individuals, it may be useful to convert the coordinates of the individual coil locations (or the locations after being projected from the coil into the cortex to a consistent, representative depth) to normalized space to allow for comparison. The same caveat as above applies: The variance in location will be a function of both the variance due to individual response (what you presumably are looking for) and the combination of the atlas registration and subject-image registration errors.

2.4.4 TMS Trial Session (e.g., Motor Mapping)

Motor mapping will be covered extensively in another chapter; however a short mention in this context is in order.

In certain cases, the actual target may not be definable using the methods described above; however the desired response may be something that can be observed during the TMS session or derived from data recorded during the session. For example, visualizing the anatomy alone cannot localize the motor "hot spot"; however it can be measured by recording the MEP. In these cases, it may be desirable to define the target as a grid that encompasses the region of interest, and the area can be covered systematically and analysed later to determine the optimal location for future TMS sessions.

2.5 Neuronavigation-Specific Preparations

In addition to selecting the targets, some housekeeping steps may be required, the details of which will be specific to the neuronavigator used.

2.5.1 Registration Landmarks

There are multiple methods to register the subject to the images at the start of the TMS session. If your system uses homologous points, where you are expected identify landmarks on the images and later on the subject, then this is probably the right time to identify them on the images.

Homologous point matching is a simple method to perform, but relies on a few key assumptions. First, that the landmarks used are unambiguous ensuring that you can select the same point on both the images and the subject. Second, it assumes that the landmark is rigid in that the landmark is in the same location w.r.t. brain at the TMS session time as it was during the scan.

The first assumption means that you need to pick easily identifiable points. The vertex on the top of the head, for example, is a poor point because its location cannot easily be identified (without resorting to using a ruler). The second assumption eliminates landmarks that may have been disturbed during the scan, such as the tragus of the ear as the subject would be wearing some sort of hearing protection.

Typical landmarks that have been used routinely in neurosurgery and are equally applicable here are the nasion, tip of the nose (if it looks unambiguous), the super-tragal notch and the cantus of the eyes (Fig. 13). The eye-based landmarks are less attractive as many are reluctant to move a pointer close to the eyes.

2.5.2 Characterizing the Coil and Tracker

The coil's location is tracked by placing a tracked object (usually called a tracker) on the coil. The position sensor reports the position of the tracker to the navigation software. The software must apply an offset from the tracker location and orientation to translate it to a location that is relevant to us (i.e., the coil's center). In some cases, the manufacturer performs this measurement and the coil is provided with the tracker permanently fixed to the coil. In other cases, a calibration tool is provided to allow the end-user to perform this calibration. This has the advantage of easily supporting the use of multiple coils from multiple manufacturers, but requires care to perform properly.

The calibration tool varies by manufacturer, but the overall process is generally the same. Using a tool with a reference pin or a plate with a cross-hairs reference, place the coil on the tool (or the tool on the coil) such that the reference point indicator is at the desired reference point on the coil. At this time, activate the calibration function in the navigator to record this reference location. The calibration tool usually has a tracker built into it (with an a-priori measurement of the reference point's location) and the navigator uses this information to determine the coil tracker to coil reference point transform (${}^{\mathrm{coil}}_{\mathrm{tr}}T$).

It is important to keep in mind that the system assumes that the tracker is rigidly fixed to the coil. The navigator never "sees" the coil, only the tracker and assumes that the tracker's location w.r.t. the coil is constant. If for some reason the tracker is moved (or you suspect to may have moved for any reason), it is important to repeat the calibration. Any error here will create a systematic error in all coil positioning.

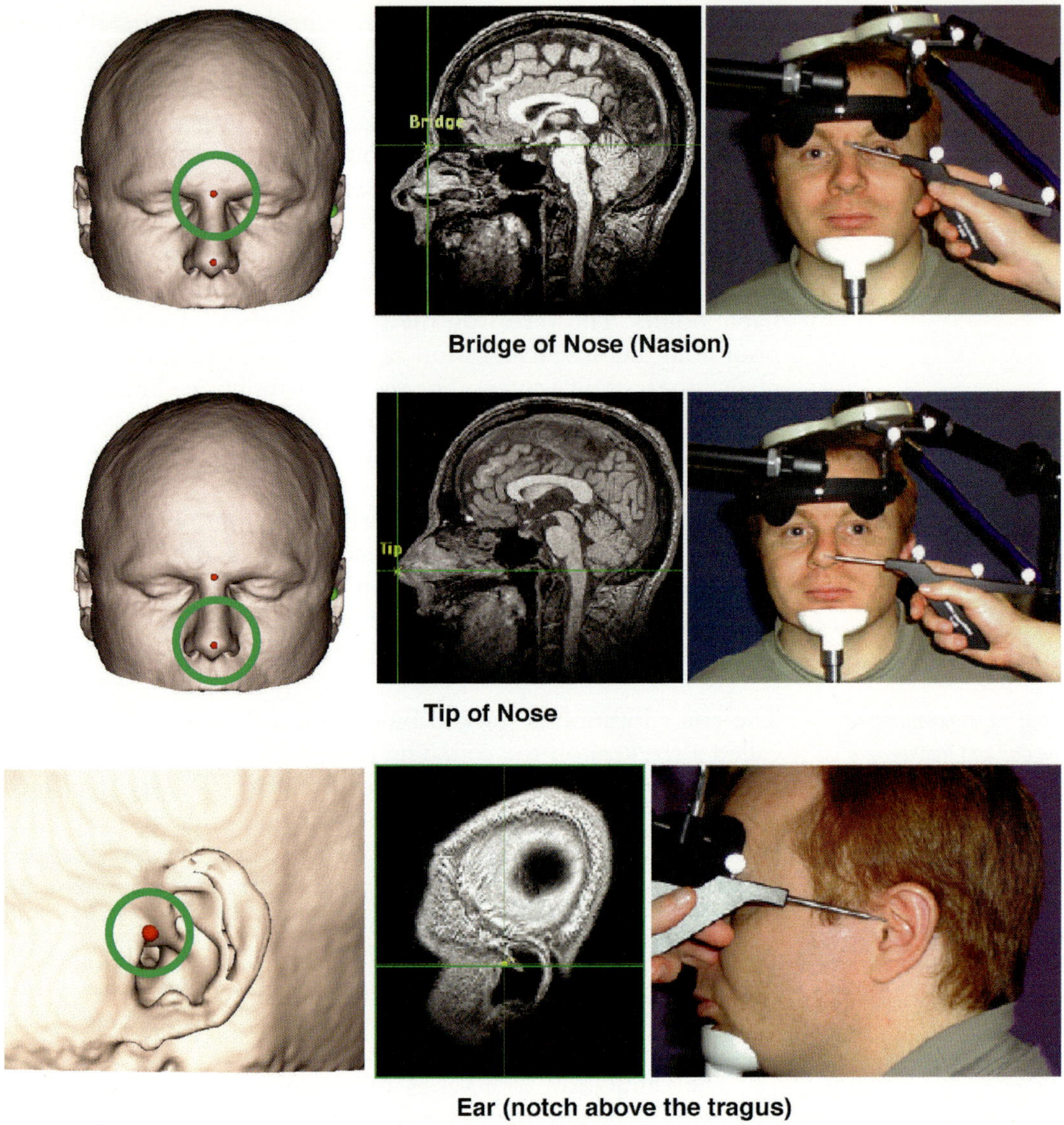

Fig. 13 Homologous points refer to the identification of the same discrete landmark in the images and on the subject. Typically the landmark is identified on the images as part of the data preparation process while the homologous landmarks are identified on the subject using a hand-held pointer tracked by the position sensor. The homologous point pairs (one in the image and the other in the subject reference space) are used to calculate the rigid body transform, usually using a least-squares method

2.6 Subject→Image Registration

Before being able to place the coil, the navigation system needs to be calibrated to be able to map the real-world coordinates of the lab to the image coordinates. This process is often referred to as subject→Image registration.

2.6.1 Registration Error

Evaluating the registration quality is difficult as the registration error varies throughout the head. Registration error is a combination of translation and rotation components. If the rotation error was 0 (only translation error), then the registration error would be a constant throughout the volume. If the translation error was 0 (only rotation), then the registration error would be 0 at the center of rotation and increase with increasing distance from that center. In practice, the error is a combination of the two, so evaluating the error can be somewhat subjective. Often a quantitative measure is given by looking at the distance between the anatomical landmarks in the images, and the coordinates of the landmarks as identified by the pointer (during the registration) after mapping them to the image space using the registration matrix itself. It is often helpful to bring the pointer back to the landmarks and visually assess consistency and to also verify by touching the skin at different locations, taking care not to use locations where the skin may have been distorted by the head fixation during the scan, or where MR image noise generated a distorted skin surface.

Different navigation systems have taken different approaches to perform the registration. In general, there are two methods, which are not necessarily mutually exclusive. First is using homologous anatomical landmarks and the second based on surface feature matching.

Homologous Point

Strengths: Relatively simple and fast to perform. When done well, it is sufficiently accurate.

Weaknesses: Relies on operator skill. Poor landmark selection on either the images or subject will lead to poor results.

With practice, registration accuracy of 3 mm or better can be achieved. Often, it is faster to repeat a poor homologous point registration than to resort to surface matching (next section).

Surface-Based Registration (Discrete Points or Line Profiles)

Surface-based registration is often used to attempt to improve the registration initially obtained by the homologous point method. In some navigation system, the step of identifying the landmarks in the images is eliminated (the system guesses approximate locations) because it assumes that any errors in the first step will be corrected using the surface registration method.

Advantages

- Potential to use more information for the registration and increase accuracy and reduce reliance on operator skill.

Disadvantages

- Increases the time to complete the registration step.
- In some situations, it can be difficult to do (particularly profile tracing), as the pointer must be traced along the skin surface. Any jittering of the pointer (hair makes this difficult to avoid) will lead to registration error. Obstacles including EEG electrodes or caps or head straps create more obstacles to work around.

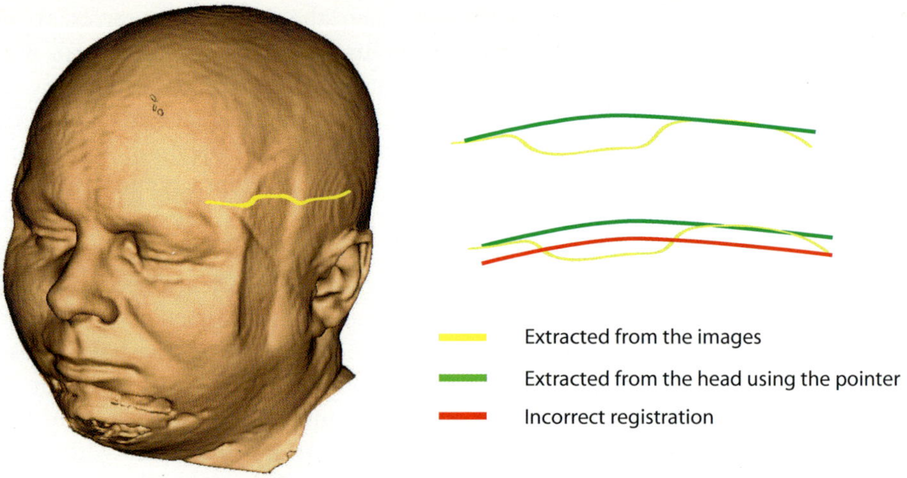

Fig. 14 Illustration of a skin deflection from the head fixation apparatus used during the MR image acquisition. Areas of significant deflection are not suitable for surface feature-based registration

- Assumes that the skin shape during the TMS session is the same as when it was scanned. This is similar to the homologous point constraint, but now applies to the skin surface regions that you trace with the pointer. If the subject's skin is distorted from the scanner head cushion, then the skin in that area is no longer a viable surface for registration.

Before tracing the skin with the pointer, take a moment to examine the skin displayed within the navigator. Look for signs of skin deflection due to the head restraint (Fig. 14). Avoid these areas because the shape you trace on the head now will be different than the model derived from the images.

Some navigation systems use discrete points on the skin as the refinement instead of tracings. These are less susceptible to jitter; however it is sometimes difficult to use as improving the fit in one area may lead to degradation in another region, and it is possible to end up "chasing your tail" as you go back and forth trying to improve regions that were degraded when you improved another region. It is important in all these cases to consider when to accept a registration. Registrations are never perfect and setting a goal of how good is good enough will save you a lot of time. Remember how big the volume that is typically stimulated and don't try to get submillimeter accuracy when a couple of mm is sufficient.

2.7 Representing the Coil Online

How the coil is represented on the screen is an active topic of research and the source of much debate. The holy grail of displays would be to display a map of the neurons that will be activated (activation map) from the stimulation given the characteristics and location of the coil, the stimulation parameters (pulse waveform, intensity), the subject's anatomy (head shape) as well as any factors that may influence the subject's stimulus threshold. The challenge is that to achieve this, several steps of modeling must be performed, each requiring certain assumptions to make the model work, and potentially many simplifications to make the model practical for use in typical TMS applications.

In order of certainty, the steps from coil position to activation map are:

Coil position→B field→E field→Neuronal activation map

Author disclosure: It should be noted that the display of the estimation of the induced electric field remains a topic of discussion and there is no consensus in both the scientific community and among navigation manufacturers as to the utility or even validity of such a display. It is important for users to understand the methodology used in any particular system and understand the pros and cons of a specific approach.

2.7.1 Position

The simplest and most reliable way to represent the coil is to represent its position and orientation. The only source of error is the error in the navigator itself, including the position sensor error and the largest component, subject-image registration error. This requires that the leap from position/orientation to neuronal activation be treated as a black box. This has been the case for TMS since it began prior to the use of neuronavigation.

2.7.2 Magnetic Field

In order to associate the magnetic field to the coil, it must be determined. This can be done using a magnetometer by stepping it to various positions in 3D space beneath the coil as it is fired to build up a volumetric field map. Alternatively, if the internal physical geometry of the coil is known, it can be modeled using a variety of 3D magnetic field modeling tools. In either case, if the field is to be used for display within a navigator, the field map must be co-registered to the physical coil so that when the coil's location is mapped to the MR space, the associated field map can also be mapped to the MR space. This is essential in order to display this field as well as if it is to be used as an input for current modeling. Currently, the hardware to perform this measurement is specialized and is performed by the coil or navigation manufacturer.

2.7.3 Online Modeling of the Electric Field

A complete review of this topic is beyond the scope of this chapter; however it is important to understand some of the fundamental aspects of these displays, as they will influence how you place your coil and thus your results.

Modeling of the electric field within the brain is an active field of research (see Wagner et. al [20] for a review). There are various techniques used to generate electric current maps from TMS, including spherical models [21, 22], Finite Element Modeling (FEM) [23–25], and Boundary Element Methods [23].

The references above are a small sample of the growing body of work in the field. While these provide examples of compelling information that can provide more insight into optimal coil placement, in the context of a real-time neuronavigator, it provides a set of challenges.

- Most of the modeling requires additional information, including segmentation of the compartments of the head derived from MRI, including scalp, skull, CSF, and brain (and often white/gray matter). As a research topic, this additional work is warranted. As a matter of practicality for the typical TMS installation, where MRI access is already a challenge, expecting the end-user to segment each brain may be ambitious.
- The computational burden for many modeling methods is well beyond the abilities of current computer hardware to calculate in an acceptable time (a few seconds per frame, let alone a few frames/second).
- Given the lack of consensus between the models (how do they compare?) and the limitations of each model, a given neuronavigator would have to pick one model over another, or implement multiple models to be unbiased. Given the challenges of any one model to implement, implementing multiple models is impractical.
- Validating the accuracy of any given model is difficult, particularly when considering the wide areas of application by the general TMS community. How would one validate a model to a degree where one can be confident of the results? How does one handle pathology? Can a typical TMS user be considered competent in segmenting and MRI for a brain lesion, or should a radiologist be consulted to help prepare the subject data? Can MRI be sufficient to delineate CSF from other soft-tissue lesions or a stroke?

Despite these challenges, multiple navigation systems offer the ability to display their model of induced electric field and provide compelling visual displays. Before relying on these displays however, it is important to understand the details of the models employed to obtain an understanding as to where they are expected to be valid and where they may not be accurate. Some of the questions one should ask are:

- What modeling method is used? Are there any peer-reviewed papers describing the methods and how they were validated?
- What are the inputs used for the coil's magnetic field? Do they simulate the field based on the coil geometry, or have they measured it? If measured, do they measure each coil individually, or a representative coil for a given model? If a representative coil was measured, has a validation been performed to ensure the consistency of the field for multiple coils of the same model?

- Does the model take into account the conductivities of the scalp, skull, CSF, Gray and White Matter? If so, are they taken from a segmentation of the subject-specific MRI, or from a standard head model fit to the individual? Do they validate their model for various pathologies?

Even if the current were modeled correctly, it is important to note that the current is not the final step. The orientation of the neuron and the current has a significant effect on the resulting stimulation. Just rotate the coil about its center over the motor cortex to see this effect. As you rotate the coil, does the change in current displayed accurately predict the change in MEP response?

2.7.4 What We Would Really Want: Predictor of Stimulated Neurons

If the induced current within the brain were accurately modeled (a big if), then the final step would be to decide how the brain would react to this current. To take this final step, the orientation of the neurons may be required (diffusion tractography), or some additional modeling to simplify the assumed interaction between the current and neurons would be required. Finally, other parameters that are not dependent on the current/neuron coupling may have to be considered (e.g., drugs that change the overall level of cortical excitability, TMS pulse shape).

2.8 Recording Experimental Data

One of the key roles of the neuronavigation system is to bring the coil information into the subject's image space, which is where all the data ultimately will end up. For example, functional activation maps from fMRI, PET, MEG, and EEG are now commonly being mapped to the anatomical image space. If one considers TMS as a probe, it is becoming more common to employ more sophisticated acquisition techniques to capture the results of the TMS pulse(s). For example, it has become commonplace to acquire EMG to perform a more objective measure of the resting motor threshold. TMS is also now being performed with simultaneous EEG or fMRI or fNIRS. As these have evolved, the role of the neuronavigator has also evolved.

Many neuronavigators support the simultaneous recording of the TMS location with EMG and EEG. This simultaneous recording requires coordination between the navigation software and the measurement equipment. Typically, the equipment is manufactured by the same company, or manufacturers have collaborated to allow for the sharing of data between their systems. In either case, the result is the ability to view and analyse more information in the same MR space, simplifying the viewing of the results and allowing for more sophisticated analysis by making the anatomical information from the MR images available for the analysis of the other data (e.g., subject-specific EEG analysis).

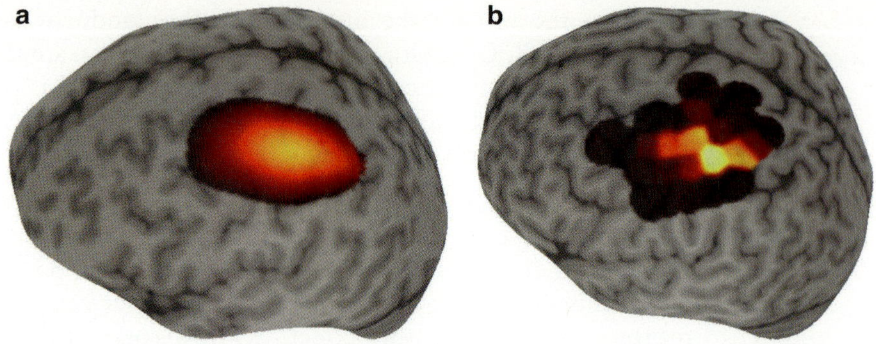

Fig. 15 (**a**) Curvilinear with motor map overlaid using a Gaussian blurring with FWHM set to twice the sample spacing. (**b**) FWHM set to 5 mm, which is smaller than the sample spacing, making the distribution difficult to interpret

2.8.1 Example: MEP Mapping

Here is one simple example of how a neuronavigator acquired both the TMS coil position and orientation and tagged it with the subsequent EMG data. MEP values were extracted from each EMG waveform. The MEP values and coil position and orientation were used to generate an interpolated MEP map that was then displayed on the 3D curvilinear display (Fig. 15). This type of display is becoming a more common feature in navigation systems as this concept of system integration continues. It is important for the end-user to understand the underlying methods used to generate these displays. In this example, the interpolation is generated by using a Gaussian smoothing algorithm. Changing the full width at half max (FWHM) of the Gaussian kernel will change the interpolated display. If one were using this technique to detect changes in the MEP map, then it would be important to maintain the same parameters in the mapping procedure for both sessions.

3 Looking Ahead: Robotic Positioning

While at the time of this writing, commercial robotic positioners were in the process of being introduced; however some were in the prototype stage. Robotic positioning of the TMS coil is not a new concept [26–30]; however with the introduction of commercially available systems, the technique will no doubt become more popular. In the context of navigation, a navigator will ultimately control any robotic system. Instead of an operator following the instructions of the computer, another computer in the robotic system will follow those same instructions. This will have the following benefits:

- Remove inter-operator variability in coil placement.
- More stable coil placement (assuming the robot can compensate for head movement).
- Potentially faster, or at least predictable coil placement (useful in clinical context).

The promise of a robot is to simply place the coil exactly where you want it and hold it there for as long as you want while compensating for head movement. As with any technology, a few caveats bear mention:

- Cost (robots are expensive).
- Safety. The human has a sense of touch and can easily ensure that the coil is gently held on the head. It is important that any robotic system has relevant safety features built-in to detect contact with the head, be it intended contact (you normally touch the head with the coil), or unwanted contact.
- Comfort. The robot should be able to hold the coil without excessive head fixation. It should also have head movement compensation. This should include the ability of the head to push against the coil and have the coil retreat as one pushes against it (to avoid the coil pinning the head in an uncomfortable location after a head twitch for example).
- Regulatory. Currently, the regulatory requirements for robotics in TMS are ambiguous and evolving. Early adopters of robotic positioning should take extra care to ensure that safety mechanisms incorporated in the robot are adequate and conform to current and potentially new regulatory requirements.

4 Summary

This chapter provided you with a small taste of the typical application of neuronavigation as applied to TMS coil positioning. Neuronavigation has evolved tremendously since its first applications over 10 years ago. With the advent of more compatible imaging modalities (e.g., EEG, EMG, fNIRS) improvements in modeling, robotic positioners, the future promises more changes and improved utility. The reader is encouraged to review the current state of the art as it is reported in the literature as well as exhibits at relevant scientific conferences.

References

1. Gugino LD et al (2001) Transcranial magnetic stimulation coregistered with MRI: a comparison of a guided versus blind stimulation technique and its effect on evoked compound muscle action potentials. Clin Neurophysiol 112:1781–1792
2. Paus T (1998) Imaging the brain before, during, and after transcranial magnetic stimulation. Neuropsychologia 37:219–224
3. Comeau R, Peters TM, Paus T (1999) Optically based frameless stereotaxy for image guided transcranial magnetic stimulation (TMS). Neuroimage 9(6):S225
4. Neggers SF et al (2004) A stereotactic method for image-guided transcranial magnetic stimulation validated with fMRI and motor-evoked potentials. Neuroimage 21:1805–1817
5. Hannula H et al (2005) Somatotopic blocking of sensation with navigated transcranial magnetic stimulation of the primary somatosensory cortex. Hum Brain Mapp 26: 100–109

6. Julkunen P et al (2009) Comparison of navigated and non-navigated transcranial magnetic stimulation for motor cortex mapping, motor threshold and motor evoked potentials. Neuroimage 44:790–795
7. Schonfeldt-Lecuona C et al (2005) Accuracy of stereotaxic positioning of transcranial magnetic stimulation. Brain Topogr 17:253–259
8. Sparing R, Buelte D, Meister IG, Paus T, Fink GR (2008) Transcranial magnetic stimulation and the challenge of coil placement: a comparison of conventional and stereotaxic neuronavigational strategies. Hum Brain Mapp 29: 82–96
9. Mazziotta JC, Toga AW, Evans A, Fox P, Lancaster J (1995) A probabilistic atlas of the human brain: theory and rationale for its development: the international consortium for brain mapping (ICBM). Neuroimage 2:89–101
10. Cantarero G et al (2011) Disrupting the ventral premotor cortex interferes with the contribution of action observation to use-dependent plasticity. J Cogn Neurosci 23:3757–3766
11. Dormal V, Andres M, Pesenti M (2012) Contribution of the right intraparietal sulcus to numerosity and length processing: an fMRI-guided TMS study. Cortex 48(5):623–629
12. Andres M, Pelgrims B, Michaux N, Olivier E, Pesenti M (2011) Role of distinct parietal areas in arithmetic: an fMRI-guided TMS study. Neuroimage 54:3048–3056
13. Beauchamp MS, Nath AR, Pasalar S (2010) fMRI-Guided transcranial magnetic stimulation reveals that the superior temporal sulcus is a cortical locus of the McGurk effect. J Neurosci 30:2414–2417
14. Bastos AC et al (1999) Diagnosis of subtle focal dysplastic lesions: curvilinear reformatting from three-dimensional magnetic resonance imaging. Ann Neurol 46:88–94
15. Talairach J, Tournoux P (1988) Co-planar stereotaxic atlas of the human brain: 3-dimensional proportional system: an approach to cerebral imaging. Thieme, New York
16. Friston KJ, Ashburner J, Kiebel SJ, Nichols TE, Penny WD (2007) Statistical parametric mapping: the analysis of functional brain images. Academic, London, p vii 647
17. Lotze M (2010) Functional lesions of the motor system with TMS–a challenge for individual functional mapping. Restor Neurol Neurosci 28:469–476
18. Thakral PP, Slotnick SD (2011) Disruption of MT impairs motion processing. Neurosci Lett 490:226–230
19. Sack AT et al (2009) Optimizing functional accuracy of TMS in cognitive studies: a comparison of methods. J Cogn Neurosci 21: 207–221
20. Wagner T, Rushmore J, Eden U, Valero-Cabre A (2009) Biophysical foundations underlying TMS: setting the stage for an effective use of neurostimulation in the cognitive neurosciences. Cortex 45:1025–1034
21. Ruohonen J, Ilmoniemi RJ (2001) Physical principles for transcranial magnetic stimulation. In: Handbook of transcranial magnetic stimulation. . Edited by A. Pascual-Leone, N.J. Davey, J. Rothwell, E.M. Wasseran, B.K. Puri, Arnold, London. ISBN 0340720093. p 17–29
22. Thielscher A, Kammer T (2002) Linking physics with physiology in TMS: a sphere field model to determine the cortical stimulation site in TMS. Neuroimage 17:1117–1130
23. Cho YS, Suh HS, Lee WH, Kim TS (2010) TMS modeling toolbox for realistic simulation. Conf Proc IEEE Eng Med Biol Soc 2010: 3113–3116
24. Salvador R, Silva S, Basser PJ, Miranda PC (2011) Determining which mechanisms lead to activation in the motor cortex: a modeling study of transcranial magnetic stimulation using realistic stimulus waveforms and sulcal geometry. Clin Neurophysiol 122:748–758
25. Windhoff M, Opitz A, Thielscher A (2013) Electric field calculations in brain stimulation based on finite elements: an optimized processing pipeline for the generation and usage of accurate individual head models. Hum Brain Mapp 34:923–935
26. Kantelhardt SR et al (2010) Robot-assisted image-guided transcranial magnetic stimulation for somatotopic mapping of the motor cortex: a clinical pilot study. Acta Neurochir (Wien) 152:333–343
27. Lancaster JL et al (2004) Evaluation of an image-guided, robotically positioned transcranial magnetic stimulation system. Hum Brain Mapp 22:329–340
28. Matthaus L, Giese A, Wertheimer D, Schweikard A (2006) Planning and analyzing robotized TMS using virtual reality. Stud Health Technol Inform 119:373–378
29. Richter L, Bruder R, Schlaefer A, Schweikard A (2010) Towards direct head navigation for robot-guided transcranial magnetic stimulation using 3D laserscans: Idea, setup and feasibility. Conf Proc IEEE Eng Med Biol Soc 2010: 2283–2286
30. Zorn L, Renaud P, Bayle B, Goffin L, Lebossé C, de Mathelin M, Foucher J (2012) Design and evaluation of a robotic system for transcranial magnetic stimulation. IEEE Trans Biomed Eng 59(3):805–815

Chapter 4

Reaching Deep Brain Structures: The H-Coils

Yiftach Roth and Abraham Zangen

Abstract

The capacity of standard transcranial magnetic stimulation (TMS) to elicit neuronal responses is limited to superficial structures. The deep TMS technology enables direct stimulation of deeper and larger brain volumes, potentially affecting extensive neuronal pathways including deeper cortical regions and fibers targeting subcortical regions, without a significant increase of the electric field induced in superficial cortical layers. This chapter includes theoretical background and design principles of deep TMS H-coils, comparison of standard TMS and deep TMS coils electric field profiles, as well as description and maps of electric field distribution of several deep TMS H-coils.

Key words Deep transcranial magnetic stimulation, Deep brain structures stimulation, Magnetic coils, Electric field

1 Background and Theoretical Foundations

Until several years ago, the capacity of transcranial magnetic stimulation (TMS) to elicit neuronal responses was limited to superficial structures. The typical coils used for TMS (such as round or figure-8 coils) induce stimulation in superficial cortical regions under the windings of the coil as the intensity of the electric field drops dramatically as a function of distance from the coil [1–4]. Therefore, to stimulate deep brain regions with such coils, a very high intensity would be required which is not feasible with standard magnetic stimulators. Moreover, the intensity required to stimulate deeper brain regions using typical coils would stimulate both cortical regions and facial nerves at high level which may lead to facial pain, facial and cervical muscle contractions, epileptic seizures, and other undesirable side effects [5].

The difficulty of efficiently activating deep neuronal structures using TMS emerges from both physical properties of the brain and physical and physiological aspects of the interaction of a TMS system with the human brain. As shown by Heller and Van Hulsteyn [6], the three-dimensional maximum of the electric field intensity

Alexander Rotenberg et al. (eds.), *Transcranial Magnetic Stimulation*, Neuromethods, vol. 89,
DOI 10.1007/978-1-4939-0879-0_4, © Springer Science+Business Media New York 2014

will always be located at the brain surface, regardless of configuration or superposition of TMS coils. However, both the TMS coil and the stimulator may be optimized for effective stimulation of deeper brain regions.

A family of coil designs for stimulation of deeper brain areas, termed H-coils, has been proposed and evaluated [5, 7, 8]. Each of these coils are based on common design principles essential for effective deep brain TMS. Yet each specific H-coil must have a unique design and configuration based on the location and size of the deep brain region/s intended to be activated and the preferred direction/s of stimulation.

The construction of deep TMS coils should meet several goals:

(a) High enough electric field intensity in the desired deep brain region that will surpass the threshold for neuronal activation.
(b) High percentage of electric field in the desired deep brain region relative to the maximal intensity in the cortex.
(c) Minimal adverse effects, such as pain, motor activation, and activation of facial muscles.

The design principles essential for effective stimulation of deeper brain regions include the following [5, 7, 8]:

1. *Summation of electric impulses.* The induced electric field in the desired deep brain regions is obtained by optimal summation of electric fields, induced by several coil elements with common directions in different locations around the skull. These coil elements are placed in various locations around the head, in such a configuration as to create high electric field intensity in the target deep brain region which, at the same time, is a high percentage of the maximal electric field at the brain surface.
2. *Minimization of non-tangential components.* Coil elements which are non-tangential to the surface induce accumulation of surface charge, which leads to the cancelation of the perpendicular component of the directly induced field at all points within the tissue and, typically, to the reduction of the electric field in all other directions (Tofts [1]; Tofts and Branston [3]; Eaton [4]; Roth et al. [7]). In order to reduce accumulation of electrostatic charge, non-tangential elements in the coils are minimized, especially around the stimulation target.
3. *A base conforming to the portion of the head on which it rests.* The base should preferably have certain degree of flexibility in order to be compatible to the head at the required position and orientation. In certain H-coils, two or more bases may be included, each conforming to the head at a certain required portion.
4. *Proper array of stimulating coil elements.* Optimal and effective summation in depth requires that stimulating coil elements

with common orientation in the base are dispersed rather than forming a dense organization. They may be organized in several groups or may be sparsely organized over the base.

5. *Proper orientation of stimulating coil elements.* Coils must be oriented such that they will produce a considerable field in a desired direction tangential to the surface, which should also be the preferable direction to activate the neuronal structures under consideration.
6. *Remote location of return paths.* The wires leading currents in a direction opposite to the preferred direction (the return paths) should be located far from the base and the desired brain region, either adjacent to secondary head regions or remote from the head. This enables a higher absolute electric field in the desired brain region.

H-coils are usually encased in a helmet. The H-coils can be positioned over the motor cortex for determination of the individual motor threshold, which is used to establish the treatment stimulation intensity. Then the coil helmet is navigated to the treatment position where the coil is attached to the head for treatment administration. The coil flexibility enables good attachment to the head and compatibility with various head sizes and shapes.

For double blind placebo-controlled (DBPC) clinical trials, the H-coils helmet can encase a sham coil which can mimic the acoustic artifact and scalp sensations of the active coil, without inducing any significant field anywhere in the brain.

2 Comparison of Deep and Standard TMS Coils Field Profiles

A comparison of the electric field profile along a line going from the coil surface into the center of a realistic phantom head model is shown in Fig. 1 for five different TMS coils: a commercial figure-8 coil (70-mm diameter of each wing), commercial double cone coil (120-mm diameter of each wing, with an opening angle of 95°), large and small custom circular coils (with diameters of 160 and 55 mm, respectively), and a version of the H-coil which was used in a previous study [8]. The electric field distribution was measured in a realistic model of the human head ($x \times y \times z = 15 \times 13 \times 18$ cm, where x, y and z are postero-anterior, right-left and inferior-superior axes, respectively), filled with physiological saline solution.

A sketch of the H-coil version is shown in Fig. 1a. The coil has 10 strips carrying a current in a common direction (postero-anterior direction) located around the desired target site (segments A–B and G–H in Fig. 1a). The average length of the strips is 110 mm. The only coil elements with radial current components are those connected to the return paths of five strips (segments C–I

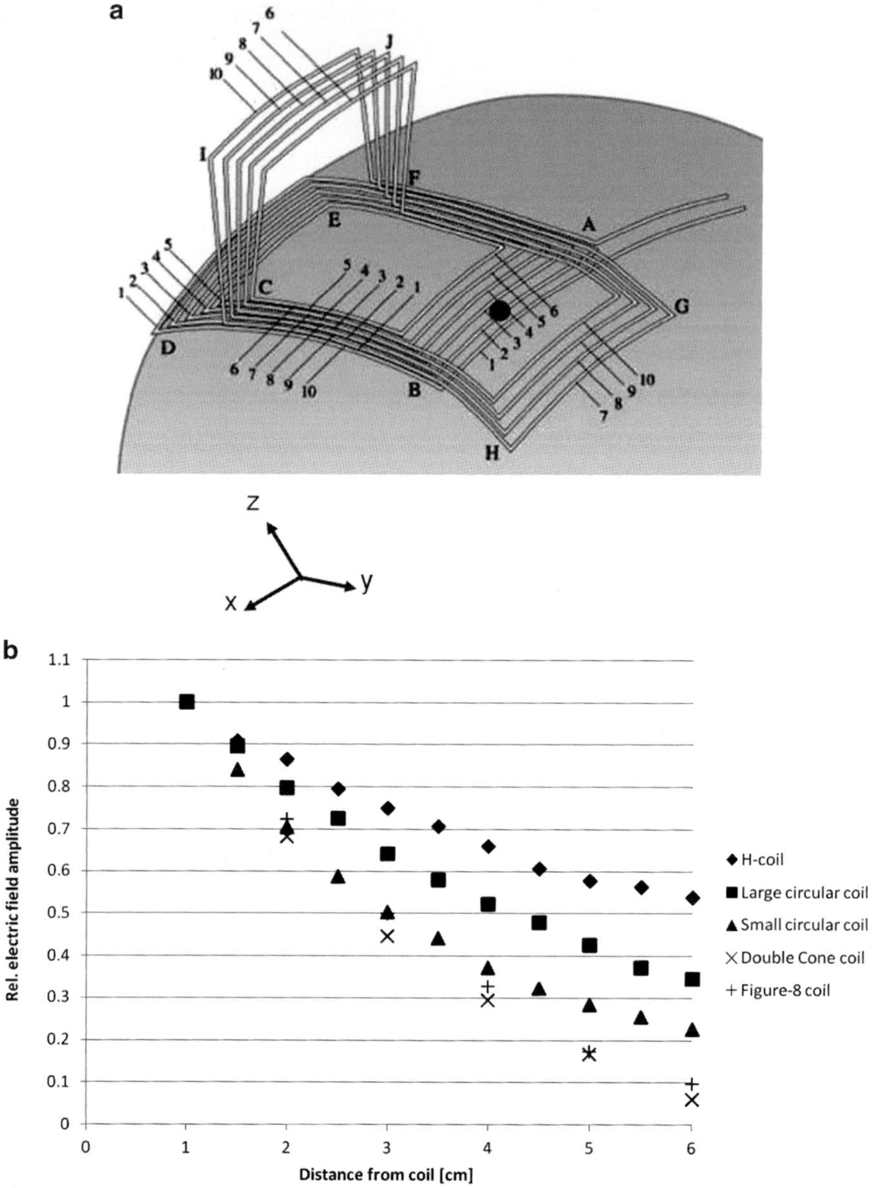

Fig. 1 (**a**) A sketch of the H-coil version used in this comparison. (**b**) Plots of the electric field amplitude induced by several TMS coils, as a function of distance from the coil, normalized to the amplitude at 1 cm distance. The electric field was calculated in a phantom head model filled with a saline solution with physiologic concentration. The coils are a H-coil version (*diamonds*), a large circular coil with 160 mm average diameter (*squares*), a small circular coil with 55 mm average diameter (*triangles*), a commercial figure-8 coil having 70 mm diameter of each wing (+'s), and a commercial double cone coil with 120 mm diameter of each wing (X's)

and J–F in Fig. 1a). The length of these wires is 80 mm. The return paths of the other five strips are placed on the contralateral hemisphere of the head (segment D–E in Fig. 1a). The wires connecting the strips and the return paths (segments B–C and F–A) are on average 90 mm long.

The electric field amplitudes for all coils were calculated along a line going downward (z axis in Fig. 1a) with coordinates of $(x, y) = (0, 3)$, (i.e., 3 cm laterally to the midline). For the figure-8 coil and the double cone coil, the line started at the coil center. For the two circular coils, the line started at the coil edge. For the H-coil, the line started at the center of elements A–B (Fig. 1a).

From the plot in Fig. 1b it can be seen that the H-coil has the most favorable field profile (i.e., field attenuation to 66 % at a distance of 4 cm). Among the rest of the coils, the large circular coil has the slowest rate of field decay with distance (i.e., field attenuation to 52 % at a distance of 4 cm). Yet the circular coil induces a nonspecific effect over a complete cortical ring underneath the coil windings.

The other three coils present a much weaker depth penetration with a strong attenuation of the electric field with depth (i.e., field attenuation to 29–37 % at a distance of 4 cm). Note that the double cone coil is much larger than the Figure-8 coil, and produces significantly higher absolute electric field amplitude at any distance. Yet the rate of decay of the field with distance from coil is similar between the two coils. This demonstrates that the coil size is not the only factor affecting the efficiency in activating deeper brain regions, and the principles detailed above must be accounted for.

3 Examples of Deep TMS H-Coils

Several examples of H-Coils are given below. The HAC coil, which is currently being used in several clinical studies around the world, is designed to stimulate medial frontal neuronal structures including the anterior cingulate cortex. The HAC coil is shown in Fig. 2a at a typical treatment position over a human head. The main stimulating elements are arranged in two groups separated by a gap and dispersed above medial brain regions in order to obtain optimal effect at deep medial cortical structures. The return paths are located over lateral regions on both sides. The coil has a flexible base intended to enable the coil elements to conform to various head shapes and sizes in order to minimize electrostatic charge accumulation. The bottom cover of the coil is made of soft biocompatible material to allow for this flexibility and attachment straps facilitate proper attachment to the head at the treatment position.. Maps of the electric field distribution induced by the HAC coil in the brain superimposed on coronal slices of anatomical MR images are shown in Fig. 2b.

The H1 coil was used in several clinical trials [9–14], including a large prospective multicenter study which included over 200 major depression patients and formed the basis for FDA approval of Deep TMS device for the treatment of major depression in the USA. The H1 coil is designed to induce activation of left and right lateral and medial prefrontal cortex structures, with a preference to

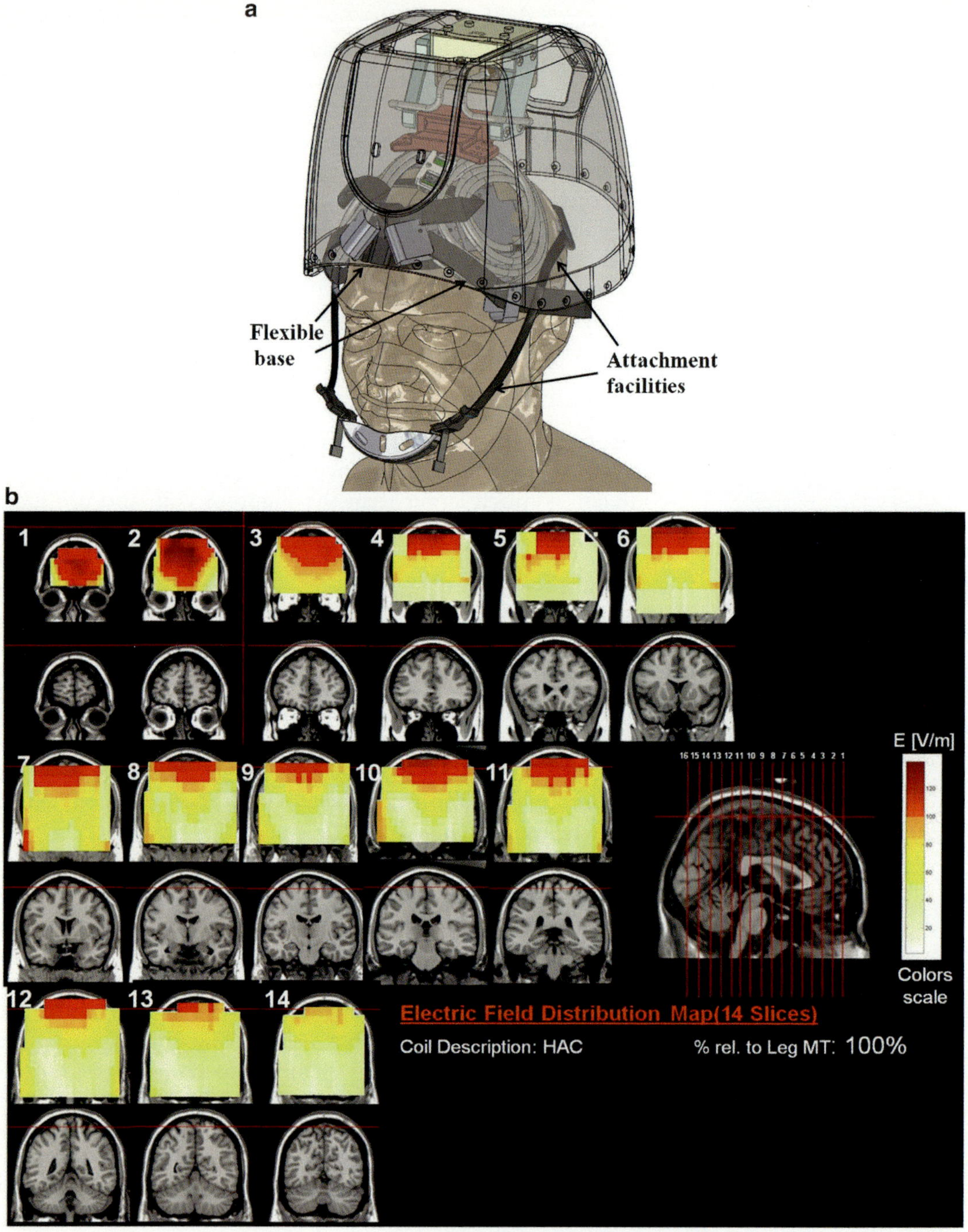

Fig. 2 (**a**) A sketch of the HAC coil at the treatment position over a human head. (**b**) Colored field maps for the HAC coil indicating the electric field absolute magnitude in each pixel over 14 coronal slices 1 cm apart. The *red pixels* indicate field magnitude above the threshold for neuronal activation, which was set to 100 V/m. The field maps are adjusted for stimulator power output level required to obtain 100 % of the leg motor threshold, at a depth of 3 cm

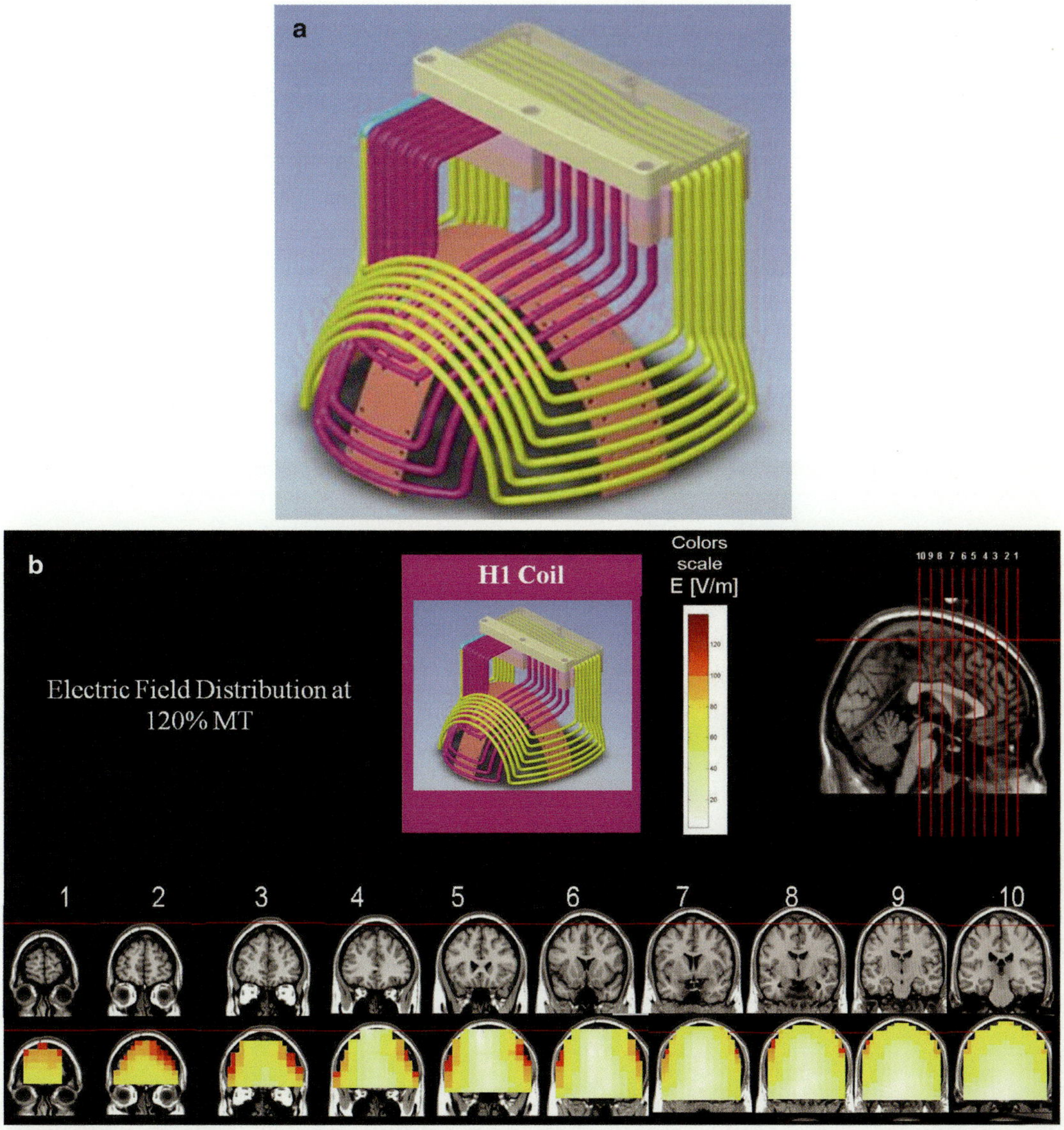

Fig. 3 (**a**) A sketch of the H1 coil. (**b**) Colored field maps for the H1 coil indicating the electric field absolute magnitude in each pixel over 10 coronal slices 1 cm apart. The *red pixels* indicate field magnitude above the threshold for neuronal activation, which was set to 100 V/m. The field maps are adjusted for stimulator power output level required to obtain 120 % of the hand motor threshold, at a depth of 1.5 cm

the left hemisphere. A sketch of the H1 coil is shown in Fig. 3a. Electric field distribution maps of the H1 coil superimposed on coronal slices of anatomical MR images are shown in Fig. 3b.

The HLRINS coil, which is currently being used in several clinical studies around the world, is designed to stimulate bilaterally neuronal structures in the insular and entorhinal cortex. A sketch of the HLRINS coil is shown in Fig. 4a. Maps of the electric

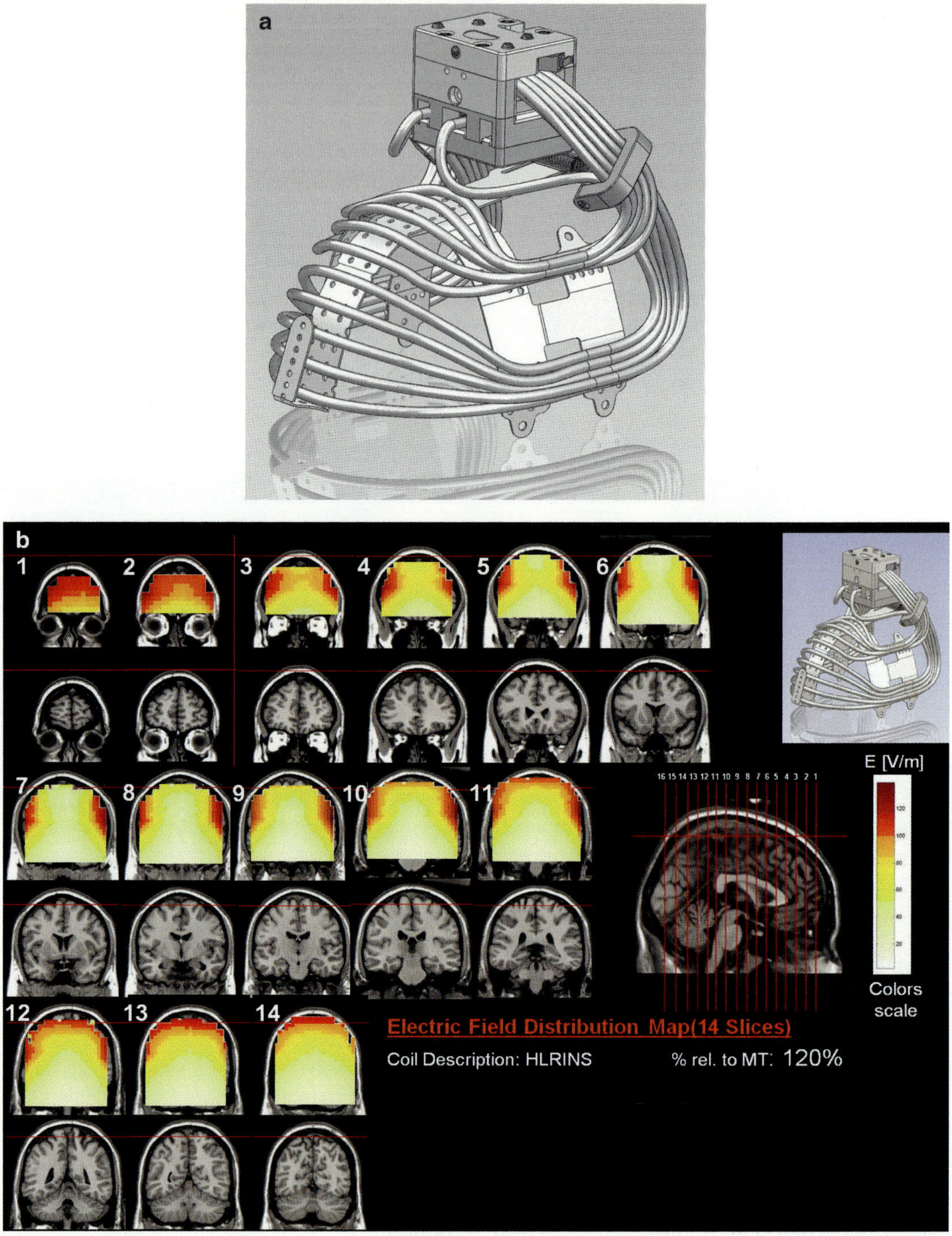

Fig. 4 (**a**) A sketch of the HLRINS coil. (**b**) Colored field maps for the HLRINS coil indicating the electric field absolute magnitude in each pixel over 14 coronal slices 1 cm apart. The *red pixels* indicate field magnitude above the threshold for neuronal activation, which was set to 100 V/m. The field maps are adjusted for stimulator power output level required to obtain 120 % of the hand motor threshold, at a depth of 1.5 cm

field distribution induced by the HLRINS coil in the brain superimposed on coronal slices of anatomical MR images are shown in Fig. 4b.

Conflict of Interest

The authors of this chapter are key inventors of deep TMS H-coil systems produced by Brainsway. The authors serve as consultants for and have financial interests in Brainsway.

References

1. Tofts PS (1990) The distribution of induced currents in magnetic stimulation of the brain. Phys Med Biol 35:1119–1128
2. Maccabee PJ et al (1990) Spatial distribution of the electric field induced in volume by round and figure '8' magnetic coils: relevance to activation of sensory nerve fibers. Electroencephalogr Clin Neurophysiol 76: 131–141
3. Tofts PS, Branston NM (1991) The measurement of electric field, and the influence of surface charge, in magnetic stimulation. Electroencephalogr Clin Neurophysiol 81:238–239
4. Eaton H (1992) Electric field induced in a spherical volume conductor from arbitrary coils: application to magnetic stimulation and MEG. Med Biol Eng Comput 30: 433–440
5. Roth Y, Padberg F, Zangen A (2007) Transcranial magnetic stimulation of deep brain regions: principles and methods. In: Marcolin M, Padberg F (eds) Transcranial stimulation as treatment in mental disorders. Advances in biological psychiatry, vol 23. Karger Publishers Zürich, Switzerland, pp 204–224
6. Heller L, van Hulsteyn DB (1992) Brain stimulation using electromagnetic sources: theoretical aspects. Biophys J 63:129–138
7. Roth Y, Zangen A, Hallett M (2002) A coil design for transcranial magnetic stimulation of deep brain regions. J Clin Neurophysiol 19: 361–370
8. Zangen A, Roth Y, Voller B, Hallett M (2005) Transcranial magnetic stimulation of deep brain regions: evidence for efficacy of the H-coil. Clin Neurophysiol 116:775–779
9. Levkovitz Y, Roth Y, Harel EV, Braw Y, Sheer A, Zangen A (2007) A randomized controlled feasibility and safety study of deep transcranial magnetic stimulation. Clin Neurophysiol 118:2730–2744
10. Levkovitz Y, Harel EV, Roth Y et al (2009) Deep transcranial magnetic stimulation of the prefrontal cortex—effectiveness in major depression. Brain Stimul 2:188–200
11. Harel EV, Zangen A, Roth Y, Reti I, Braw Y, Levkovitz Y (2011) H-coil repetitive transcranial magnetic stimulation for the treatment of bipolar depression: an add-on, safety and feasibility study. World J Biol Psychiatry 12: 119–126
12. Isserles M, Rosenberg O, Dannon P, Lerer B, Zangen A (2011) Cognitive emotional reactivation during deep transcranial magnetic stimulation over the prefrontal cortex of depressive patients affects antidepressant outcomes. J Affect Disord 128:235–242
13. Levkovitz Y, Rabany L, Harel EV, Zangen A (2011) Deep transcranial magnetic stimulation add-on for treatment of negative symptoms and cognitive deficits of schizophrenia: a feasibility study. Int J Neuropsychopharmacol 14:991–996
14. Harel EV, Rabany L, Deutsch L, Bloch Y, Zangen A, Levkovitz Y (2012) H-coil repetitive transcranial magnetic stimulation for treatment resistant major depressive disorder: an 18-week continuation safety and feasibility study. World J Biol Psychiatry 2012 Feb 7. [Epub prior to print]

Part II

Transcranial Magnetic Stimulation Methods

Chapter 5

Single-Pulse Transcranial Magnetic Stimulation (TMS) Protocols and Outcome Measures

Faranak Farzan

Abstract

The integrity of the brain circuitries can be quantified through application of single pulse transcranial magnetic stimulation (TMS) to the cortex and concurrent recording of the neurophysiological response through electromyography at the periphery, and more recently centrally through electroencephalography and functional magnetic resonance imaging. In this chapter, we first introduce the neurophysiological techniques and the corresponding outcome measures that have been employed to capture the neural response to a single TMS pulse. We then describe the physical and physiological factors that can affect the TMS outcome measures and should be carefully controlled for. We then introduce single pulse TMS protocols motor threshold, contralateral silent period, ipsilateral silent period, and the input/output curve. We describe the neurobiological mechanisms and brain circuitries that each protocol examines, and provide a step-by-step guideline for conducting each protocol. Finally, we provide an overview of the application of single pulse TMS protocols in basic and cognitive neuroscience, and clinical investigations.

Key words Motor threshold, Contralateral silent period, Ipsilateral silent period, Input/output curve, Electromyography, Electroencephalography, Inhibition, Excitation, Motor cortex

1 Introduction

The assessment and modification of brain dynamics can be achieved through single pulse transcranial magnetic stimulation (TMS) in combination with electrophysiological and/or neuroimaging techniques. Single pulse TMS was introduced in the 1980s to study the motor system. Historically, therefore, single pulse TMS has been applied to the motor system at stimulation intensities that produce a motor-evoked response (MEP) at the periphery measurable by surface electromyography (EMG). To date, based on the EMG outcomes, several single pulse TMS protocols have been designed to investigate the neurophysiology of the motor system. Among these protocols, motor threshold (MT), contralateral silent period (cSP), ipsilateral silent period (iSP), and the input/output (I/O) curve are the focus of our discussion in this chapter.

Alexander Rotenberg et al. (eds.), *Transcranial Magnetic Stimulation*, Neuromethods, vol. 89,
DOI 10.1007/978-1-4939-0879-0_5, © Springer Science+Business Media New York 2014

In order to utilize single pulse TMS protocols to study the brain, it is important to understand the mechanisms underlying a biological response to TMS. While the effects of TMS at the cellular level are not yet completely understood, the biological correlates of the EMG outcomes in single pulse TMS protocols have been investigated in vivo via pharmacological intervention. In addition, neuroimaging modalities, such as positron electron tomography (PET) and near-infrared spectroscopy (NIRS), may be interleaved with TMS to more directly investigate the contribution of certain neurotransmitter systems to TMS-induced responses. Beyond the biological correlates of the EMG outcomes, the impact of different TMS parameters on tissue and subsequent EMG response, as well as the effect of different biological parameters on the TMS electrical field distribution have been investigated to better characterize the interaction between a single TMS pulse and the underlying tissue. Collectively, the results of these endeavors have demonstrated that various stimulation parameters and biological properties can influence TMS outcome measures and, thus, can influence the results of single pulse TMS protocols.

Today, due in large part to technological advancements, several other electrophysiological and neuroimaging techniques can be combined with single pulse TMS to examine brain areas and neural processes outside the motor system. For example, the assessment of the prefrontal cortex can now be achieved through quantifying TMS evoked potentials (TEPs) by interleaving TMS with electroencephalography (EEG). In the motor cortex, this provides the advantage of concurrently recording EMG and EEG measures (e.g., MEP and TEP) and thereby better disentangling the spinal and cortical mechanisms that underlie the neural responses to TMS. This further allows for examining the motor system at intensities that do not result in a detectable motor response at the periphery. Additionally, functional magnetic resonance imaging (fMRI) interleaved with single pulse TMS can allow for the exploration of TMS effect on subcortical structures. The recordings of the brain response through EEG and fMRI also further facilitate the effort of characterizing TMS mechanism of actions. Therefore, these advances have expanded the application of single pulse TMS beyond the assessment of the motor system and local processes.

The aim of this chapter is twofold. The main purpose is to present the theoretical and practical grounds required to understand and utilize the single pulse protocols of MT, cSP, iSP, and I/O curve for clinical and experimental purposes. We will also discuss the application of single pulse TMS in the assessment and modulation of non-motor brain systems and functions. Therefore, in this chapter: (1) We introduce the neurophysiological techniques and the corresponding outcome measures that have been employed to capture the neural response to a single TMS pulse; (2) We describe the physical and physiological factors that can alter the TMS interaction with tissue and, thereby, the outcome measures of a single TMS pulse; (3) We describe the procedures, neurobiological correlates, and technical considerations associated with the single pulse TMS protocols of MT, cSP, iSP, and I/O curve for the motor system, and provide examples of single pulse TMS in examination of non-motor regions; and (4) We review several applications of single pulse TMS protocols in research and clinical practice.

2 Single Pulse TMS: From Electrical Energy to Biological Response

As described in previous chapters, the activation of neurons through TMS technique follows Faraday's principle of induction of electric current by a time-varying magnetic field. A simplified schematic of a TMS machine is an oscillating RLC circuit with a combination of low resistance cables (R), a bank of energy storing capacitors (C), and an inductor (L). The inductor is the TMS coil that the TMS operator holds over the subject's scalp. The TMS coil is typically made of one or two wound loops of copper wires housed in a plastic or ceramic casing. TMS coils are available in different geometry and sizes and each can produce electric fields of different pattern and focality. To apply a single pulse of TMS to the brain, the operator places the TMS coil tangential to the subject's scalp. Depending on the brain target and several other factors, such as the coil shape and type, the coil handle is oriented at a specific angle with respect to an underlying biological landmark.

When the operator delivers a pulse, the stored energy in the capacitors is released and an intense time-varying current pulse of several kilo Amperes runs though the coil. The time-varying current in the coil induces a short pulse of time-varying magnetic field perpendicular to the plane of the coil and the surface of the brain tissue. This time-varying magnetic field, in turn, induces an electric field parallel to the plain of the coil in nearby conductors, including the brain. This field generates a current in the tissue and can result in depolarization of cell membranes and initiation of action potentials. Therefore, the stored electrical energy in the TMS machine capacitors is ultimately transformed into an induced current in the brain without significant attenuation by tissues that have high electrical impedance (e.g., skull, dura).

While the cellular effect of TMS and the microscopic interaction between the electrical energy and the underlying biological tissue are not fully understood, the net macroscopic outcome can be captured at peripheral, subcortical, and cortical regions (Fig. 1). Since the 1980s, various electrophysiological and neuroimaging approaches have been utilized in online (simultaneous stimulation and recording) and off-line (stimulation followed by a delayed recording) manners to examine the local and remote effects of the

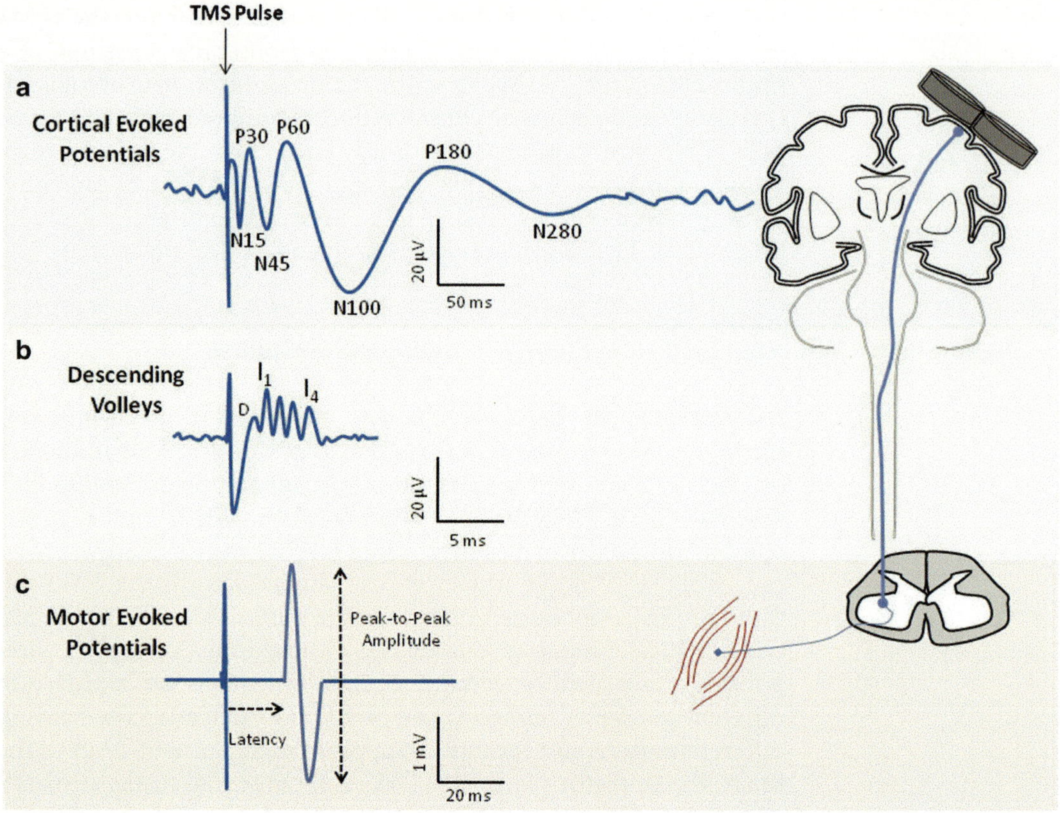

Fig. 1 Some outcome measures of a single pulse TMS in the motor cortex. (**a**–**c**) Represent the EEG, epidural, and EMG outcomes of a single suprathreshold TMS pulse applied to the primary motor cortex—as depicted on the right. In (**a**), N15, P30, N45, P60, N100, P180, and N280 indicate the positive and negative deflection at 15, 30, 45, 60, 100, 180, and 280 ms following the TMS pulse. Some of these peaks have been associated with cortical excitatory (N15, P30) and inhibitory mechanisms (N100). (**b**) Demonstrates the D- and I-waves in response to the TMS pulse that are recorded from the epidural space in patients with implanted electrodes, and are suggested to reflect the direct and transsynaptic activation of the corticospinal tract. (**c**) Demonstrates the latency and peak-to-peak amplitude of a motor-evoked potential (MEP) recorded from the target muscle following the TMS pulse

TMS-induced current. In this section, we will introduce some of the neuroimaging and electrophysiological tools that can capture the TMS-induced responses. Some of the neuroimaging approaches are discussed in more details in subsequent chapters.

3 TMS Outcome Measures

3.1 Motor Response at the Periphery Through Electromyography

When a TMS pulse is applied to the motor cortex, the effect of the pulse can be characterized by examining the properties of the compound muscle action potential (CMAP), recorded by surface EMG, which may be visually observed as a twitch in the target muscle (Fig. 1c). Conventionally, a muscle twitch that is produced as a result of motor cortex stimulation is referred to as motor-evoked potential (MEP). However, a TMS-induced MEP has more variability than a CMAP obtained through peripheral nerve stimulation. Also, a TMS-induced MEP has a longer duration and latency of onset and may have a more polyphasic shape compared to a peripherally induced CMAP. These differences have been attributed, in part, to intrinsic fluctuations of cortical excitability.

Distinct features of TMS-induced MEPs, such as the latency, magnitude, and area under the rectified MEP, can be used in single pulse TMS protocols to derive unique information about the integrity of corticospinal pathways. Latency refers to the time between the delivery of a single TMS pulse over the scalp and the appearance of an MEP at the periphery (Fig. 1c). Latency is mainly reflective of the number of synapses between the stimulation site and the peripheral target muscle, as well as the integrity of the white matter fibers, such as the diameter and the myelin sheath thickness. Latency is measured in units of milliseconds and varies across muscles, being shorter in the muscles of the upper extremities (such as facial muscles) and longer in the muscles of the lower extremities (such as leg muscles). MEP latency can be used to approximate the speed of the combined central and spinal conduction time.

The magnitude of an MEP is by far the most utilized outcome measure of single TMS pulse protocols to the motor cortex. MEP amplitude is often measured from the negative trough to the positive peak, a measure referred to as the *peak-to-peak amplitude* and represented in units of micro- or millivolts (Fig. 1c). MEP amplitude can also be measured from baseline EMG level to the first positive or negative deflection. Furthermore, the magnitude of an MEP can be measured as the area under the rectified MEP curve, measured in units of micro- or millivolts × milliseconds. Area under the curve has the advantage of distinguishing two MEPs of the same peak-to-peak amplitude but varying durations.

Other EMG characteristics include the frequency spectrum of the MEP, EMG, and the mean level of the background EMG activity during the voluntary contraction of the target muscle. A limited number of studies have also quantified the TMS-induced motor response through intramusculature EMG, which permits recording from single motor units through insertion of needle electrodes into the muscle tissue. The interested readers may refer to earlier works to learn more about this method and results [1, 2].

3.2 Spinal Responses Through Peripheral Electrical Stimulation

While TMS-induced motor responses provide valuable information about the overall integrity of the corticospinal pathways, they cannot be used to differentiate between the integrity of spinal and cortical mechanisms. Therefore, to further evaluate the effect of TMS on spinal pathways and to better disentangle cortical from spinal lesions, the amplitude and latency of spinal reflexes can be examined before and after administration of a single TMS pulse.

Spinal reflexes may include the H-reflex, or Hoffmann's reflex, which is a muscle reaction in response to the electrical stimulation of the nerve innervating a particular muscle. The H-reflex has a latency of 25–35 ms with respect to electrical nerve stimulation onset and is observed after the M-wave (a short response with a latency of 3–6 ms) and prior to the F-wave (a long latency response at high intensity). As the intensity of peripheral nerve stimulation increases, the amplitude of H-reflexes is reduced and M-waves are increased [3]. Changes in spinal reflexes, therefore, provide a means by which the modulation of spinal inhibitory and excitatory mechanisms could be systematically studied [3]. A change in spinal reflexes following a TMS pulse (compared to prior) might be an indication of modulation of spinal mechanism, while an absence of change might be indicative of TMS acting primarily on supraspinal pathways.

3.3 Descending Volleys Through Epidural Recordings

In order to more directly assess the effect of a single TMS pulse on the corticospinal tract, a limited number of studies have examined TMS-induced descending volleys. Descending volleys are often measured at the cervical level of the spinal cord, and are recorded during or after surgery in patients in whom spinal stimulators are implanted in the dura space for treatment of medical conditions such as pain and scoliosis [4–14]. The first epidural recordings were conducted in anesthetized humans, while later studies included conscious humans. Several of these studies have examined the descending volleys produced by both TMS and transcranial electric stimulation (TES), providing further insight into the TMS mechanism of actions as compared to TES.

Collectively, the results of these studies have shown that single pulse TMS or TES results in a series of traveling waves in the spinal cord which are separated from one another by an inter-peak interval of about 1.5 ms (~600 Hz frequency). In TES and with specific TMS coil orientation and intensity, an early descending volley is recorded which resembles the results of animal experiments involving the direct electrical stimulation of exposed motor cortex and direct recordings form the pyramidal tract neurons [15]. This early wave is often referred to as the direct-wave (D-wave). Following TMS, the first descending wave typically has a longer latency than the D-wave and is often referred to as the Indirect-wave (I-wave). It has been suggested that these volleys are generated by synchronized orthodromic action potentials originating from fast conducting corticospinal neurons which monosynoptically connect with spinal motorneurons [16]. Each TMS-induced descending volley triggers the release of glutamate which results in depolarization of the postsynaptic cellular membrane. If the spatial and temporal summation of the glutamate release exceeds the firing threshold, an action potential is induced in the spinal motorneuron which then propagates down the peripheral motor axons and evokes a motor response (i.e., MEP).

Based on the results of the epidural studies and computer simulations, several models have been proposed to explain the mechanisms of TMS activation and the origin of the D- and I-waves. For example, it has been postulated that I-waves may reflect the indirect activation of pyramidal neurons through interneuron circuitries of the gray matter or the cortico-cortical fibers connecting the primary motor cortex with adjacent regions [17]. For example, the late onset latency of I-waves compared to D-waves has been associated with the transsynaptic activation of corticospinal neurons through activation of horizontally oriented excitatory interneurons located one or more synapses away from the pyramidal neurons (for a review, refer to: [18]). I-waves may also originate from the activation of cortico-cortical fibers between the motor cortex and the neighboring regions, such as premotor and somatosensory areas [17]. This is consistent with the horizontal orientation of the TMS-induced current in gray matter tissue which may favor the activation of horizontally oriented neurons as compared to pyramidal neurons (which are largely perpendicular to the surface of the scalp and, hence, the orientation of the TMS-induced current).

3.4 Local Field Potentials Through Electro-encephalography

While epidural recordings have provided valuable insight about TMS mechanisms of action, such recordings are invasive, limited to a selective population of patients, and are an indirect method for evaluating the effect of TMS on cortical tissues. Due to commercialization of TMS compatible EEG systems, it has become possible to record TEPs shortly following a TMS pulse. TMS combined with EEG more closely resembles animal experiments in which an electrical pulse is delivered to the cortex and the local field potentials are simultaneously recorded from the stimulated cortical tissue. The current induced by a TMS pulse can initiate an action potential in the cortical tissue giving rise to numerous excitatory and inhibitory postsynaptic potentials. EEG recordings following a TMS pulse reflect the temporal and spatial summation of the TMS-induced excitatory and inhibitory postsynaptic potentials.

Following initial attempts to combine TMS with EEG [19, 20], a growing number of studies have characterized the EEG correlates of a TMS pulse [21–29]. To date, the EEG traces that are recorded following a TMS pulse have been quantified temporally, spectrally, and spatially giving rise to a variety of TMS probed EEG outcome measures that can be used to delineate the effect of TMS with more precision than the TMS-EMG approach. The TMS-EEG technique and its application will be fully described in future chapters. Here, we will briefly describe some of the EEG characteristics referenced in this chapter.

The temporal characteristics of EEG measurements include characterizing the amplitude, latency [26], and/or slope [30, 31] of TEP peaks (i.e., negative or positive deflection from baseline).

For example, a series of studies have documented the generation of several peaks in the TEPs within the first 300 ms of a single TMS pulse applied to the motor cortex. The latency of the TEPs replicated across studies include a negativity at 15 ms (N15), a positivity at 30 ms (P30) followed by N45, P55, N100, P180, and N280 [26]. A few studies have also examined the neurophysiological mechanisms that underlie these TEP peaks. To date, the earlier peaks have been suggested to be associated with fast excitatory and inhibitory mechanisms whereas the later peaks, such as the N100 response, have been associated with slow acting inhibitory mechanisms [27].

EEG outcome can also be characterized spectrally through examining the frequency of cortical oscillatory activities following a TMS pulse, as compared to a control or a baseline condition. Furthermore, EEG spatial outcome measures may include the characteristics of TEP propagation from the stimulation site to intra and interhemispheric regions. The test/retest reliability of several EEG measures of single pulse TMS have been demonstrated for the motor and prefrontal cortices [32, 33].

3.5 Whole Brain Response Through Neuroimaging

Several neuroimaging modalities, such as fMRI, PET, SPECT, and NIRS, have been interleaved with TMS in an online (simultaneous recording) and/or off-line (with a temporal delay) manner to more fully characterize the neurophysiological effects of TMS on the whole brain [34–38].

3.5.1 Positron Emission Tomography

In PET, a positron-emitting radioisotope (also called radionuclide) such as carbon-11, oxygen-15, or flourine-18, are added to a biologically active molecule (such as fluorodeoxy glucose, oxygen, or a specific neurotransmitter) creating a radioligand which is injected to the bloodstream. The PET scanner detects the pair of gamma rays emitted by the radioligand as it interacts with biological tissue. A three-dimensional map of the radioligand distribution and concentration in biological tissues is then produced. Therefore, by examining the radioligand distribution following a TMS pulse, the interleaved TMS-PET can be used to study the effect of a TMS pulse on glucose metabolism, cerebral blood flow, amino acid uptake, activity of a unique cell population (such as microglial cells), and activity of a specific neurotransmitter or receptor system [35, 38]. Similar to PET, SPECT is based on imaging the distribution of radioligands in the biological tissue. In SPECT, the half-life of the radioisotopes is longer, thereby, eliminating the need for an on-site cyclotron, making SPECT a significantly less expensive approach than PET.

3.5.2 Functional Magnetic Resonance Imaging

In fMRI, the outcome measure is change in the blood flow that is detected by generating a strong magnetic field (~3 T) around the head and imaging the magnetized neuronal tissue as it returns to its baseline state. Due to the differential magnetic properties of oxygenated and deoxygenated hemoglobin, a blood oxygen level-dependant (BOLD) contrast is created. When TMS is interleaved with fMRI, changes in the BOLD pattern can be used to examine the TMS-induced modification in neuronal metabolic activity and energy usage. Compared to PET and SPECT, fMRI is noninvasive and, compared to EEG, fMRI has a high spatial resolution thereby, making it a powerful tool for examining the effect of TMS on remote cortical and subcortical structures and the cerebellum. However, compared to EEG, fMRI has a poor temporal resolution. Although the hemodynamic changes in fMRI are on the order of seconds, hemodynamic lag is constant and, with appropriate designs, it may be possible to associate the activity of neuronal populations to events that last a few 100 ms [39]. A TMS-fMRI study has demonstrated that a TMS pulse that generates an MEP results in modifications of BOLD pattern in the ipsilateral as well as contralateral motor cortex, and also bilaterally in the temporal cortex [34]. As the mechanism of action of fMRI and TMS is based on the induction of a magnetic field, the online combination of the two techniques, although possible, is challenging. In particular, the accurate placement and online monitoring of the TMS coil inside the scanner is one of the biggest challenges yet to be fully resolved.

3.5.3 Near-Infrared Spectroscopy

In NIRS, near-infrared light is emitted by a transmitter positioned around the head and the concentration of oxygenated or deoxygenated hemoglobin is measured by quantifying the intensity and wavelength of the near-infrared light absorption in different tissues. Therefore, in interleaving TMS with NIRS, changes in the concentration of oxygenated and deoxygenated hemoglobin are the outcome measures. In addition to being noninvasive (unlike PET and SPECT), NIRS can more easily be interleaved with TMS than fMRI in an online approach as the magnetic field induced by TMS does not interfere with NIRS. However, limitations include the sensitivity of NIRS to environmental light and the limited depth of infrared penetration which results in poor spatial resolution compared to fMRI.

Each electrophysiological and neuroimaging technique has its strengths and weaknesses and the information provided by these modalities is complementary. While TMS interleaved with EEG and EMG can capture the neurophysiological effect of TMS with a high temporal resolution (i.e., milliseconds), TMS interleaved with fMRI, PET and SPECT have a temporal resolution on the order of seconds to minutes and are less ideal for examining the fast and transient neurophysiological response to a single pulse of TMS. Instead, such approaches may be more valuable for examining the effect of trains of TMS pulses that can induce neurophysiological modification outlasting the duration of TMS delivery. On the other hand, fMRI, PET, and SPECT provide superior spatial resolution compared to EEG, in which local field potentials are spatially dispersed by the skull. Using high density EEG, mathematical modeling (such as source localization and minimum norm estimation) has been applied to overcome the relatively weak spatial resolution of EEG. This is important as, for the purpose of examining the physiological effect of single pulse TMS (which is less likely to induce any long lasting modification of brain dynamics), EEG and EMG appear to be the two more plausible and practical methods for capturing induced response.

3.6 The Association Between Different TMS Outcome Measures

The association between cortical, subcortical, spinal, and peripheral outcomes of a single TMS pulse has been studied by several investigators, often with the aim of mapping the neural pathways affected by a TMS pulse and identifying the origin of TMS-induced MEPs. Although the strength of the descending cortical output and the excitability of spinal motorneurons determine the size of an MEP at the periphery, not all cortical-evoked activities or descending volleys are destined for the muscle whose cortical representation has been targeted by TMS. Therefore, TEPs, descending volleys, and MEPs may not linearly correlate.

The association between the TMS-induced MEPs and the descending volleys has been explored in a few studies [2, 13]. For example, the latencies of the D- and I-waves have been compared with single motor unit recordings from the target muscle using needle EMG. It has been illustrated that, similar to epidural recordings, there are several peaks in the firing pattern of the motor units and these peaks separated by an inter-peak interval of approximately 1.5 ms; similar to the inter-peaks interval of the descending volleys [2]. Moreover, a growing line of research is examining the correlation between the temporal and spectral characteristics of EEG and EMG measures of single pulse TMS. One such investigations demonstrated a correlation between MEP amplitude and the peak-to-peak amplitude of the TEP N15-P30 component [40]. In a similar study, the power of specific cortical oscillatory activity prior to a TMS pulse was shown to be weaker when TMS induced larger MEPs [41]. More studies are needed to fully map the neural pathways that are directly and indirectly activated by single TMS pulses.

4 Factors Affecting the Outcome Measures of a TMS Pulse

Several physical and biological factors have been found to influence the outcome measures of single pulse TMS. Research conducted to date has provided clear evidence for significant contribution of several of these factors, while the effect of several others is currently being investigated. In a recent study, using the Delphi technique and participation of 42 TMS experts in online surveys, a consensus-based methodological checklist was developed listing 27 physical and biological factors that should be reported and/or controlled for in single pulse TMS studies of the motor cortex to improve the quality of data collection and reporting [42]. In this chapter, we will review the mechanisms underlying the modulatory effects of some of these factors. We will refer to these factors as *physical*, if they are related to the TMS equipment or stimulation parameters, and *physiological*, if they are related to the properties or state of biological tissues.

4.1 Physical Factors

Several physical properties of the currently available magnetic stimulators can influence neuronal activation. Among these are: (1) pulse shape, such as the duration and rise time of the induced magnetic field; (2) coil orientation and, thus, the direction of the induced current in the tissue [43]; and (3) the shape, size and material of the TMS coil.

4.1.1 Pulse Shape and Intensity

The intensity of a TMS pulse is determined by the capacitance voltage of the TMS device, which can be easily adjusted. The capacitance voltage affects the initial steepness of the induced magnetic field and, therefore, affects the amplitude of the electrical current induced in the brain. It has been observed that by increasing the intensity of TMS, one can increase MEP amplitudes, modify the pattern of descending volleys (such as increasing the number of I-waves and generating a D-wave), and also increase the amplitude of several EEG peaks [1, 13, 44].

The shape (including duration) of the magnetic pulse, on the other hand, is determined by the circuitry of the TMS stimulator, and cannot be readily adjusted in most commercially available stimulators. The pulse shape is determined by the resonant frequency of the RLC circuit ($f_0 = \frac{1}{2\pi} \times \sqrt{LC}$) and is, thus, dependant on the capacitor and inductor of each TMS device. In some devices, the pulse shape is adjusted through a shunting diode that can dampen quarter cycles of the oscillation period producing TMS pulses of several different shapes. An undampened pulse resembles a *polyphasic* sinusoidal waveform. A pulse that looks like a cosine function is referred to as a *biphasic* pulse. A pulse containing a half-cycle is referred to as a *half-sine* pulse while a pulse with approximately a quarter-cycle is referred to as a *monophasic* pulse. Monophasic pulses were the first pulse shapes applied to the motor cortex in the 1980s [45]. The monophasic pulse has a rapid rise time and a slow decay toward zero and a short duration of hyperpolarization. A monophasic pulse has more than one phase, despite what the name implies, but also has an asymmetrical shape. It is the initial phase that depolarizes the neurons. Pulse duration is highest for the monophasic pulse (~600 μs) compared to the biphasic pulse

(~300 μs). Currently available magnetic stimulators may produce monophasic, biphasic, and/or half-sine waveforms; although future devices may allow for additional adjustable pulse shapes [46].

The effect of pulse shape on tissue has been investigated by comparing the MEP characteristics across different pulse shapes [47]. Some studies have also evaluated the effect of pulse shape on the size and pattern of epidural volleys [48]. The duration of TMS pulse has been shown to influence the physiological effect of TMS [49]. In general, biphasic TMS pulses are suggested to be more powerful than monophasic pulses and, accordingly, generate a more complex pattern of neural activation [48]. The effects of TMS pulse shape on other TMS outcome measures, such as EEG, are currently being investigated.

4.1.2 Coil Physical Property

The focality and pattern of the induced current in tissue varies as a function of TMS coil geometry. The geometry of a TMS coil is often described by its shape and the size of the outer diameter. In circular coils, the magnitude of the induced electrical field is maximal under the rim of the circular coil and reduces in the center. To achieve more focal stimulation, flat figure-of-eight coils are used which are made by attaching two flat circular coils. In the figure-of-eight coil, the currents in the two wings are in opposite directions, resulting in summation of the currents at the junction of the wings (Fig. 2b) generating a maximum field in the center of the coil. Unlike flat coils, the circular wings of double-cone coils are bent, making them ideal for stimulating the posterior lobes. However, by changing the angle between the wings of the bent coil, the focality of the coil may be shifted from the center.

Descending volleys have been shown to be different for figure-of-eight versus circular coils [50]. Circular coils have been shown to produce long latency D-waves, perhaps suggestive of activation of pyramidal neurons closer to the cell body [50]. In addition to the coil geometry, the tightness of the coil wire windings and the properties of the medium in the core of the coil (e.g., air versus iron) can affect TMS output. Therefore, similar coil shapes of different manufacturers may produce slightly different results. For example, the iron used in iron-core coils serves to funnel away heat which may modify the output relative to air-core coils. The effect of coil shape on other TMS outcome measures, such as physiological measures obtained through EEG, have yet to be elucidated.

4.1.3 Coil Orientation

Coil orientation is reported in term of the current in the coil or thc induced current in the tissue. It should be noted that the direction of the current in the coil is opposite to the direction of the induced current immediately underneath the coil. In circular coils, the current direction can be anticlockwise or clockwise (Fig. 2a). When a circular coil utilizing an anticlockwise current is centered over the vertex, the direction of the current in the left wing of the coil will

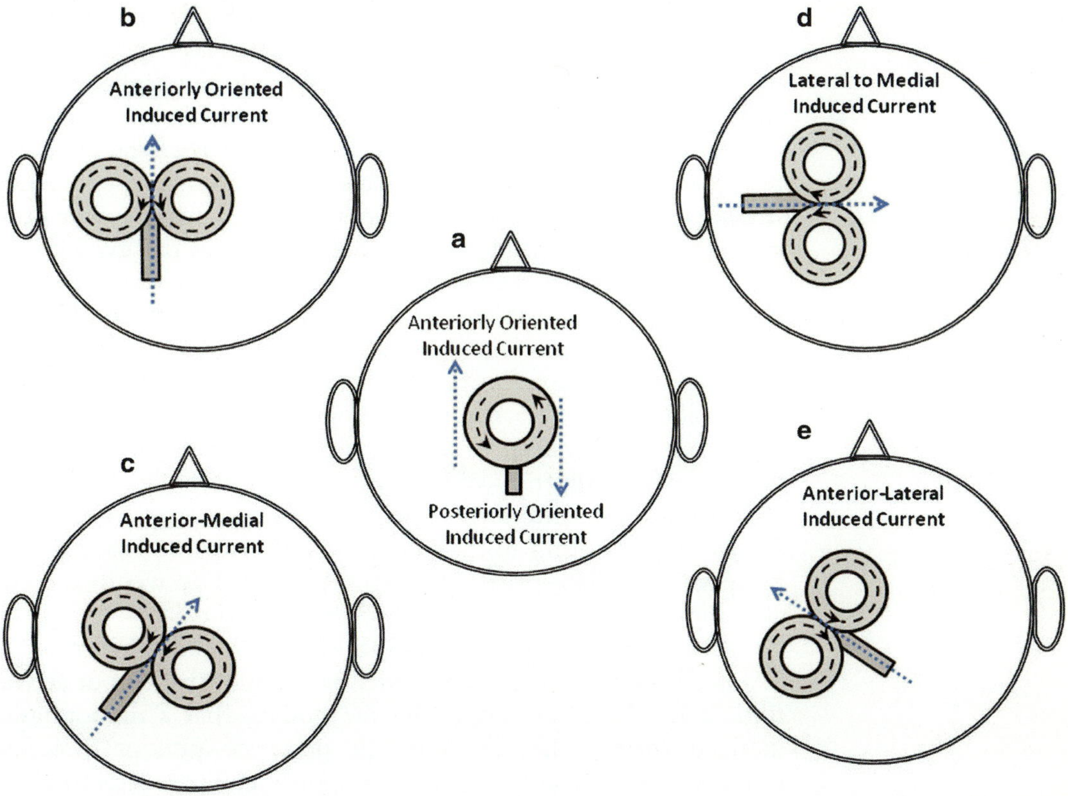

Fig. 2 Coil orientation. (**a**) Illustrates a round coil and (**b–e**) depict a figure of eight coil. Each panel illustrates different coil orientations as positioned on the head. The coil currents are depicted by *dashed black arrows* drawn inside the coil rim, and the direction of the induced currents in the tissue are illustrated by *blue dashed arrows* along the axis of the coil handle

be posteriorly oriented thereby inducing an anteriorly oriented current in the left hemisphere. In the right wing of the coil, however, the current will be directed anteriorly inducing a posteriorly oriented current in the right hemisphere. Similarly, anterior–medial, lateral–medial, and anterior–lateral currents can be induced in tissue by positioning the coil handle at an appropriate angle with respect to the midsagittal line (Fig. 2b–e).

The direction of the induced current can significantly influence the mechanisms of neuronal activation, such as the type of neurons recruited (e.g., interneuron *versus* pyramidal) as well as the site of the neuronal depolarization (e.g., soma *versus* axon hillock: [51–54]). The effect of coil orientation on TMS outcomes has been primarily examined in the motor cortex. For example, it has been shown that posterior-anterior, anterior–posterior, and lateral–medial-induced currents in the brain produce different patterns of descending volleys (for review: [17]). For example, the lateral–medial-induced current can more easily create a D-wave as compared to posteriorly or anteriorly oriented current. The effect of coil orientation also varies across different TMS pulse shapes, with monophasic pulses being more affected than biphasic pulses [55], particularly at high stimulation intensities [56].

As described in Sect. 3.3, several theories have been put forth to explain how a TMS pulse activates tissue. It has been postulated that TMS may more readily activate the gyral crown because of its proximity to the coil. However, since neurons in the gyral crown are perpendicular to the surface of the cortex, it has been suggested that the activation of cortical columns must be achieved indirectly via synaptic inputs through an initial activation of horizontally oriented excitatory interneurons [2, 57]. TMS-induced generation of I-waves is also considered to be evidence for the transsynaptic activation of pyramidal neurons compared to TES, which more readily induces D-waves. Using this model, it has been postulated that anteriorly oriented currents produce indirect I-waves by activating horizontal interneurons with higher probability while medially oriented currents produce D-waves by directly activating pyramidal tracts with higher probability.

Although transsynaptic activation of pyramidal neurons [2, 57] is one of the most cited model of TMS mechanism of activation, it has been challenged by the finding that a monophasic-induced current that flows in the posterior–anterior direction requires less intensity to evoke a response compared to an anterior–posterior-induced current. This is while the first-order interneurons that synapse with the pyramidal cells are located in the anterior bank of the sulcus, as suggested by some animal and neuroimaging studies [58]. TMS combined with PET has also localized the site of TMS to the sulcal banks [54]. Therefore, these discrepancies have led to other plausible hypotheses for the TMS site of action. For example, Sommer et al., have proposed that an anteriorly directed monophasic pulse may activate cortical columns at the anterior bank of the central sulcus and may be soma depolarizing, while a posteriorly oriented monophasic pulse may be soma hyperpolarizing and dendrite-depolarizing causing an antidromic current flow in the cortical column [58].

Some studies have also examined the effect of coil orientation on non-motor regions. In the visual cortex, lateral–medial-oriented current has been found to be optimal for induction of phosphenes by single pulses of TMS [59, 60]. Additionally, in the visual cortex, induced current perpendicular to the stimulated gyrus is suggested to be optimal [61]. One study demonstrated that coil orientation over the prefrontal cortex influences cognitive performance, with anterior–lateral-induced current being the most effective [62]. Therefore, to optimally target underlying tissue, the orientation between the induced current and the underlying tissue should be taken into consideration, especially when using monophasic pulses.

4.2 Physiological Factors

In addition to various TMS parameters, several tissue properties may also modify the TMS-induced electrical field distribution [63]. To date, the effect of coil-to-cortex distance, gyral geometry, fiber orientation, and corticospinal fluid have been examined.

4.2.1 Coil-to-Cortex Distance

The strength of the induced magnetic field falls exponentially with increasing distance from the coil surface [64]. For example, TMS delivered through the figure-of-eight coil over the motor cortex may directly activate the outermost 2–2.5 cm of the brain tissue. Intersubject variability in TMS-induced responses may be partly related to differences in the coil-to-cortex distance (rather than underlying neurophysiological mechanisms). Since coil-to-cortex is not homogenous across cortical and cerebellar regions, it should be taken into consideration when non-motor regions are stimulated at an intensity determined based on the excitability of motor regions. In addition, there is a small distance between the wound copper wires and the surface of the coil. This distance, which varies across coil manufactures, may affect the depth and pattern of TMS-induced cortical activation.

4.2.2 Tissue Types and Fiber Orientation

As discussed in Sect. 4.1.3, coil orientation can alter physiological response to a TMS pulse. This is partly due to the geometry, spatial distribution, and electrical properties of different classes of neurons that are stimulated. Through finite element analysis, accurate modeling of head geometry, and high resolution diffusion weighted imaging, it has been demonstrated that the strength of the induced electric field may vary depending on gyral orientation and white matter anisotropy [63]. For example, it has been found that maximum TMS-induced electrical field strength can be obtained when induced currents are parallel to gyral orientation. It has been further shown that the spatial distribution of the electric field differs in white and gray matter. In gray matter, the field strength has been found to be largest at the gyral crown, while in white matter, field strength has been found to be largest deep in the fiber and is related to white matter anisotropy [63].

It has been suggested that a TMS pulse can modify the charge on the inside and outside of cell membranes thereby leading to depolarization or hyperpolarization of neurons. The transfer of charge across the cell membrane depends on fiber diameter, length, and orientation. In vitro studies and computer simulations suggest that TMS depolarizes neurons where the induced current has its maximum peak [65–67]. The results of in vitro nerve stimulation, however, suggests depolarization may also occur where the negative-going first spatial derivative of the electric field is maximal (i.e., where neurons bend away from the direction of the induced current) and hyperpolarization may occur at the positive going spatial derivative of the current [65, 66]. In the future, finite element calculation and high resolution diffusion tensor imaging (DTI) may be used to monitor and navigate coil placement with respect to underlying brain tissue.

4.2.3 Corticospinal Fluid

The time-varying magnetic field produced with each TMS pulse induces current in any medium that has nonzero conductivity and contains a proportion of the magnetic flux bound within it [68]. The layers of biological tissue that are located underneath the TMS coil, such as the skin, skull, cerebrospinal fluid (CSF), and gray and white matter brain tissue, each have distinct electrical properties. This leads to an inhomogeneous dispersion of the electrical field in each layer and a buildup of electrostatic charge at the boundaries. In a finite element model (FEM) of electrical stimulation in the spinal cord, it was shown that when the conductivity of the CSF is taken into account, the induced current may be shunted away from the spinal cord [69]. Similarly, through real head model simulation of electrical field distribution, it has been demonstrated that the electrical properties (such as the conductivity and permittivity) of the biological tissues underneath the TMS coil can alter the direction of the induced current and the site of the maximum activation. This, quite predictably, might affect physiological response to TMS. Wagner et al. [70] simulated a real head model to study the induced current density of a TMS pulse taking into account the geometrical asymmetry and the electrical properties of skin, skull, CSF, and gray and white matter. Several results were obtained: First, depending on the geometry of the head, the induced current had components not only parallel to the plane of the coil but also components normal to the head surface. Furthermore, the maximum induced current was produced in the CSF and at the boundary surface of the CSF and gray matter. Of note, adjusting the geometry and electrical properties of tissue resulted in modification of the direction and location of the maximum induced current [70].

4.2.4 Additional Factors

Several additional factors, such as the duration of time awake, circadian rhythm, hormonal levels, and menstrual cycle, have been shown to influence the physiological effects of TMS. In addition, it has now become possible to monitor the physiological state of the brain immediately before the delivery of TMS through methods such as EEG (as described in Sect. 3.4). The results of these endeavors are generating a list of parameters, such as phase or power of specific cortical oscillatory activity at the time of TMS delivery, which could aid in the prediction of physiological response to subsequent TMS pulses.

In general, when comparing TMS-induced outcome measures across patients, studies, and laboratories, careful attention should be paid toward two groups of factors: (1) factors that can alter TMS response without modifying the neurobiological processes that are being investigated (e.g., the effect of coil geometry and coil-to-cortex distance on TMS indices of cortical excitability) and (2) factors than can transiently alter the brain state and, hence, temporarily modify the neurobiological processes that are meant to be investigated by TMS (e.g., transient effect of menstrual cycle on cortical excitability and therefore a modification of the TMS indices of cortical excitability). The state-dependency of TMS makes it a powerful tool for the investigation of subtle changes in biological tissues but, at the same time, demands meticulous monitoring and reporting of transient changes in brain state in order to eliminate any misinterpretation of outcome measures. In this regard, readers are recommended to refer to the recommended guidelines that are regularly presented and updated in the consensus manuscripts (for example: [42]).

5 Single Pulse TMS Protocols

In this section, we will describe four single pulse TMS-EMG protocols: (a) motor threshold (MT), (b) input/output (I/O) curve, (c) cSP, and (d) iSP. For each protocol, we present several guidelines provided by the committee of the international federation of clinical neurophysiology (IFCN). Furthermore, we provide a review of the neurobiological underpinnings of the outcome measures in each protocol. Finally, we outline technical details that need to be considered when undertaking each protocol, such as factors affecting measurement reliability. Guidelines for acquisition of EMG and quantification of MEP can be found elsewhere [16]. In addition to the conventional TMS-EMG protocols, we provide experimental evidence to demonstrate that advancements in neuroimaging technology, as described in Sects. 3.4 and 3.5, have made it possible to expand the utility of single pulse TMS to assess non-motor regions of the cortex and brain dynamics.

5.1 Motor Threshold

In TMS, the corticospinal excitability of human subjects is often determined through identifying the motor threshold (MT). MT has been used to characterize each individual's level of excitability to extract information about the integrity of the corticospinal pathway for experimental and clinical purposes. It often also serves as a baseline measure of excitability to guide the intensity of other TMS protocols (e.g., paired pulse and repetitive TMS). MT can be identified with the target muscle at rest, referred to as resting MT (rMT), or during voluntary contraction of the target muscle, referred to as active MT (aMT).

5.1.1 Protocol

Resting Motor Threshold

RMT is often defined as the minimum stimulation intensity needed to produce MEPs with peak-to-peak amplitude of ≥50 μV in 50 % of trials when ten consecutive single pulses are applied with the TMS coil fixed on the hot spot of the target muscle [43]. This method has been referred to as the *relative frequency* method. A few alternative approaches have also been proposed for rMT determination including: (1) visual method: visualization of TMS-induced twitches in the target muscle without EMG recording [71], (2) two-threshold method: determination of the highest intensity that is required to produce no MEP responses in ten trials, and lowest intensity that is required to produce an MEP in ten consecutive trials. The MT is then the average of these two intensities [72], (3) adaptive method: an automated threshold-tracking technique based on parameter estimation by sequential testing and maximum likelihood regression [73], and (4) supervised parametric estimation technique: mathematical identification of threshold by fitting a sigmoid curve to the probability of response as a function of intensity [74]. Depending on the utilization of rMT and the availability of technical resources, any of these approaches may be used. However, it should be noted that, while visualization has not been completely invalidated, the use of EMG is highly recommended [75]. It has been further suggested that the adaptive methods provide the most accurate MT while requiring the same or even fewer stimuli compared to other methods [76].

Step-by-step procedure. **(1)** Before identifying the rMT, the hot spot of the target muscle has to be identified with the target muscle at rest. The hot spot can be obtained using a similar approach as the motor map protocol (detailed in other chapters of this book). In short, coil placement is determined by delivering suprathreshold single pulse TMS of a fixed intensity to several cortical spots contralateral to the target muscle and identifying the coil placement that produces the largest response in the target muscle. For example, for hand muscles, the axis of the a figure-of-eight coil is held approximately at a 45° angle with respect to the midsagittal line [77] thereby inducing a current approximately perpendicular to the motor strip and the central sulcus.

(2) Once the hot spot is determined, the position, angle, and orientation of the coil should be kept constant throughout rMT determination. This can be achieved by two methods. First, the apex of the coil can be marked with a felt pen on the scalp or a tightly fitting swim cap. Second, the coil can be tracked in space with respect to the subject's head position utilizing varied neuronavigation technologies..

(3) Once the hot spot is marked, rMT can be obtained through any of the protocols described above. Many laboratories use the relative frequency protocol. However, this protocol has been criticized due to several limitations, such as not taking into account the "probabilistic nature" of the MT and lack of more specific criterion for identifying the threshold. To address this, a more detailed version of relative frequency method has been proposed by the IFCN: **(3a)** Single pulses of TMS are applied starting at a subthreshold intensity (e.g., 35 % of maximum stimulator output) that does not produce an MEP. The interstimulus interval between consecutive pulses should be longer than 5 s. **(3b)** Intensity is increased in steps of 5 % until an intensity is reached at which TMS consistently produces MEPs with peak-to-peak amplitude of 50 μV or higher. **(3c)** Intensity is reduced by 1 % steps until an intensity is reached at which less than 5 out of 10 pulses produce an MEP of 50 μV or higher. **(3d)** RMT is defined as this intensity +1 % [16].

Another limitation of relative frequency method is that it requires a large number of pulses before the MT is reached. To overcome this limitation, any of the above mentioned adaptive methods may be employed.

Active Motor Threshold

Step-by-step procedure. Determination of aMT is similar to rMT, except that there is a voluntary contraction of the target muscle. In determining aMT:

(1) The hot spot of the target muscle is identified (as described above). **(2)** Coil placement is marked using a felt pen or tracking technique (as described above). **(3)** Subjects are instructed to maintain a voluntary contraction of the target muscle at about 20 % of their maximum force. To maintain a fixed force, a gauge meter may be used that subjects squeeze with the target muscle. The pressure exerted on the meter is displayed as a percentage or maximum value, and participants are asked to maintain that same level of force throughout the procedure.

(4) Single pulses of TMS, separated by 6–7 s, are applied to the hot spot and aMT is determined using the same methods described above (for instance, by employing the modified relative frequency method). The only difference is that aMT is defined as the minimum intensity required to produce an MEP of 200 μV peak-to-peak amplitude (instead of 50 μV in RMT) while subjects maintain a voluntary contraction of the target muscle. This higher MEP amplitude (i.e., 200 μV) is used to minimize the error of identifying background EMG activity as a TMS-related MEP.

5.1.2 Neurobiological Correlates of the MT

Resting Motor Threshold

Several lines of evidence suggest that rMT reflects the excitability of elements activated by TMS. These may include the ion-dependant neuronal membrane and cortico-cortical axons and their excitatory synaptic connections with corticospinal neurons [78]. For instance, the blockade of voltage-gated sodium and calcium channels, which are essential for regulating axonal excitability, is shown to increase rMT [78, 79]. Furthermore, NMDA antagonists, such as ketamine, have been shown to reduce rMT, while GABAergic drugs, such as lorazepam, vigabatrine, and baclofen, display no influence on rMT [80] (for a review: [81]). DTI has demonstrated that rMT may also reflect the microstructure of white matter tracts [82]. For example, rMT was associated with the integrity of the white matter tracts of the primary motor cortex, as well as the premotor cortex, internal capsule, cerebral peduncles, corpus callosum, and corona radiata [82].

RMT varies across muscles and this variance can be attributed to the strength of corticospinal projections [83], size of the cortical representation of the muscle, and accessibility of the muscle by TMS. A lower rMT is observed for thoracic compared to pelvic limbs [84]. The rMT is also lower for hand muscles compared to proximal arm, lower limb, and truncal muscles [85]. Muscles most commonly studied are the intrinsic hand muscles; including the abductor pollicis brevis (APB), the first dorsal interosseous (FDI), and the abductor digiti minimi (ADM) muscles.

Active Motor Threshold

Active motor threshold is almost always lower than the rMT. At a fixed stimulation intensity, muscle contraction increases MEP amplitude compared to rest. This has been attributed to the elevation of excitability of spinal motorneurons as well as cortical neurons [8]. Epidural recordings have revealed that contraction of hand muscle has no effect on the D-wave produced by TMS-induced lateral–medial current. However, the amplitude of the TMS-induced I-waves is increased by as much as 50 %. Facilitation of I-waves has also been observed during the contraction of lower leg muscles [10]. It has been proposed that voluntary muscle contraction reduces the temporal dispersion of TMS-induced descending volleys, leading to increased MEP amplitude and a shortened MEP latency compared to rest [16]. However, following a period of muscle contraction, MEP amplitudes and D- and I-waves are reduced in response to TMS at rest [13, 48], reviewed in [50]. This has been associated with reduction of corticospinal excitability following a period of repetitive pyramidal neuron firing [50].

5.1.3 Technical Considerations

Resting Motor Threshold

RMT has a high degree of intersubject variability and several factors can play a role in elevating the inter- and intra-subject variability. Intersubject variability in rMT can be due to variance in the integrity and neurobiology of motor pathways. However, several other factors, including confounding factors related to stimulation parameters as well as transient changes in brain state, contribute to inter- and intra-subject rMT variability. As examples, rMT is positively related to the skull-to-cortex distance [86, 87]. Herbsman et al., demonstrated that skull-to-cortex distance and a DTI measure of motor bundle fiber orientation accounted for more than 80 % of the intersubject variance observed in rMT [88]. Despite this finding, the magnitude of the induced electric field has been found to suffer from a similar degree of intersubject variability [89]. For example, in a recent study, Danner et al. [89] compared the intersubject variability in rMT with the variance in the intensity required to induce a fixed peak current in tissue. It was concluded that, while induced currents may provide a better baseline measure to estimate the excitability of non-motor regions, it may not significantly reduce the degree of intersubject variability [89].

The state of the underlying biological tissue may also affect rMT. For example, imagining hand movement during determination may affect rMT [90]. Furthermore, the characteristics of underlying cortical oscillations have been shown to modify rMT [91]. While the latter may provide further insight into the neurobiological mechanism underlying rMT and cortical excitability, the former should be controlled for by instructing subjects to not attend to the twitches produced in their muscle, and it may be best if the subjects are seated facing away from the computer screen that displays the TMS-induced MEPs.

Moreover, rMT is not an absolute value and is reported as a percentage of the maximum stimulator output. Therefore, rMT depends on the maximum energy of the TMS stimulator, which may vary slightly across different device manufacturers. Furthermore, other factors, such as the pulse shape, coil orientation, and geometry, can play a role in increasing the inter- and intra-subject variability of rMT [92]. For example, an individual's rMT increases as a function of pulse shape as follows: monophasic > half sine > biphasic. That is, a lower rMT may be obtained with a biphasic than a monophasic pulse [47]. It has also been reported that rMT is lower and MEP latency is shorter when the induced current is oriented anteriorly to the central sulcus (posterior–anterior direction) during monophasic stimulation and when the current in the first half of the cycle is posteriorly oriented [47] during biphasic stimulation. Therefore, slight changes in coil orientation may also affect rMT. To minimize coil displacement during rMT determination, the use of MRI-guided neuronavigation has been recommended [93]. Although some studies have shown that rMT does not differ when determined with neuronavigation than without, MEPs have been found to have larger amplitudes and shorter latencies when neuronavigation was used to deliver consecutive pulses [94].

Finally, as will be discussed in Sect. 6, any disease that leads to alteration of cortical or spinal excitability can influence rMT. Increases in rMT are typically observed in diseases that are associated with the damage of a significant portion of corticospinal tract, such as in stroke and spinal cord injury. Decreases in rMT are typically observed in diseases associated with hyperexcitability of corticospinal pathway, such as in untreated epilepsy, Alzheimer's disease, and early stages of amyotrophic lateral sclerosis (ALS). Medications that modify cortical excitability, such as ion-channel blockers, can modify rMT and their effects should be taken into account when comparing subjects. In general, since rMT has high degree of intersubject variability and is affected by several technical and physiological factors, the diagnostic application of rMT should be carried out with care and by reporting and controlling for the confounding factors or coexisting medical conditions.

A number of studies have investigated the test/retest reliability of the rMT within subjects and it has been shown that, across three sessions, rMT has moderate to good reliability when measured from the ABP, FDI, extensor digitorum communis (EDC), and flexor carpi radialis (FCR) muscles [95]. Both the inter- and intra-investigator reliability of rMT for the tibialis anterior (TA) muscle has also been demonstrated [96].

Active Motor Threshold

All technical considerations pertaining to rMT apply to aMT protocol. AMT is prone to a larger variability across subjects and studies. First, a broader range of MEP size criterion (e.g., 50–400 μV) has been used across studies. Also, the methods for monitoring the magnitude of voluntary contraction, and the extent of voluntary muscle contraction varies across studies, ranging from 10 to 100 % of maximum force. To minimize variability, readers are suggested to use surface EMG and a gauge meter to systematically quantify EMG activity and the amount of force, and also to follow the most up-to-date guidelines and methodologies put forth by experts in the field [16].

5.2 Input/Output Curve

5.2.1 Protocol

The input/output (I/O) curve, also referred to as the *stimulus-response* (S-R) curve or *recruitment* curve, is obtained by measuring MEP amplitudes across a wide range of intensities as single TMS pulses are applied over the cortical hot spot of the target muscle at rest or during tonic activation. The MEP sizes obtained in the I/O protocol are then used to model the I/O function by fitting the data to a Boltzmann equation [16].

Step-by-step procedure. **(1)** The hot spot of the target muscle is identified as described above. In short, TMS coil placement is determined by delivering suprathreshold single pulse TMS of a fixed intensity to several cortical spots contralateral to the target muscle and identifying the coil placement that produces the largest response in the target muscle. **(2)** Once the hot spot is determined, similar to the previous protocols, the position, angle and orientation of the coil should be kept the same throughout the I/O curve determination. This can be achieved by marking the apex of the coil with a felt pen or using a tracking device.

(3) Starting at subthreshold intensities, 5–10 pulses, separated by 6–7 s, are applied as the TMS coil is fixed on the hot spot. Repeat this step again in increasing increments (e.g., 5 %) up to an intensity equivalent to 100 % of the maximum stimulator output.

(4) Mean peak-to-peak amplitude of the MEPs are calculated for each intensity and plotted as a function of the % maximum stimulator output (Fig. 3). For muscles that have relatively low rMT (and can, therefore, be fully characterized within the available TMS stimulator output), the data will have a sigmoid shape and can be fitted by a Boltzmann equation. For example, if a wide range of intensities are tested starting from an intensity below the

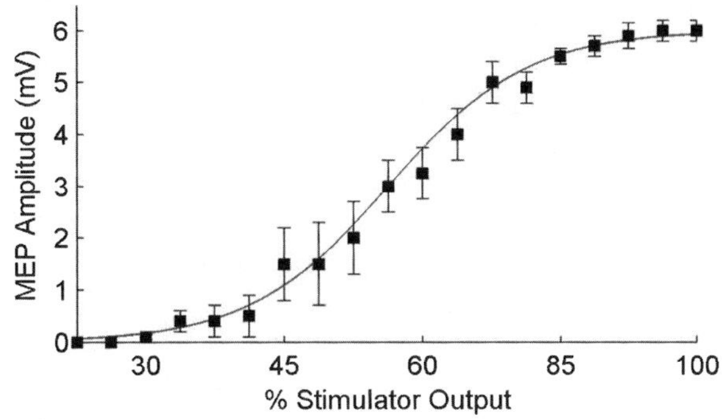

Fig. 3 Input/output curve. The input/output curve is obtained by fitting a Boltzmann equation (*sigmoid curve*) to the data

rMT to maximum stimulator output, the initial part of the curve will be a flat line indicating the "zero" MEP peak, then start to increase linearly with intensity, and finally reach a plateau after which increases in intensity no longer increase MEP amplitude. The I/O curve has a sigmoid shape for the intrinsic hand muscles (Fig. 3) [1, 97–100]. **(5)** The slope (shape) and the intensity after which MEP amplitude plateaus are used as the outcome measures of this single pulse protocol.

If the I/O curve is measured for an active target muscle, the plateau phase is reached at a lower intensity. In a preactivated hand muscle, the transition from the linearly increasing phase to the plateau phase occurs at around 140 % of rMT, corresponding to about 170 % aMT [16]. For a muscle at rest, the turning point may be reached at higher intensities [16]. It has been suggested that the intensity corresponding to this turning point to the plateau phase may be an optimal intensity for application in clinical diagnosis (e.g., in assessment of central motor conduction time) as it likely activates all the underlying neural elements while introducing less subject discomfort than maximum stimulator output [16].

5.2.2 Neurobiological Correlates of the Input/Output Curve

It has been suggested that the I/O curve assesses the activity of neurons that are less excitable or further away from the TMS center of maximum activation [83]. The slope and the plateau phase of the I/O curve vary across different muscles and are thought to be related to the strength of corticospinal projections [98, 101]. For example, the slope is steeper for intrinsic hand muscles that have lower rMT. The slope of the I/O has been shown to become less steep by sodium and calcium channels blockers (e.g., lamotrigine) and $GABA_A$ receptor agonist (e.g., lorazepam) [100]. The effect of lorazepam has been found to be maximal at the highest intensity of stimulation [100]. As lorazepam has been shown to have no effect on rMT [100], modulation of the I/O curve with lorazepam further indicates that rMT and I/O curve provide complementary information about the motor system, with the I/O curve also providing information about the GABAergic circuitry. Furthermore, Boroojerdi et al. have found that lorazepam and lamotrigine have no effect on the spinal M- and F-reflexes [100], demonstrating that at least some characteristics of the I/O curve seem to be primarily modulated by cortical mechanisms. In addition to GABA, the administration of an indirect dopaminergic–adrenergic agonist, D-amphetamine, has been found to increase the I/O curve slope [100]. Finally, a recent magnetic resonance spectroscopy (MRS) study has demonstrated a significant correlation between the slope of the I/O curve and glutamate concentration within the primary motor cortex. It has been proposed that a higher MRS-glutamate suggests a greater amount of presynaptic glutamate is stored and available for release with increasing stimulation intensity [102].

5.2.3 Technical Considerations

In determining the I/O curve, slightly different methods have been used. In one method, stimulations are presented in a ramping order, starting with the lowest intensity then incrementally increasing to the highest intensity. In an alternate method, the order of intensities is randomized. The rationale behind this method is to avoid artificially increasing the MEP amplitudes at higher intensities due to a serial order effect. A recent study, however, has compared the I/O curves obtained by both methods and found no differences [103]. In addition to stimulation order, the increments between the tested intensities may vary across studies. While some studies have used 1 % increments to obtain a more accurate estimate of the curve, others have used 5 % increments. It has been demonstrated that the I/O curve obtained with 5 % increment method can also accurately predict rMT obtained with 1 % increments [104].

Several factors have also been found or suggested to influence I/O curve and should be controlled for. For example, the level of background EMG activity, presentation of the stimulus at the start or during constant muscle contraction, interstimulus interval, coil orientation, coil size, and pulse shape may alter the I/O curve. The test/retest reliability of the I/O curve of a few muscles have been examined [105], and the I/O curves of APB, FDI, EDC, and FCI muscles have exhibited moderate to good test/retest reliability [95].

5.3 Silent Period

A single TMS pulse applied to the motor cortex has both an excitatory and an inhibitory effect. To examine the excitatory effect with EMG, the motor cortex is stimulated at intensities equal to or higher than MT and the MEP in the contralateral hand muscle is measured. To examine the inhibitory effect of a TMS pulse using EMG, the TMS pulse is applied to the motor cortex during voluntary contraction of the target muscle. The inhibitory effect of TMS is represented by a suppression of the background EMG activity that follows the MEP. This period of EMG silence, that may last up

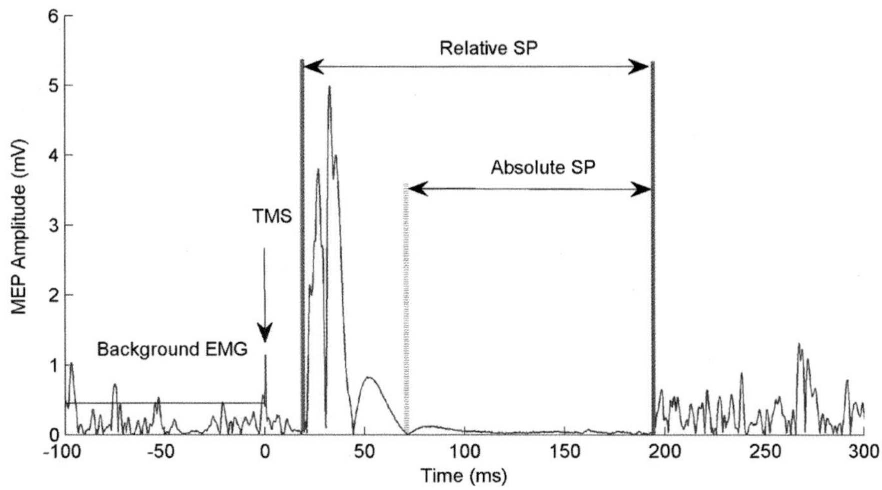

Fig. 4 Contralateral silent period. The *x*-axis represents time in milliseconds with respect to the TMS onset. The *y*-axis represents MEP amplitude in millivolt (mV). The *solid black line* represents the mean EMG activity. The first *solid pink line* illustrates the iSP onset latency, and the second *solid pick line* marks the iSP offset latency. The duration of iSP is measured by subtracting the onset latency from the offset latency

to hundreds of milliseconds, can be induced in both the contralateral and ipsilateral muscles and are referred to as contralateral and ipsilateral SP, respectively.

5.3.1 Protocol

Contralateral Silent Period

Contralateral silent period (cSP) involves a transient suppression of EMG activity following the delivery of a suprathreshold TMS pulse to the motor cortex during voluntary contraction of a contralateral target muscle [106] (Fig. 4).

Step-by-step procedure. **(1)** The hot spot of the target muscle is identified. In short, coil placement is determined by delivering suprathreshold single pulse TMS of a fixed intensity to several cortical spots contralateral to the target muscle and identifying the coil placement that produces the largest response in the target muscle. **(2)** Once the target is determined, the same coil placement should be maintained throughout the procedure by marking the coil placement on the scalp or using tracking methods. **(3)** MT (active or resting) is identified as described in Sect. 5.1. **(4)** Subjects are instructed to contract their target muscles by exerting their maximum force on a gauge meter. Based on the meter output, participants are instructed to maintain 20 % of their maximum force throughout the experiment.

(5a) The intensity of stimulation is set at 140 % of rMT (or a higher percentage of aMT) and ten single TMS pulses, separated out by about 10 s, are applied to the hot spot as subjects maintain the voluntary hand contraction. **(5b)** In healthy subjects, the intensity for inducing a period of EMG silence is almost the same as the intensity for inducing an MEP. However, in pathological conditions, the

SP may be observed without a preceding MEP, and in some conditions SP may be shortened. An alternative approach is to obtain the I/O curve of the SP similar to the procedures described for the I/O curve of the MEP amplitude in Sect. 5.2. In short, step 5a is repeated for a range of intensities to obtain the I/O curve across a wide range of intensities.

(6a) Data analysis is performed (Fig. 4). The suprathreshold TMS pulse induces an MEP followed by a period of suppressed background EMG activity. Commonly reported outcomes of a cSP protocol are the peak-to-peak amplitude of the initial MEP and the duration of the relative and absolute SP. The mean level of background EMG activity can also be quantified and reported. *Relative* SP is measured from MEP onset to the onset of reemergence of EMG background activity. MEP onset is chosen as the initial start point of SP as the true onset of SP is not known [107]. *Absolute* cSP can be measured from the end of the MEP to the onset of reemergence of EMG background activity. The offset of cSP can be identified manually or by thresholding through automated algorithms [108]. **(6b)** An I/O curve can also be obtained for cSP by obtaining duration of cSP against incremental increases in intensity of stimulation. The cSP I/O curve can be fitted by a sigmoid function, such as the Boltzmann equation [109]. The duration of cSP increases with stimulation intensity and plateaus at high intensities. The slope of the cSP I/O curve has been shown to vary between proximal arm versus distal hand muscles [110].

Ipsilateral Silent Period

iSP is suppression of background EMG activity during voluntary contraction of the target muscle obtained when a suprathreshold TMS pulse is applied to the ipsilateral motor cortex [111, 112].

Step-by-step Procedure. **(1)** The cortical stimulation site is the hot spot of the muscle contralateral to the target muscle [113]. In short, TMS coil placement is determined by delivering suprathreshold single pulse TMS of a fixed intensity to several cortical spots ipsilateral to the target muscle (for which iSP is determined) and identifying the coil placement that produces the largest MEP in the muscle contralateral to the stimulation site (muscle on the opposite side of the target muscle). **(2)** Once the target is determined, the same coil placement should be maintained throughout the procedure by marking the coil placement on the scalp or using coil tracking technologies. **(3)** The participant is instructed to maintain voluntary contraction of the target muscle (ipsilateral to the stimulation site) at 20 % of his/her maximum force, using a similar procedure as described for the cSP. **(4)** Similar to the cSP protocol, stimulation intensity is set at 140 % rMT. Alternatively, a better estimation of iSP measures can be obtained by measuring the I/O curve as described for the cSP protocol.

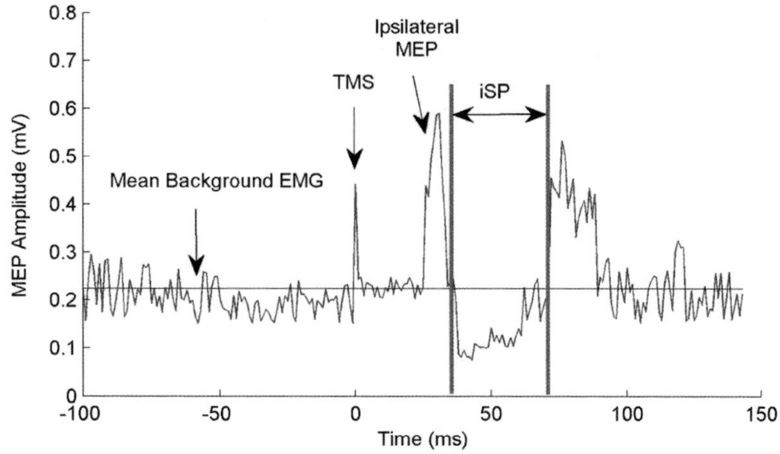

Fig. 5 Ipsilateral silent period. The *x*-axis represents time in milliseconds with respect to the TMS onset. The *y*-axis represents MEP amplitude in millivolt (mV). The *solid horizontal line* represents the mean EMG activity. The first *solid vertical line* illustrates the iSP onset latency, and the second *vertical line* marks the iSP offset latency. The duration of iSP is measured by subtracting the onset latency from the offset latency

(5) Data analysis is performed. The EMG traces obtained in the iSP protocol can be described by the onset latency, duration, and offset latency of the SP (Fig. 5). The SP may be preceded by an MEP. However, the hot spot for generating an ipsilateral MEP (iMEP) has been shown to be more lateral than the hot spot for generating an ipsilateral iSP or a contralateral MEP [113]. The transcallosal transmission time (TCT) can also be measured by subtracting the onset latency of the contralaterally induced MEP from the onset latency of the ipsilateral-induced SP for a given target muscle. The onset latency of MEP and SP are delayed in the ipsilateral (i.e., iSP protocol) compared to the contralateral muscle (i.e., cSP protocol).

5.3.2 Neurobiological Correlates of the Silent Period

Contralateral Silent Period

Several lines of evidence suggest that the first 50 ms of the cSP is due to spinal mechanisms while the later portion is mediated by cortical inhibition [107, 114]. Siebner et al. reported a significant prolongation of the cSP duration following a continuous intrathecal administration of baclofen in a patient with generalized dystonia [115]. Also, tiagabine, a GABA uptake inhibitor, has been shown to prolong cSP duration [116]. Furthermore, administration of vigabatrin, a selective GABAergic drug that increases the availability of GABA in the brain, has also been shown to enhance cSP [117]. Several studies have suggested that cSP is associated with $GABA_B$ receptor-mediated inhibitory neurotransmission [114–116, 118]. Furthermore, $GABA_A$ receptor agonists, such as lorazepam, can prolong cSP. Dopaminergic drugs may also prolong cSP duration [119].

Ipsilateral Silent Period

It has been suggested that iSP is mediated by transcallosal pathways between the targeted hemisphere and the contralateral hemisphere [111], such as inhibitory control of the contralaterally stimulated primary motor cortex. This suggestion is partly based on the impairment of iSP duration in patients with lesions of corpus callosum [112]. Spinal reflexes have been found to remain unchanged during iSP, a finding that further suggests iSP is primarily mediated by supraspinal mechanisms. However, iSP has also been produced through stimulation of the corticospinal tract caudal to the corpus callosum, suggesting that some non-callosal mechanism may also generate iSP [120]. In addition, a second phase of iSP is observed in FDI muscles but not in ABP muscles, and it has been suggested this second phase is *not* mediated by transcallosal pathways [121]. Furthermore, abnormalities of iSP, such as prolongation of iSP offset, is often accompanied by abnormalities of central motor conduction time from the non-stimulated motor cortex to the target muscle. Therefore, it has been proposed that iSP also reflects the integrity of the corticospinal tract from the non-stimulated motor cortex to the muscle ipsilateral to the stimulated motor cortex [121].

5.3.3 Technical Considerations

Contralateral Silent Period

As briefly discussed above, cSP duration can increase with stimulus intensity [121]. Interestingly, cSP is little affected by the amount of voluntary contraction of the target muscle [121]. While the I/O curves of MEP amplitudes and cSP duration are characterized by a linear increase in the stimulation intensity and a plateau phase at high intensities, both I/O curves have unique slopes and may reach plateau at different intensities. As such, some investigators have suggested that, instead of evaluating the cSP at an intensity determined by a percentage of MT (e.g., 140 % of RMT) CSP should be obtained for an intensity at which the I/O curve of cSP reaches a plateau [109].

In general, as will be discussed in Sect. 6, any medical condition or intervention that alters cortical inhibition may affect cSP duration. Furthermore, cSP duration can be affected by TMS coil geometry and has been reported to be longer for circular as compared to figure-of-eight coils [121]. It has also been shown that cSP duration is shortest for monophasic pulses, longest for biphasic pulses, and intermediate for half-sine pulses [47]. Hyperventilation may also influence cSP and, therefore, should be monitored and avoided [122].

Ipsilateral Silent Period

Several factors have been shown to modulate iSP parameters. First, similar to cSP, to compare iSP across subjects, it is ideal to identify the iSP at an intensity that produces the maximum SP duration. Second, it has been reported that an anterior–medial-induced current orientation engenders the longest iSP durations [123]. Furthermore, a recent study has shown that several characteristics of iSP are age- and sex-dependant. For instance, it has been shown that the iSP onset and offset latencies and TCT are prolonged in older adults [124] while MEP onset latency is unchanged. Moreover, women have been shown to have shorter iSP onset and offset latencies regardless of age or menstrual cycle [124]. Interhemispheric asymmetry in iSP duration has also been reported in some, but not all, studies [124]. Therefore, to ensure interpretation accuracy and result reproducibly, these potential confounding factors should be reported and controlled for.

5.4 Single Pulse TMS Protocols in Non-motor Regions of Cortex

In recent years, several investigators have embarked on employing single pulse TMS to study the neurophysiology of non-motor cortical regions. This has been, in part, made possible by advances in neuroimaging methodologies, such as the combination of concurrent TMS and EEG and fMRI recording as discussed in Sect. 3.5. In addition, these neuroimaging modalities have allowed for better elucidation of the neurophysiological underpinnings of EMG response with higher temporal (through EEG) and spatial (through fMRI) precision. In this section, we will provide some examples of single pulse TMS applied in combination with other imaging modalities. The comprehensive discussion of TMS combined with EEG and fMRI and the guidelines are provided in subsequent chapters.

One major advantage of interleaved TMS-EEG (and TMS-fMRI/NIRS) is that it permits for the examination of the effect of a broader range of TMS intensities in motor and non-motor regions. This feature of TMS-EEG may ultimately lead to the design of a more optimal MT and I/O curve protocols for both motor and non-motor regions. For instance, in a series of TMS-EEG studies, the effect of single pulse TMS of sub- and supra-threshold intensities are examined in motor and prefrontal cortices [24, 44]. As discussed in Sect. 3.4, the TMS outcome in EEG studies can be the amplitude of TMS-induced TEPs in local and remote regions. Results of recent studies have revealed that, even at an intensity of 60 % of rMT applied at rest, TMS produces distinct and measureable EEG responses in motor and prefrontal cortices (EEG peaks identified at about 15, 30, 45, 100, 180, and 280 ms following the TMS pulse) [24, 44]. Similar approaches may be used to identify the I/O curve and excitability threshold for several other brain areas. In addition to EEG peaks, the intensity dependence of several other EEG parameters, such as frequency of oscillation, propagation extent of TMS-induced TEPs, and TCT time, can be obtained which may prove valuable in research and clinical investigations.

Moreover, the correlation between EEG and EMG outcomes of TMS can be examined to identify the origin of TMS-induced EMG responses in single pulse protocols [40, 41]. EEG correlates of neural inhibition, excitation, interhemispheric connectivity and several other brain dynamics could ultimately be added to EMG measures to more precisely identify TMS impact on different cortical tissues, which can lead to an expansion of TMS utility for examining a broader range of human brain dynamics [125]. Thus far, the sensitivity, reliability, and validity of several EEG measures have been investigated and promising results have been reported [32, 125, 126].

Hemodynamic response is also being investigated to accurately characterize the impact of single pulse TMS at various intensities in non-cortical brain regions. Although the number of these studies is still far fewer than the TMS-EEG approach, partly due to the technical limitations and poor temporal resolution, some investigators have explored the hemodynamic correlates of the excitatory and inhibitory response to TMS. For instance, the results of one TMS-NIRS study suggest the concentration of oxygenated hemoglobin is increased in response to single pulse TMS administered to the motor cortex at intensities of 90 or 110 % of aMT during voluntary muscle contraction. By contrast, TMS administered at 120 or 140 % of aMT leads to a decrease in deoxygenated hemoglobin [127]. It has been suggested that this may reflect the induction of a long lasting inhibition by suprathreshold TMS pulses [38].

In summary, the single pulse TMS protocols of MT, cSP, iSP, and I/O curve can be employed to examine corticospinal excitability and cortical excitatory, inhibitory and interhemispheric processes in the motor system. Additionally, single pulse TMS can be interleaved with several other neuroimaging modalities and novel outcome measures, such as local field potentials and hemodynamic responses, expand the application of single pulse TMS to evaluate brain dynamics in motor and non-motor brain regions.

6 Clinical Applications

Single pulse TMS protocols can be used to (a) assess the presence and severity of neurological impairments, (b) investigate the pathophysiology of the disease, (c) monitor disease progression, and (d) evaluate the mechanism of actions of various therapeutic approaches. To date, the single pulse TMS protocols of MT, I/O curve, cSP, and iSP have been used for clinical assessments, and the sensitivity of some of these protocols for diagnostic utility has been examined. In order to obtain a precise clinical assessment, often several TMS protocols and neuroimaging measures must be integrated.

6.1 Applications in Stroke Assessment and Follow-up

Changes in MT, I/O curve, and SP following a brain lesion have been used to characterize lesion location and severity [83, 128]. In stroke, EMG measures can be employed immediately after the stroke to examine excitability of the corticospinal pathway in the affected and unaffected hemispheres in order to characterize lesion severity and the integrity of the unaffected regions with the aim of recovery prediction. Furthermore, these measures can be reassessed longitudinally over weeks and months to monitor functional status and recovery. It has been demonstrated that observation of MEPs in the affected limb shortly after stroke predicts good recovery [83, 129, 130]. Conversely, the absence of MEPs in the affected hand and an increase of MEP amplitudes in the unaffected hand may predict poor recovery [83, 131].

The excitatory imbalance between the affected and unaffected hemisphere has been correlated with lesion severity and has been shown to decrease over the recovery period [83, 131]. Since rMT has been found not to be significantly different between homologous motor areas in healthy subjects [132], any interhemispheric asymmetry of rMT can be used as a prognostic measure in stroke. However, it is still not clear to what extent left–right asymmetry may be an indication of an underlying pathological condition. An increase in excitability at the unaffected site is thought to reflect a reduction in interhemispheric inhibition from the affected hemisphere [133]. Thus, an interhemispheric difference in rMT may be an indication of unilateral damage to the corticospinal pathway. Prolongation of cSP has also been observed in cerebral stroke [134, 135].

Single pulse protocols can also be employed to investigate stroke in non-cortical regions. For instance, MT has been shown to be altered in cerebellar stroke. It has been reported that aMT and rMT increase in the motor cortex contralateral to the affected cerebellar hemisphere [136]. This might be associated with deep cerebellar nuclei (e.g., dentate nucleus) which have been shown, through animal studies, to exert excitatory control over the contralateral motor cortex via disynaptic connections through the ventral thalamus. Therefore, MT may also be used to assess the modification of excitability in brain regions remote to the site of stroke. Prolongation of cSP has also been reported in unilateral cerebellar stroke with damage of dentate nucleus [136].

6.2 Applications in TBI Assessment and Follow-up

Traumatic brain injury (TBI) has been identified as a major risk factor for age-related neurological impairments, such as cognitive decline and Alzheimer's disease. As such, TMS measures in TBI assessment involve identifying biomarkers that may identify individuals at higher risk of developing neurological impairments later in life. In particular, this is of interest in sports injuries, which could ultimately lead to preventive strategies for athletes more prone to head injuries or in whom previous incidents have modified brain circuitry making the athlete more susceptible to neurological impairment. In this regard, several investigators have reported prolongation of cSP following sports-related concussion [137–139].

Beaumont et al. have examined several single pulse and paired pulse TMS indices of cortical excitability and inhibitory circuits along with performance on several neuropsychological and movement tasks in athletes who had suffered a sports-related concussion approximately 30 years prior. It was reported that, while rMT, I/O curve, and paired pulse measures of inhibition and excitation were normal, cSP was prolonged compared to athletes without a history of head injury. Furthermore, although daily function had not yet been affected, athletes who had a history of head injury were significantly slower in performing a movement task, and this slowness correlated strongly with cSP prolongation. Collectively, these results suggest that cSP prolongation may be a marker of movement slowness later in life in athletes who suffered from head injuries earlier in their career. As suggested by Beaumont et al., longitudinal monitoring of athletes may help identify those who might suffer from cognitive or movement impairments.

This line of research could be expanded to include TMS-EEG approaches to examine cortical reactivity in non-motor regions which may prove more relevant to the pathophysiology of cognitive impairment. For example, single pulse TMS of prefrontal and parietal brain regions may be optimal in investigating risk factors associated with TBI-related cognitive decline.

6.3 Applications in Neurodegenerative Diseases Assessment and Follow-up

Single pulse TMS protocols can also be used to assess disease progression and functional outcome in neurodegenerative disorders, such as Parkinson's disease, Alzheimer's disease, multiple sclerosis (MS), and ALS. The findings of several clinical assessments suggest that changes in corticospinal excitability vary with illness severity, making TMS a potentially powerful tool for evaluating functional status and, ultimately, predicting functional outcome in certain patients. Here, we will provide some evidence from clinical TMS investigations into a few several neurodegenerative disorders.

In patients with *Parkinson's disease*, shortening of cSP has been repeatedly reported, in particular in unmedicated patients [119, 140]. Furthermore, several interventions, such as repetitive TMS [141], 8 weeks of intense exercise [142], and dopaminergic treatments [143], have been shown to normalize cSP duration in this patient population.

In *Alzheimer's disease*, reduced rMT [144, 145] and, in some studies, reduced cSP duration [144] have been reported. Furthermore, it has been demonstrated that spinal reflexes (H- and F-reflex) are normal in patients, providing evidence that rMT reduction is likely caused by an increase in cortical excitability. Moreover, a progressive increase in rMT has been reported in patients whose cognitive deficits worsened over time despite medication. In a recent pilot study, both cognitive abilities and rMT were obtained before and 1 year after patients received acetylcholinesterase inhibitor therapy. Interestingly, in patients who displayed a stabilized cognitive performance, TMS measures of cortical excitability (including the rMT) were not changed. As such, rMT may be used to examine the efficacy of different therapeutic approaches and monitor disease progression. However, to examine the excitability of non-motor regions, a line of research investigating EEG measures of cortical excitability in response to single pulses of TMS applied to prefrontal and parietal regions is emerging. Preliminary findings suggest that early EEG response to single pulse TMS (e.g., EEG peak at 30 ms post-TMS) may be used to distinguish patients with Alzheimer's disease and mild cognitive impairment from normal aging [146]. These clinical investigations may lead to identification of predictive measures of cognitive decline which could be employed to identify and treat high-risk individuals at early stages of the disease.

In patients with *Amyotrophic lateral sclerosis*, an increase in MT has been demonstrated [83, 147, 148] which has been found to be preceded by a period of decreased MT during early disease phases [149]. Furthermore, increases in rMT and aMT have been found to correlate with the severity of ALS [150]. In addition, shortening of cSP duration, in particular during early disease stages, has been reported [83, 147–149]. Reduction of MT early in the disease has been suggested to correlate with the glutamate-induced excitatory neurotoxicity in this illness. Moreover, some investigators have observed a correlation between disease progression and increased MT [147]. Neurophysiological measures that can be obtained from a combination of single pulse TMS protocols of corticospinal excitability (e.g., rMT, motor mapping, and central motor conduction time) may be integrated to identify the subtle presence of upper motor neuron impairments that are not easily diagnosed through EMG examination [148]. This has the potential of improving ALS diagnosis in patients with motor neuron diseases [148]. The combination of rMT and central motor conduction time has also proven useful as a follow-up assessment measure to monitor functional status and disease progression in this population [151]. It has been suggested that the combination of several TMS parameters and neuroimaging modalities will improve differentiating ALS from other motor neuron diseases will also provide more insight into the pathophysiological underpinnings of this neurodegenerative disorder.

In *multiple sclerosis*, impairments in the single pulse TMS protocols of iSP, cSP, and central motor conduction time have been identified and their potentialities as sensitive diagnostic measures are being investigated. Prolongation of central motor conduction time has been proposed to be a sensitive index of impairments and disease progression in MS. In addition, abnormalities of iSP, such as prolonged duration and offset latency, have been reported [152] consistent with the atrophy of the corpus callosum in MS. However, it is not clear whether iSP and central motor conduction time impairments are due to two different abnormal neural mechanisms or are both related to corticospinal tract degradation [152]. In addition, some studies have reported abnormal elevation in rMT in relapsing patients which has been found to be accompanied by a shortened cSP duration. Interestingly, rMT has been found to be normal and cSP duration prolonged in remitting patients [153].

Therefore, single pulse TMS protocols have applications in monitoring disease progression. Future longitudinal assessments of TMS-induced EMG and EEG outcome measures in patients with neurodegenerative disorders *versus* age-matched normal aging subjects will permit identifying indices that can be used to assess the integrity of the neurophysiological processes throughout disease progression to predict functional outcome and refine treatment plans.

6.4 Applications in Epilepsy

Findings of TMS studies in epileptic patients have revealed that changes in cortical excitability varies across different types of epilepsy, medicated *versus* untreated patients, and time of testing relative to a prior or future seizure episode [154]. For example, interhemispheric differences in rMT were shown to be exaggerated in patients with idiopathic generalized epilepsy (IGE) who displayed versive or circling seizures as compared to IGE patients without versive or circling seizures and normal subjects [155]. In another study, rMT was tested 48 h after unmedicated patients had their first grand mal seizure, and it was shown, probably due to protective and compensatory mechanisms, rMT was increased without any change in MEP amplitudes. This is suggestive of a decrease in cortical excitability [156]. The rMT has also been employed to examine the effect of anticonvulsant medications on cortical excitability. For example, it has been shown that rMT is reduced in unmedicated patients with IGE compared to healthy subjects. Following the administration of an antiepileptic medication (e.g., valporate), rMT becomes significantly higher in patients compared to healthy controls [157]. In addition, prolongation of cSP has been observed in IGE patients [158] and in partial seizure during periods with high chance of relapse [159]. As it has been reported that spinal excitability (H- and F-reflexes) are normal in patients, it has been argued that increases in cSP may reflect an increase in the excitability of cortical $GABA_B$ receptor-mediated inhibition [159].

In addition to the above-mentioned TMS measures of excitability in the motor system, a new line of research is employing other multimodal interleaved approaches to characterize modifications of non-motor systems in various types of epilepsy and to evaluate the effect/s of antiepileptic medications across brain networks. For example, single pulse TMS interleaved with fMRI has been employed to investigate the effect of valporate in the brain of healthy subjects when the motor or the prefrontal cortex is stimulated with single pulse of TMS [160, 161]. Similarly, single pulse TMS applied to the temporal cortex can be used to identify EEG correlates of modified excitability in regions that might be more closely implicated in some types of epilepsy.

6.5 Applications in Pediatric Population

The safety of application of several TMS protocols in children has been investigated and a growing number of studies are now utilizing single pulse TMS to evaluate TMS measures of cortical excitation, inhibition, and functional connectivity in healthy children. The results of these endeavors are paralleled by clinical investigations in children with neurological disorders, such as Tourette syndrome [162], autism, attention deficit hyperactive disorder [163], and Periventricular leukomalacia [164]. Converging lines of evidence suggest that single pulse TMS protocols, such as MT, iSP, and I/O curve, go through maturational changes. For example, as reviewed by Garvey et al. [165], rMT is hard to identify, even at 100 % of TMS machine output, in children younger than 6 years old. ISP is absent in children 4–6 years of age and has a longer latency and shorter duration until in the age of 12 [166]. Moreover, the slope of the I/O curve becomes steeper with age and MEP onset latency shortens. Given the age-dependant changes in TMS measures of cortical reactivity, single pulse TMS can be used to monitor delayed maturation in children and may be used for prognosis in children with neurodevelopment disorders [165].

In addition, the noninvasive nature of EEG and its combination with TMS has permitted to further examine maturation and impairment of cortical reactivity in children. For instance, EEG measures of single pulse TMS can be used to investigate neurophysiological impairments that may underlie the pathophysiology of attention deficit hyperactive disorder. To this end, the N100 EEG response (negative deflection at latency of 100 ms following TMS) following a single TMS pulse (Fig. 1a), which has been associated with inhibitory cortical processes in adults, has been found to be larger in healthy children compared to adults. Moreover, a negative correlation has been found between the N100 response and age in children, suggesting that N100 response may also undergo maturation [167]. Recently, a TMS-EEG study in children with attention deficit hyperactive disorder illustrated that the N100 may be used as a marker of abnormal brain maturation in younger populations [163].

6.6 Other Clinical Applications

The diagnostic and prognostic values of single pulse TMS protocols have been evaluated for several other neurological and psychiatric populations that, due to space limitations, cannot be fully reviewed here. For instance, reduction of cSP duration has been found in several psychiatric illnesses, such as schizophrenia [168–170], bipolar disorder [171], unipolar major depression [172], obsessive compulsive disorder [173], and borderline personality disorder [174]. Single pulse protocols have also been used in pain research to investigate the pathophysiology of fibromyalgia [175] and chronic neuropathic pain [176]. Additionally, impairments in TMS measures have also been investigated for cerebellar diseases, such as spinocerebellar and Fredrick's ataxia, in which prolongation of cSP and central motor conduction time have been observed [177–179].

Collectively, this growing body of clinical results may lead to the discovery of common and distinct neural mechanisms that underlie various disorders of the human nervous system and may aid the development of general and disease-specific biomarkers of neural deficits and disease progression.

7 Experimental Applications

Single pulse TMS has been widely used in fundamental science, cognitive neuroscience, and social psychology. The experimental application of single pulse TMS may be categorized by: (1) Studies that use single pulse TMS to investigate neural dynamics that underlie various brain states and functions (in these studies, single pulse TMS may be used to interrupt the neural mechanisms of specific brain regions to identify their roles in function and behavior); and (2) studies that use expected brain response to single pulse TMS to investigate the effect of an intervention on brain dynamics.

7.1 Single Pulse TMS in Investigating the Brain Dynamics and the Brain–Behavior Relationship

Single pulse TMS protocols can be used to examine the state of corticospinal excitability, cortical inhibition, and interhemispheric and functional connectivity during sleep, wakeful resting, and task performance. For example, in sleep research [31, 180, 181], single pulse TMS may be used to evaluate cortical excitability through comparing TMS measures obtained during rest to measures obtained after waking from REM *versus* NREM sleep [180] and during sleep deprivation [181]. In one TMS-EEG study, single pulse TMS was utilized to identify the effect of "time-awake" on excitability of the prefrontal cortex. Indeed, it was shown the EEG measure of cortical excitability increases as a function of wake time [31]. Furthermore, EEG and EMG response to single pulse TMS applied during movement, such as a visually triggered hand movement [27], can be used to identify the subtle changes in cortical reactivity during motor execution.

Single pulse TMS may also be used to probe region-specific properties of cortical tissues. To this end, the natural frequency of cortical oscillations in the frontal, occipital, and parietal cortices has been investigated [182]. In this experiment, single pulse TMS was separately applied to each of these brain regions and the frequency of the TMS-induced TEP, as measured using EEG, was evaluated for each region. Results demonstrated that each region had a unique prominent frequency of oscillation.

The role of specific neural dynamics or networks in cognitive function can be studied by interfering with normal neural reactivity using a TMS pulse and evaluating the impact of this interference on behavior and task performance [183]. For example, in a study that examined the neural dynamics of sensorimotor integration in grasp planning, single pulse TMS applied to the anterior–intraparietal sulcus within 200 ms of object presentation reduced the electrophysiological correlates of enhanced motor planning and impaired subjects' ability to use learned object knowledge in movement planning [183].

7.2 Single Pulse TMS in Assessing the Impact of an Intervention

Single pulse TMS can be used to evaluate intervention-specific modifications of brain dynamics. To achieve this, single pulse TMS protocols are assessed prior to the administration of an intervention and are reassessed following the course of an intervention. The time interval between the intervention and the follow-up assessments should be carefully determined based on the lasting effect of the intervention. For instance, MT and cSP have been extensively used to evaluate the effect of various pharmacological agents on the motor cortex neurophysiology (for a review, refer to [81]). For example, cSP is prolonged in response to alcohol intake, while rMT and aMT remain unchanged. This illustrates that ethanol potentiates intracortical inhibitory mechanism in the motor cortex. Prolongation of cSP has been observed after administration of several antiepileptic drugs, such as gabapentin and carbamazepine [80]. In addition, the effects of drugs on non-motor regions can be studied using the TMS-EEG approach. For example, single pulse TMS was used to assess the effect of alcohol on the prefrontal cortex [23, 184]. Moreover, TMS combined with fMRI has been used to investigate the effect of psychoactive drugs on subcortical structures and network connectivity. In one TMS-fMRI study, TMS was applied to either the motor or the prefrontal cortex in healthy subjects before and after oral administration of two psychoactive drugs (valporate and lamotrigine) [160, 161]. An assessment of changes in functional connectivity suggested these drugs had region- and network-specific effects when administered in combination as opposed to alone.

Single pulse TMS can also be used to investigate the effect of repetitive TMS on motor cortical physiology. As will be discussed in the future chapters, repetitive TMS can induce neurophysiological and behavioral changes that may outlast the duration of stimulation by minutes to hours. The neurophysiological impact of rTMS can be studied by examining TMS measures of cortical excitability before and after rTMS application. Single pulse TMS can also be used to evaluate how long the rTMS-induced effects last. This can be achieved by obtaining a TMS-induced response (e.g., MEP) to single pulses of fixed suprathreshold intensity at multiple time points prior to and following rTMS delivery (e.g., 5 min, 15 min, ... 90 min) and identifying the time point at which the MEP responses return to baseline level. Similarly, the lasting effect of rTMS on non-motor regions of the cortex can be explored through monitoring the changes of TMS-induced EEG responses [185]. Moreover, rMT is often used to adjust the intensity of rTMS interventions based on each individual's cortical excitability.

8 Summary

The building block of all TMS protocols is a single pulse of TMS which by its interaction with the brain permits studying and modulating the underlying tissue, neural pathways, and brain networks that the stimulated area is functionally and structurally connected to. Application of a single TMS pulse of sufficient intensity to the motor cortex can activate the corticospinal tract either directly or transynaptically through intracortical and cortico-cortical circuitries. This activation can result in a discharge of corticospinal neurons at about 600 Hz frequency for about 10 ms which can lead to the generation of MEPs in the target muscle. This period of excitability is followed by an inhibitory period that may last up to several 100 ms and results in the suppression of background EMG activity when TMS is applied during voluntary muscle contraction of an ipsilateral or contralateral muscle. A ramping increase in TMS intensity leads to an approximately linear increase in the magnitude and duration of TMS-induced MEPs and SPs, which reach a plateau at high intensities. In this chapter, we described how these TMS-induced indices of brain reactivity can be employed in clinics and research laboratories to characterize brain dynamics and various function in health and disease.

However, the cascade of events generated following a single TMS pulse can also have an impact on neural dynamics of areas remote to the stimulation site. Conversely, the neural reactivity of nonstimulated regions may impact the peripheral motor response to TMS. To further evaluate the impact of motor cortical stimulation on other regions, and to assess functional connectivity between brain areas, paired pulse TMS studies have been designed which will be discussed in later chapters. However, an alternative, and perhaps complementary approach, to paired pulse technique is the combination of TMS with other neuroimaging modalities, such as EEG and fMRI. This approach further allows for the assessment of non-motor regions of the cortex in which TMS application does not generate a measurable response at the periphery. We introduced several neuroimaging modalities that have been interleaved with TMS, and the results of several studies have provided us with further insight into TMS mechanisms of action. As the number of interleaved studies grows, it is expected the field will refine existing single pulse protocols and establish several others that, together, may lead to optimal utility of the TMS single pulse approach in both experimental and clinical applications.

References

1. Hess CW, Mills KR, Murray NM (1987) Responses in small hand muscles from magnetic stimulation of the human brain. J Physiol 388:397–419
2. Day BL, Dressler D, Maertens de Noordhout A, Marsden CD, Nakashima K, Rothwell JC, Thompson PD (1989) Electric and magnetic stimulation of human motor cortex: surface EMG and single motor unit responses. J Physiol 412:449–473
3. Knikou M (2008) The H-reflex as a probe: pathways and pitfalls. J Neurosci Methods 171:1–12
4. Boyd SG, Rothwell JC, Cowan JM, Webb PJ, Morley T, Asselman P, Marsden CD (1986) A method of monitoring function in corticospinal pathways during scoliosis surgery with a note on motor conduction velocities. J Neurol Neurosurg Psychiatry 49:251–257
5. Burke D, Hicks R, Gandevia SC, Stephen J, Woodforth I, Crawford M (1993) Direct comparison of corticospinal volleys in human subjects to transcranial magnetic and electrical stimulation. J Physiol 470:383–393
6. Nakamura H, Kitagawa H, Kawaguchi Y, Tsuji H (1996) Direct and indirect activation of human corticospinal neurons by transcranial magnetic and electrical stimulation. Neurosci Lett 210:45–48
7. Di Lazzaro V, Oliviero A, Profice P, Saturno E, Pilato F, Insola A, Mazzone P, Tonali P, Rothwell JC (1998) Comparison of descending volleys evoked by transcranial magnetic and electric stimulation in conscious humans. Electroencephalogr Clin Neurophysiol 109: 397–401
8. Di Lazzaro V, Restuccia P, Oliviero A, Profice P, Ferrara L, Insola A, Mazzone P, Tonali P, Rothwell JC (1998) Effects of voluntary contraction on descending volleys evoked by transcranial stimulation in conscious humans. J Physiol 508:625–633
9. Di Lazzaro V, Oliviero A, Profice P, Insola A, Mazzone P, Tonali P, Rothwell JC (1999) Direct recordings of descending volleys after transcranial magnetic and electric motor cortex stimulation in conscious humans. Electroencephalogr Clin Neurophysiol Suppl 51: 120–126
10. Di Lazzaro V, Oliviero A, Profice P, Meglio M, Cioni B, Tonali P, Rothwell JC (2001) Descending spinal cord volleys evoked by transcranial magnetic and electrical stimulation of the motor cortex leg area in conscious humans. J Physiol 537:1047–1058
11. Di Lazzaro V, Oliviero A, Saturno E, Pilato F, Insola A, Mazzone P, Profice P, Tonali P, Rothwell JC (2001) The effect on corticospinal volleys of reversing the direction of current induced in the motor cortex by transcranial magnetic stimulation. Exp Brain Res 138: 268–273
12. Di Lazzaro V, Oliviero A, Pilato F, Saturno E, Insola A, Mazzone P, Tonali PA, Rothwell JC (2002) Descending volleys evoked by transcranial magnetic stimulation of the brain in conscious humans: effects of coil shape. Clin Neurophysiol 113:114–119
13. Di Lazzaro V, Oliviero A, Pilato F, Mazzone P, Insola A, Ranieri F, Tonali PA (2003) Corticospinal volleys evoked by transcranial stimulation of the brain in conscious humans. Neurol Res 25:143–150
14. Di Lazzaro V, Oliviero A, Pilato F, Saturno E, Dileone M, Meglio M, Cioni B, Papacci F, Tonali PA, Rothwell JC (2004) Comparison of descending volleys evoked by transcranial and epidural motor cortex stimulation in a conscious patient with bulbar pain. Clin Neurophysiol 115:834–838
15. Amassian VE, Stewart M, Quirk GJ, Rosenthal JL (1987) Physiological basis of motor effects of a transient stimulus to cerebral cortex. Neurosurgery 20:74–93
16. Groppa S, Oliviero A, Eisen A, Quartarone A, Cohen LG, Mall V, Kaelin-Lang A, Mima T, Rossi S, Thickbroom GW, Rossini PM, Ziemann U, Valls-Sole J, Siebner HR (2012) A practical guide to diagnostic transcranial magnetic stimulation: report of an IFCN committee. Clin Neurophysiol 123:858–882
17. Di Lazzaro V, Profice P, Ranieri F, Capone F, Dileone M, Oliviero A, Pilato F (2012) I-wave origin and modulation. Brain Stimul 5(4):512–525
18. Ziemann U, Rothwell JC (2000) I-waves in motor cortex. J Clin Neurophysiol 17: 397–405
19. Cracco RQ, Amassian VE, Maccabee PJ, Cracco JB (1989) Comparison of human transcallosal responses evoked by magnetic coil and electrical stimulation. Electroencephalogr Clin Neurophysiol 74:417–424
20. Ilmoniemi RJ, Virtanen J, Ruohonen J, Karhu J, Aronen HJ, Naatanen R, Katila T (1997) Neuronal responses to magnetic stimulation reveal cortical reactivity and connectivity. Neuroreport 8:3537–3540
21. Paus T, Sipila PK, Strafella AP (2001) Synchronization of neuronal activity in the

human primary motor cortex by transcranial magnetic stimulation: an EEG study. J Neurophysiol 86:1983–1990

22. Kahkonen S, Kesaniemi M, Nikouline VV, Karhu J, Ollikainen M, Holi M, Ilmoniemi RJ (2001) Ethanol modulates cortical activity: direct evidence with combined TMS and EEG. Neuroimage 14:322–328
23. Kahkonen S, Wilenius J, Nikulin VV, Ollikainen M, Ilmoniemi RJ (2003) Alcohol reduces prefrontal cortical excitability in humans: a combined TMS and EEG study. Neuropsychopharmacology 28:747–754
24. Kahkonen S, Komssi S, Wilenius J, Ilmoniemi RJ (2005) Prefrontal transcranial magnetic stimulation produces intensity-dependent EEG responses in humans. Neuroimage 24: 955–960
25. Komssi S, Aronen HJ, Huttunen J, Kesaniemi M, Soinne L, Nikouline VV, Ollikainen M, Roine RO, Karhu J, Savolainen S, Ilmoniemi RJ (2002) Ipsi- and contralateral EEG reactions to transcranial magnetic stimulation. Clin Neurophysiol 113:175–184
26. Komssi S, Kahkonen S (2006) The novelty value of the combined use of electroencephalography and transcranial magnetic stimulation for neuroscience research. Brain Res Brain Res Rev 52:183–192
27. Nikulin VV, Kicic D, Kahkonen S, Ilmoniemi RJ (2003) Modulation of electroencephalographic responses to transcranial magnetic stimulation: evidence for changes in cortical excitability related to movement. Eur J Neurosci 18:1206–1212
28. Thut G, Northoff G, Ives JR, Kamitani Y, Pfennig A, Kampmann F, Schomer DL, Pascual-Leone A (2003) Effects of single-pulse transcranial magnetic stimulation (TMS) on functional brain activity: a combined event-related TMS and evoked potential study. Clin Neurophysiol 114:2071–2080
29. Esser SK, Huber R, Massimini M, Peterson MJ, Ferrarelli F, Tononi G (2006) A direct demonstration of cortical LTP in humans: a combined TMS/EEG study. Brain Res Bull 69:86–94
30. Vyazovskiy VV, Olcese U, Lazimy YM, Faraguna U, Esser SK, Williams JC, Cirelli C, Tononi G (2009) Cortical firing and sleep homeostasis. Neuron 63:865–878
31. Huber R, Maki H, Rosanova M, Casarotto S, Canali P, Casali AG, Tononi G, Massimini M (2013) Human cortical excitability increases with time awake. Cereb Cortex 23(2): 332–338
32. Lioumis P, Kicic D, Savolainen P, Makela JP, Kahkonen S (2009) Reproducibility of TMS-Evoked EEG responses. Hum Brain Mapp 30:1387–1396
33. Farzan F, Barr MS, Levinson AJ, Chen R, Wong W, Fitzgerald PB, Daskalakis ZJ (2010) Reliability of long interval cortical inhibition in healthy human subjects: a TMS-EEG study. J Neurophysiol 104(3):1339–1346
34. Bohning DE, Shastri A, Wassermann EM, Ziemann U, Lorberbaum JP, Nahas Z, Lomarev MP, George MS (2000) BOLD-f MRI response to single-pulse transcranial magnetic stimulation (TMS). J Magn Reson Imaging 11:569–574
35. Paus T, Jech R, Thompson CJ, Comeau R, Peters T, Evans AC (1997) Transcranial magnetic stimulation during positron emission tomography: a new method for studying connectivity of the human cerebral cortex. J Neurosci 17:3178–3184
36. Pogarell O, Koch W, Popperl G, Tatsch K, Jakob F, Zwanzger P, Mulert C, Rupprecht R, Moller HJ, Hegerl U, Padberg F (2006) Striatal dopamine release after prefrontal repetitive transcranial magnetic stimulation in major depression: preliminary results of a dynamic [123I] IBZM SPECT study. J Psychiatr Res 40:307–314
37. Pogarell O, Koch W, Popperl G, Tatsch K, Jakob F, Mulert C, Grossheinrich N, Rupprecht R, Moller HJ, Hegerl U, Padberg F (2007) Acute prefrontal rTMS increases striatal dopamine to a similar degree as D-amphetamine. Psychiatry Res 156:251–255
38. Siebner HR, Bergmann TO, Bestmann S, Massimini M, Johansen-Berg H, Mochizuki H, Bohning DE, Boorman ED, Groppa S, Miniussi C, Pascual-Leone A, Huber R, Taylor PC, Ilmoniemi RJ, De Gennaro L, Strafella AP, Kahkonen S, Kloppel S, Frisoni GB, George MS, Hallett M, Brandt SA, Rushworth MF, Ziemann U, Rothwell JC, Ward N, Cohen LG, Baudewig J, Paus T, Ugawa Y, Rossini PM (2009) Consensus paper: combining transcranial stimulation with neuroimaging. Brain Stimul 2:58–80
39. Bestmann S, Ruff CC, Blankenburg F, Weiskopf N, Driver J, Rothwell JC (2008) Mapping causal interregional influences with concurrent TMS-fMRI. Exp Brain Res 191: 383–402
40. Maki H, Ilmoniemi RJ (2010) The relationship between peripheral and early cortical activation induced by transcranial magnetic stimulation. Neurosci Lett 478(1):24–28

41. Maki H, Ilmoniemi RJ (2010) EEG oscillations and magnetically evoked motor potentials reflect motor system excitability in overlapping neuronal populations. Clin Neurophysiol 121: 492–501
42. Chipchase L, Schabrun S, Cohen L, Hodges P, Ridding M, Rothwell J, Taylor J, Ziemann U (2012) A checklist for assessing the methodological quality of studies using transcranial magnetic stimulation to study the motor system: an international consensus study. Clin Neurophysiol 123(9):1698–1704
43. Rossini PM, Barker AT, Berardelli A, Caramia MD, Caruso G, Cracco RQ, Dimitrijevic MR, Hallett M, Katayama Y, Lucking CH et al (1994) Non-invasive electrical and magnetic stimulation of the brain, spinal cord and roots: basic principles and procedures for routine clinical application. Report of an IFCN committee. Electroencephalogr Clin Neurophysiol 91:79–92
44. Komssi S, Lonen S, Ilmoniemi RJ (2004) The effect of stimulus intensity on brain responses evoked by transcranial magnetic stimulation. Hum Brain Mapp 21:154–164
45. Barker AT, Jalinous R, Freeston IL (1985) Non-invasive magnetic stimulation of human motor cortex. Lancet 1:1106–1107
46. Gattinger N, Moessnang G, Gleich B (2012) flexTMS—a novel repetitive transcranial magnetic stimulation device with freely programmable stimulus currents. IEEE Trans Biomed Eng 59:1962–1970
47. Sommer M, Alfaro A, Rummel M, Speck S, Lang N, Tings T, Paulus W (2006) Half sine, monophasic and biphasic transcranial magnetic stimulation of the human motor cortex. Clin Neurophysiol 117:838–844
48. Di Lazzaro V, Oliviero A, Mazzone P, Insola A, Pilato F, Saturno E, Accurso A, Tonali P, Rothwell JC (2001) Comparison of descending volleys evoked by monophasic and biphasic magnetic stimulation of the motor cortex in conscious humans. Exp Brain Res 141: 121–127
49. Rothkegel H, Sommer M, Paulus W, Lang N (2010) Impact of pulse duration in single pulse TMS. Clin Neurophysiol 121:1915–1921
50. Di Lazzaro V, Oliviero A, Pilato F, Saturno E, Dileone M, Mazzone P, Insola A, Tonali PA, Rothwell JC (2004) The physiological basis of transcranial motor cortex stimulation in conscious humans. Clin Neurophysiol 115: 255–266
51. Wagner T, Gangitano M, Romero R, Theoret H, Kobayashi M, Anschel D, Ives J, Cuffin N, Schomer D, Pascual-Leone A (2004) Intracranial measurement of current densities induced by transcranial magnetic stimulation in the human brain. Neurosci Lett 354: 91–94
52. Pell GS, Roth Y, Zangen A (2010) Modulation of cortical excitability induced by repetitive transcranial magnetic stimulation: influence of timing and geometrical parameters and underlying mechanisms. Prog Neurobiol 93: 59–98
53. Thielscher A, Opitz A, Windhoff M (2011) Impact of the gyral geometry on the electric field induced by transcranial magnetic stimulation. Neuroimage 54:234–243
54. Fox PT, Narayana S, Tandon N, Sandoval H, Fox SP, Kochunov P, Lancaster JL (2004) Column-based model of electric field excitation of cerebral cortex. Hum Brain Mapp 22:1–14
55. Niehaus L, Meyer BU, Weyh T (2000) Influence of pulse configuration and direction of coil current on excitatory effects of magnetic motor cortex and nerve stimulation. Clin Neurophysiol 111:75–80
56. Brasil-Neto JP, Cohen LG, Panizza M, Nilsson J, Roth BJ, Hallett M (1992) Optimal focal transcranial magnetic activation of the human motor cortex: effects of coil orientation, shape of the induced current pulse, and stimulus intensity. J Clin Neurophysiol 9: 132–136
57. Day BL, Rothwell JC, Thompson PD, Dick JP, Cowan JM, Berardelli A, Marsden CD (1987) Motor cortex stimulation in intact man. 2. Multiple descending volleys. Brain 110(Pt 5):1191–1209
58. Sommer M, Norden C, Schmack L, Rothkegel H, Lang N, Paulus W (2013) Opposite optimal current flow directions for induction of neuroplasticity and excitation threshold in the human motor cortex. Brain Stimul 6(3): 363–370
59. Meyer BU, Diehl R, Steinmetz H, Britton TC, Benecke R (1991) Magnetic stimuli applied over motor and visual cortex: influence of coil position and field polarity on motor responses, phosphenes, and eye movements. Electroencephalogr Clin Neurophysiol Suppl 43:121–134
60. Kammer T, Beck S, Erb M, Grodd W (2001) The influence of current direction on phosphene thresholds evoked by transcranial magnetic stimulation. Clin Neurophysiol 112: 2015–2021
61. Kammer T, Vorwerg M, Herrnberger B (2007) Anisotropy in the visual cortex investigated by neuronavigated transcranial magnetic stimulation. Neuroimage 36: 313–321

62. Hill AC, Davey NJ, Kennard C (2000) Current orientation induced by magnetic stimulation influences a cognitive task. Neuroreport 11:3257–3259
63. Opitz A, Windhoff M, Heidemann RM, Turner R, Thielscher A (2011) How the brain tissue shapes the electric field induced by transcranial magnetic stimulation. Neuroimage 58:849–859
64. Rudiak D, Marg E (1994) Finding the depth of magnetic brain stimulation: a re-evaluation. Electroencephalogr Clin Neurophysiol 93: 358–371
65. Amassian VE, Eberle L, Maccabee PJ, Cracco RQ (1992) Modelling magnetic coil excitation of human cerebral cortex with a peripheral nerve immersed in a brain-shaped volume conductor: the significance of fiber bending in excitation. Electroencephalogr Clin Neurophysiol 85:291–301
66. Maccabee PJ, Amassian VE, Eberle LP, Cracco RQ (1993) Magnetic coil stimulation of straight and bent amphibian and mammalian peripheral nerve in vitro: locus of excitation. J Physiol 460:201–219
67. Nagarajan SS, Durand DM, Warman EN (1993) Effects of induced electric fields on finite neuronal structures: a simulation study. IEEE Trans Biomed Eng 40:1175–1188
68. Roth BJ, Saypol JM, Hallett M, Cohen LG (1991) A theoretical calculation of the electric field induced in the cortex during magnetic stimulation. Electroencephalogr Clin Neurophysiol 81:47–56
69. Scivill I, Barker AT, Freeston IL (1996) Finite element modeling of magnetic stimulation of the spine. In: Proc. 18th Annu. Int. Conf. IEEE Engineering in Medicine and Biology Society. pp 393–394
70. Wagner TA, Zahn M, Grodzinsky AJ, Pascual-Leone A (2004) Three-dimensional head model simulation of transcranial magnetic stimulation. IEEE Trans Biomed Eng 51: 1586–1598
71. Pridmore S, Fernandes Filho JA, Nahas Z, Liberatos C, George MS (1998) Motor threshold in transcranial magnetic stimulation: a comparison of a neurophysiological method and a visualization of movement method. J ECT 14:25–27
72. Mills KR, Nithi KA (1997) Corticomotor threshold to magnetic stimulation: normal values and repeatability. Muscle Nerve 20: 570–576
73. Awiszus F (2003) TMS and threshold hunting. Suppl Clin Neurophysiol 56:13–23
74. Tranulis C, Gueguen B, Pham-Scottez A, Vacheron MN, Cabelguen G, Costantini A, Valero G, Galinovski A (2006) Motor threshold in transcranial magnetic stimulation: comparison of three estimation methods. Neurophysiol Clin 36:1–7
75. Rossi S, Hallett M, Rossini PM, Pascual-Leone A (2009) Safety, ethical considerations, and application guidelines for the use of transcranial magnetic stimulation in clinical practice and research. Clin Neurophysiol 120: 2008–2039
76. Awiszus F (2011) Fast estimation of transcranial magnetic stimulation motor threshold: is it safe? Brain Stimul 4:58–59; discussion 60–63
77. Brasil-Neto JP, McShane LM, Fuhr P, Hallett M, Cohen LG (1992) Topographic mapping of the human motor cortex with magnetic stimulation: factors affecting accuracy and reproducibility. Electroencephalogr Clin Neurophysiol 85:9–16
78. Chen R, Samii A, Canos M, Wassermann EM, Hallett M (1997) Effects of phenytoin on cortical excitability in humans. Neurology 49:881–883
79. Ziemann U, Lonnecker S, Steinhoff BJ, Paulus W (1996) Effects of antiepileptic drugs on motor cortex excitability in humans: a transcranial magnetic stimulation study. Ann Neurol 40:367–378
80. Ziemann U, Lonnecker S, Steinhoff BJ, Paulus W (1996) The effect of lorazepam on the motor cortical excitability in man. Exp Brain Res 109:127–135
81. Ziemann U (2004) TMS and drugs. Clin Neurophysiol 115:1717–1729
82. Kloppel S, Baumer T, Kroeger J, Koch MA, Buchel C, Munchau A, Siebner HR (2008) The cortical motor threshold reflects microstructural properties of cerebral white matter. Neuroimage 40:1782–1791
83. Chen R, Cros D, Curra A, Di Lazzaro V, Lefaucheur JP, Magistris MR, Mills K, Rosler KM, Triggs WJ, Ugawa Y, Ziemann U (2008) The clinical diagnostic utility of transcranial magnetic stimulation: report of an IFCN committee. Clin Neurophysiol 119:504–532
84. Rossini PM, Rosinni L, Ferreri F (2010) Brain-behavior relations: transcranial magnetic stimulation: a review. IEEE Eng Med Biol Mag 29:84–95
85. Chen R, Tam A, Butefisch C, Corwell B, Ziemann U, Rothwell JC, Cohen LG (1998) Intracortical inhibition and facilitation in different representations of the human motor cortex. J Neurophysiol 80:2870–2881
86. Kozel FA, Nahas Z, deBrux C, Molloy M, Lorberbaum JP, Bohning D, Risch SC, George MS (2000) How coil-cortex distance relates

to age, motor threshold, and antidepressant response to repetitive transcranial magnetic stimulation. J Neuropsychiatry Clin Neurosci 12:376–384

87. McConnell KA, Nahas Z, Shastri A, Lorberbaum JP, Kozel FA, Bohning DE, George MS (2001) The transcranial magnetic stimulation motor threshold depends on the distance from coil to underlying cortex: a replication in healthy adults comparing two methods of assessing the distance to cortex. Biol Psychiatry 49:454–459
88. Herbsman T, Forster L, Molnar C, Dougherty R, Christie D, Koola J, Ramsey D, Morgan PS, Bohning DE, George MS, Nahas Z (2009) Motor threshold in transcranial magnetic stimulation: the impact of white matter fiber orientation and skull-to-cortex distance. Hum Brain Mapp 30:2044–2055
89. Danner N, Kononen M, Saisanen L, Laitinen R, Mervaala E, Julkunen P (2012) Effect of individual anatomy on resting motor threshold—computed electric field as a measure of cortical excitability. J Neurosci Methods 203:298–304
90. Izumi S, Findley TW, Ikai T, Andrews J, Daum M, Chino N (1995) Facilitatory effect of thinking about movement on motor-evoked potentials to transcranial magnetic stimulation of the brain. Am J Phys Med Rehabil 74:207–213
91. Rossini PM, Desiato MT, Lavaroni F, Caramia MD (1991) Brain excitability and electroencephalographic activation: non-invasive evaluation in healthy humans via transcranial magnetic stimulation. Brain Res 567:111–119
92. Kammer T, Beck S, Thielscher A, Laubis-Herrmann U, Topka H (2001) Motor thresholds in humans: a transcranial magnetic stimulation study comparing different pulse waveforms, current directions and stimulator types. Clin Neurophysiol 112:250–258
93. Herwig U, Schonfeldt-Lecuona C, Wunderlich AP, von Tiesenhausen C, Thielscher A, Walter H, Spitzer M (2001) The navigation of transcranial magnetic stimulation. Psychiatry Res 108:123–131
94. Julkunen P, Saisanen L, Danner N, Niskanen E, Hukkanen T, Mervaala E, Kononen M (2009) Comparison of navigated and non-navigated transcranial magnetic stimulation for motor cortex mapping, motor threshold and motor evoked potentials. Neuroimage 44:790–795
95. Malcolm MP, Triggs WJ, Light KE, Shechtman O, Khandekar G, Gonzalez Rothi LJ (2006) Reliability of motor cortex transcranial magnetic stimulation in four muscle representations. Clin Neurophysiol 117:1037–1046
96. Cacchio A, Cimini N, Alosi P, Santilli V, Marrelli A (2009) Reliability of transcranial magnetic stimulation-related measurements of tibialis anterior muscle in healthy subjects. Clin Neurophysiol 120:414–419
97. van der Kamp W, Zwinderman AH, Ferrari MD, van Dijk JG (1996) Cortical excitability and response variability of transcranial magnetic stimulation. J Clin Neurophysiol 13:164–171
98. Devanne H, Lavoie BA, Capaday C (1997) Input-output properties and gain changes in the human corticospinal pathway. Exp Brain Res 114:329–338
99. Ridding MC, Rothwell JC (1997) Stimulus/response curves as a method of measuring motor cortical excitability in man. Electroencephalogr Clin Neurophysiol 105:340–344
100. Boroojerdi B, Battaglia F, Muellbacher W, Cohen LG (2001) Mechanisms influencing stimulus-response properties of the human corticospinal system. Clin Neurophysiol 112:931–937
101. Brouwer B, Ashby P (1990) Corticospinal projections to upper and lower limb spinal motoneurons in man. Electroencephalogr Clin Neurophysiol 76:509–519
102. Stagg CJ, Bestmann S, Constantinescu AO, Moreno LM, Allman C, Mekle R, Woolrich M, Near J, Johansen-Berg H, Rothwell JC (2011) Relationship between physiological measures of excitability and levels of glutamate and GABA in the human motor cortex. J Physiol 589:5845–5855
103. Pearce AJ, Clark RA, Kidgell DJ (2013) A comparison of two methods in acquiring stimulus-response curves with transcranial magnetic stimulation. Brain Stimul 6(3):306–309
104. Ray J, McNamara B, Boniface S (2002) Acquisition and expression of proximal and distal upper limb stimulus-response curves to transcranial magnetic stimulation. Muscle Nerve 25:202–206
105. Carroll TJ, Riek S, Carson RG (2001) Reliability of the input-output properties of the cortico-spinal pathway obtained from transcranial magnetic and electrical stimulation. J Neurosci Methods 112:193–202
106. Fuhr P, Agostino R, Hallett M (1991) Spinal motor neuron excitability during the silent period after cortical stimulation. Electroencephalogr Clin Neurophysiol 81:257–262
107. Inghilleri M, Berardelli A, Cruccu G, Manfredi M (1993) Silent period evoked by transcranial stimulation of the human cortex and cervicomedullary junction. J Physiol 466:521–534

108. Daskalakis ZJ, Molnar GF, Christensen BK, Sailer A, Fitzgerald PB, Chen R (2003) An automated method to determine the transcranial magnetic stimulation-induced contralateral silent period. Clin Neurophysiol 114: 938–944
109. Kimiskidis VK, Papagiannopoulos S, Sotirakoglou K, Kazis DA, Kazis A, Mills KR (2005) Silent period to transcranial magnetic stimulation: construction and properties of stimulus-response curves in healthy volunteers. Exp Brain Res 163:21–31
110. van Kuijk AA, Anker LC, Pasman JW, Hendriks JC, van Elswijk G, Geurts AC (2009) Stimulus-response characteristics of motor evoked potentials and silent periods in proximal and distal upper-extremity muscles. J Electromyogr Kinesiol 19:574–583
111. Wassermann EM, Fuhr P, Cohen LG, Hallett M (1991) Effects of transcranial magnetic stimulation on ipsilateral muscles. Neurology 41:1795–1799
112. Meyer BU, Roricht S, Grafin von Einsiedel H, Kruggel F, Weindl A (1995) Inhibitory and excitatory interhemispheric transfers between motor cortical areas in normal humans and patients with abnormalities of the corpus callosum. Brain 118(Pt 2):429–440
113. Wassermann EM, Pascual-Leone A, Hallett M (1994) Cortical motor representation of the ipsilateral hand and arm. Exp Brain Res 100:121–132
114. Chen R, Lozano AM, Ashby P (1999) Mechanism of the silent period following transcranial magnetic stimulation. Evidence from epidural recordings. Exp Brain Res 128:539–542
115. Siebner HR, Dressnandt J, Auer C, Conrad B (1998) Continuous intrathecal baclofen infusions induced a marked increase of the transcranially evoked silent period in a patient with generalized dystonia. Muscle Nerve 21:1209–1212
116. Werhahn KJ, Kunesch E, Noachtar S, Benecke R, Classen J (1999) Differential effects on motorcortical inhibition induced by blockade of GABA uptake in humans. J Physiol 517 (Pt 2):591–597
117. Pierantozzi M, Marciani MG, Palmieri MG, Brusa L, Galati S, Caramia MD, Bernardi G, Stanzione P (2004) Effect of Vigabatrin on motor responses to transcranial magnetic stimulation: an effective tool to investigate in vivo GABAergic cortical inhibition in humans. Brain Res 1028:1–8
118. Nakamura H, Kitagawa H, Kawaguchi Y, Tsuji H (1997) Intracortical facilitation and inhibition after transcranial magnetic stimulation in conscious humans. J Physiol 498(Pt 3): 817–823
119. Priori A, Berardelli A, Inghilleri M, Accornero N, Manfredi M (1994) Motor cortical inhibition and the dopaminergic system. Pharmacological changes in the silent period after transcranial brain stimulation in normal subjects, patients with Parkinson's disease a nd drug-induced parkinsonism. Brain 117 (Pt 2):317–323
120. Compta Y, Valls-Sole J, Valldeoriola F, Kumru H, Rumia J (2006) The silent period of the thenar muscles to contralateral and ipsilateral deep brain stimulation. Clin Neurophysiol 117:2512–2520
121. Saisanen L, Pirinen E, Teitti S, Kononen M, Julkunen P, Maatta S, Karhu J (2008) Factors influencing cortical silent period: optimized stimulus location, intensity and muscle contraction. J Neurosci Methods 169:231–238
122. Priori A, Berardelli A, Mercuri B, Inghilleri M, Manfredi M (1995) The effect of hyperventilation on motor cortical inhibition in humans: a study of the electromyographic silent period evoked by transcranial brain stimulation. Electroencephalogr Clin Neurophysiol 97:69–72
123. Chen R, Yung D, Li JY (2003) Organization of ipsilateral excitatory and inhibitory pathways in the human motor cortex. J Neurophysiol 89:1256–1264
124. Petitjean M, Ko JY (2013) An age-related change in the ipsilateral silent period of a small hand muscle. Clin Neurophysiol 124(2):346–353
125. Farzan F, Barr MS, Fitzgerald PB, Daskalakis ZJ (2011) Combination of transcranial magnetic stimulation with electromyography and electroencephalography: application in diagnosis of neuropsychiatric disorders. In: EMG methods for evaluating muscle and nerve function. Edited by Schwartz M: InTech Open Access
126. Voineskos AN, Farzan F, Barr MS, Lobaugh NJ, Mulsant BH, Chen R, Fitzgerald PB, Daskalakis ZJ (2010) The role of the corpus callosum in transcranial magnetic stimulation induced interhemispheric signal propagation. Biol Psychiatry 68:825–831
127. Mochizuki H, Ugawa Y, Terao Y, Sakai KL (2006) Cortical hemoglobin-concentration changes under the coil induced by single-pulse TMS in humans: a simultaneous recording with near-infrared spectroscopy. Exp Brain Res 169:302–310
128. Liepert J, Restemeyer C, Kucinski T, Zittel S, Weiller C (2005) Motor strokes: the lesion

location determines motor excitability changes. Stroke 36:2648–2653
129. Hendricks HT, Pasman JW, van Limbeek J, Zwarts MJ (2003) Motor evoked potentials in predicting recovery from upper extremity paralysis after acute stroke. Cerebrovasc Dis 16:265–271
130. Hendricks HT, Pasman JW, van Limbeek J, Zwarts MJ (2003) Motor evoked potentials of the lower extremity in predicting motor recovery and ambulation after stroke: a cohort study. Arch Phys Med Rehabil 84:1373–1379
131. Trompetto C, Assini A, Buccolieri A, Marchese R, Abbruzzese G (2000) Motor recovery following stroke: a transcranial magnetic stimulation study. Clin Neurophysiol 111:1860–1867
132. Cicinelli P, Traversa R, Bassi A, Scivoletto G, Rossini PM (1997) Interhemispheric differences of hand muscle representation in human motor cortex. Muscle Nerve 20:535–542
133. Niehaus L, Bajbouj M, Meyer BU (2003) Impact of interhemispheric inhibition on excitability of the non-lesioned motor cortex after acute stroke. Suppl Clin Neurophysiol 56:181–186
134. Classen J, Schnitzler A, Binkofski F, Werhahn KJ, Kim YS, Kessler KR, Benecke R (1997) The motor syndrome associated with exaggerated inhibition within the primary motor cortex of patients with hemiparetic. Brain 120(Pt 4):605–619
135. Braune HJ, Fritz C (1995) Transcranial magnetic stimulation-evoked inhibition of voluntary muscle activity (silent period) is impaired in patients with ischemic hemispheric lesion. Stroke 26:550–553
136. Battaglia F, Quartarone A, Ghilardi MF, Dattola R, Bagnato S, Rizzo V, Morgante L, Girlanda P (2006) Unilateral cerebellar stroke disrupts movement preparation and motor imagery. Clin Neurophysiol 117:1009–1016
137. De Beaumont L, Lassonde M, Leclerc S, Theoret H (2007) Long-term and cumulative effects of sports concussion on motor cortex inhibition. Neurosurgery 61:329–336; discussion 336–337
138. De Beaumont L, Theoret H, Mongeon D, Messier J, Leclerc S, Tremblay S, Ellemberg D, Lassonde M (2009) Brain function decline in healthy retired athletes who sustained their last sports concussion in early adulthood. Brain 132:695–708
139. Tremblay S, de Beaumont L, Lassonde M, Theoret H (2011) Evidence for the specificity of intracortical inhibitory dysfunction in asymptomatic concussed athletes. J Neurotrauma 28:493–502
140. Nakashima K, Wang Y, Shimoda M, Sakuma K, Takahashi K (1995) Shortened silent period produced by magnetic cortical stimulation in patients with Parkinson's disease. J Neurol Sci 130:209–214
141. Siebner HR, Mentschel C, Auer C, Lehner C, Conrad B (2000) Repetitive transcranial magnetic stimulation causes a short-term increase in the duration of the cortical silent period in patients with Parkinson's disease. Neurosci Lett 284:147–150
142. Fisher BE, Wu AD, Salem GJ, Song J, Lin CH, Yip J, Cen S, Gordon J, Jakowec M, Petzinger G (2008) The effect of exercise training in improving motor performance and corticomotor excitability in people with early Parkinson's disease. Arch Phys Med Rehabil 89:1221–1229
143. Lefaucheur JP (2005) Motor cortex dysfunction revealed by cortical excitability studies in Parkinson's disease: influence of antiparkinsonian treatment and cortical stimulation. Clin Neurophysiol 116:244–253
144. Alagona G, Bella R, Ferri R, Carnemolla A, Pappalardo A, Costanzo E, Pennisi G (2001) Transcranial magnetic stimulation in Alzheimer disease: motor cortex excitability and cognitive severity. Neurosci Lett 314: 57–60
145. de Carvalho M, de Mendonca A, Miranda PC, Garcia C, Luis ML (1997) Magnetic stimulation in Alzheimer's disease. J Neurol 244:304–307
146. Julkunen P, Jauhiainen AM, Kononen M, Paakkonen A, Karhu J, Soininen H (2011) Combining transcranial magnetic stimulation and electroencephalography may contribute to assess the severity of Alzheimer's disease. Int J Alzheimers Dis 2011:654794
147. Attarian S, Azulay JP, Lardillier D, Verschueren A, Pouget J (2005) Transcranial magnetic stimulation in lower motor neuron diseases. Clin Neurophysiol 116:35–42
148. Triggs WJ, Menkes D, Onorato J, Yan RS, Young MS, Newell K, Sander HW, Soto O, Chiappa KH, Cros D (1999) Transcranial magnetic stimulation identifies upper motor neuron involvement in motor neuron disease. Neurology 53:605–611
149. Mills KR (2003) The natural history of central motor abnormalities in amyotrophic lateral sclerosis. Brain 126:2558–2566
150. Khedr EM, Ahmed MA, Darwish ES, Ali AM (2011) The relationship between motor cortex excitability and severity of Alzheimer's

disease: a transcranial magnetic stimulation study. Neurophysiol Clin 41:107–113

151. de Carvalho M, Scotto M, Lopes A, Swash M (2003) Clinical and neurophysiological evaluation of progression in amyotrophic lateral sclerosis. Muscle Nerve 28:630–633
152. Jung P, Beyerle A, Humpich M, Neumann-Haefelin T, Lanfermann H, Ziemann U (2006) Ipsilateral silent period: a marker of callosal conduction abnormality in early relapsing-remitting multiple sclerosis? J Neurol Sci 250:133–139
153. Caramia MD, Palmieri MG, Desiato MT, Boffa L, Galizia P, Rossini PM, Centonze D, Bernardi G (2004) Brain excitability changes in the relapsing and remitting phases of multiple sclerosis: a study with transcranial magnetic stimulation. Clin Neurophysiol 115: 956–965
154. Tassinari CA, Cincotta M, Zaccara G, Michelucci R (2003) Transcranial magnetic stimulation and epilepsy. Clin Neurophysiol 114:777–798
155. Aguglia U, Gambardella A, Quartarone A, Girlanda P, Le Piane E, Messina D, Oliveri RL, Zappia M, Quattrone A (2000) Interhemispheric threshold differences in idiopathic generalized epilepsies with versive or circling seizures determined with focal magnetic transcranial stimulation. Epilepsy Res 40:1–6
156. Delvaux V, Alagona G, Gerard P, De Pasqua V, Delwaide PJ, Maertens de Noordhout A (2001) Reduced excitability of the motor cortex in untreated patients with de novo idiopathic "grand mal" seizures. J Neurol Neurosurg Psychiatry 71:772–776
157. Reutens DC, Berkovic SF, Macdonell RA, Bladin PF (1993) Magnetic stimulation of the brain in generalized epilepsy: reversal of cortical hyperexcitability by anticonvulsants. Ann Neurol 34:351–355
158. Macdonell RA, King MA, Newton MR, Curatolo JM, Reutens DC, Berkovic SF (2001) Prolonged cortical silent period after transcranial magnetic stimulation in generalized epilepsy. Neurology 57:706–708
159. Cincotta M, Borgheresi A, Benvenuti F, Liotta P, Marin E, Zaccara G (2002) Cortical silent period in two patients with meningioma and preoperative seizures: a pre- and postsurgical follow-up study. Clin Neurophysiol 113: 597–603
160. Li X, Teneback CC, Nahas Z, Kozel FA, Large C, Cohn J, Bohning DE, George MS (2004) Interleaved transcranial magnetic stimulation/functional MRI confirms that lamotrigine inhibits cortical excitability in healthy young men. Neuropsychopharmacology 29:1395–1407
161. Li X, Ricci R, Large CH, Anderson B, Nahas Z, Bohning DE, George MS (2010) Interleaved transcranial magnetic stimulation and fMRI suggests that lamotrigine and valproic acid have different effects on corticolimbic activity. Psychopharmacology (Berl) 209:233–244
162. Jackson SR, Parkinson A, Manfredi V, Millon G, Hollis C, Jackson GM (2013) Motor excitability is reduced prior to voluntary movements in children and adolescents with Tourette syndrome. J Neuropsychol 7(1):29–44
163. Bruckmann S, Hauk D, Roessner V, Resch F, Freitag CM, Kammer T, Ziemann U, Rothenberger A, Weisbrod M, Bender S (2012) Cortical inhibition in attention deficit hyperactivity disorder: new insights from the electroencephalographic response to transcranial magnetic stimulation. Brain 135(Pt 7): 2215–2230
164. Vry J, Linder-Lucht M, Berweck S, Bonati U, Hodapp M, Uhl M, Faist M, Mall V (2008) Altered cortical inhibitory function in children with spastic diplegia: a TMS study. Exp Brain Res 186:611–618
165. Garvey MA, Mall V (2008) Transcranial magnetic stimulation in children. Clin Neurophysiol 119:973–984
166. Heinen F, Fietzek UM, Berweck S, Hufschmidt A, Deuschl G, Korinthenberg R (1998) Fast corticospinal system and motor performance in children: conduction proceeds skill. Pediatr Neurol 19:217–221
167. Bender S, Basseler K, Sebastian I, Resch F, Kammer T, Oelkers-Ax R, Weisbrod M (2005) Electroencephalographic response to transcranial magnetic stimulation in children: evidence for giant inhibitory potentials. Ann Neurol 58:58–67
168. Fitzgerald PB, Brown TL, Daskalakis ZJ, Kulkarni J (2002) A transcranial magnetic stimulation study of inhibitory deficits in the motor cortex in patients with schizophrenia. Psychiatry Res 114:11–22
169. Fitzgerald PB, Brown TL, Marston NA, Oxley T, De Castella A, Daskalakis ZJ, Kulkarni J (2004) Reduced plastic brain responses in schizophrenia: a transcranial magnetic stimulation study. Schizophr Res 71:17–26
170. Daskalakis ZJ, Christensen BK, Chen R, Fitzgerald PB, Zipursky RB, Kapur S (2002) Evidence for impaired cortical inhibition in schizophrenia using transcranial magnetic stimulation. Arch Gen Psychiatry 59:347–354

171. Levinson AJ, Young LT, Fitzgerald PB, Daskalakis ZJ (2007) Cortical inhibitory dysfunction in bipolar disorder: a study using transcranial magnetic stimulation. J Clin Psychopharmacol 27:493–497
172. Bajbouj M, Lisanby SH, Lang UE, Danker-Hopfe H, Heuser I, Neu P (2006) Evidence for impaired cortical inhibition in patients with unipolar major depression. Biol Psychiatry 59:395–400
173. Richter MA, de Jesus DR, Hoppenbrouwers S, Daigle M, Deluce J, Ravindran LN, Fitzgerald PB, Daskalakis ZJ (2012) Evidence for cortical inhibitory and excitatory dysfunction in obsessive compulsive disorder. Neuropsychopharmacology 37: 1144–1151
174. Barnow S, Volker KA, Moller B, Freyberger HJ, Spitzer C, Grabe HJ, Daskalakis ZJ (2009) Neurophysiological correlates of borderline personality disorder: a transcranial magnetic stimulation study. Biol Psychiatry 65:313–318
175. Salerno A, Thomas E, Olive P, Blotman F, Picot MC, Georgesco M (2000) Motor cortical dysfunction disclosed by single and double magnetic stimulation in patients with fibromyalgia. Clin Neurophysiol 111:994–1001
176. Lefaucheur JP, Drouot X, Menard-Lefaucheur I, Keravel Y, Nguyen JP (2006) Motor cortex rTMS restores defective intracortical inhibition in chronic neuropathic pain. Neurology 67:1568–1574
177. Oechsner M, Zangemeister WH (1999) Prolonged postexcitatory inhibition after transcranial magnetic stimulation of the motor cortex in patients with cerebellar ataxia. J Neurol Sci 168:107–111
178. Tamburin S, Fiaschi A, Andreoli A, Marani S, Manganotti P, Zanette G (2004) Stimulus-response properties of motor system in patients with cerebellar ataxia. Clin Neurophysiol 115:348–355
179. Teo JT, Schneider SA, Cheeran BJ, Fernandez-del-Olmo M, Giunti P, Rothwell JC, Bhatia KP (2008) Prolonged cortical silent period but normal sensorimotor plasticity in spinocerebellar ataxia 6. Mov Disord 23: 378–385
180. Bertini M, Ferrara M, De Gennaro L, Curcio G, Fratello F, Romei V, Pauri F, Rossini PM (2004) Corticospinal excitability and sleep: a motor threshold assessment by transcranial magnetic stimulation after awakenings from REM and NREM sleep. J Sleep Res 13:31–36
181. Kreuzer P, Langguth B, Popp R, Raster R, Busch V, Frank E, Hajak G, Landgrebe M (2011) Reduced intra-cortical inhibition after sleep deprivation: a transcranial magnetic stimulation study. Neurosci Lett 493:63–66
182. Rosanova M, Casali A, Bellina V, Resta F, Mariotti M, Massimini M (2009) Natural frequencies of human corticothalamic circuits. J Neurosci 29:7679–7685
183. Verhagen L, Dijkerman HC, Medendorp WP, Toni I (2012) Cortical dynamics of sensorimotor integration during grasp planning. J Neurosci 32:4508–4519
184. Kahkonen S, Wilenius J (2007) Effects of alcohol on TMS-evoked N100 responses. J Neurosci Methods 166:104–108
185. Thut G, Pascual-Leone A (2010) A review of combined TMS-EEG studies to characterize lasting effects of repetitive TMS and assess their usefulness in cognitive and clinical neuroscience. Brain Topogr 22:219–232

Chapter 6

Paired-Pulse Transcranial Magnetic Stimulation (TMS) Protocols

Andrew Vahabzadeh-Hagh

Abstract

Transcranial magnetic stimulation applied in a paired-pulse sequence (ppTMS) provides a noninvasive method to evaluate the excitability of corticocortical connections. It is most commonly used to interrogate motor cortical pathways, but literature has also demonstrated its applicability in non-motor cortical regions. Commonly used protocols include short-interval intracortical inhibition, long-interval intracortical inhibition, intracortical facilitation, and short-interval intracortical facilitation. The similarities among these protocols allow for the implementation of multiple measures within a given study. This provides a much greater breadth of data collection compared to traditional single-pulse TMS studies. Although the mechanisms of these circuits of inhibition or facilitation remain only partially understood, their reproducibility has encouraged their widespread use in the TMS community. Additional methods of paired-pulse stimulation include interhemispheric and interregional ppTMS such as via peripheral nerve or cerebellar stimulation. More newly developed techniques have incorporated the use of real-time EEG to further develop its applications beyond motor cortical regions. ppTMS has found exciting applications in various neuropathologies, especially in the fields of epilepsy and stroke, as a means of individualized prognostic and therapeutic optimization. In the field of neurophysiology, ppTMS has a bright future.

Key words Facilitation, Inhibition, Interhemispheric, Interregional, Intracortical, Paired pulse

1 Background

Transcranial magnetic stimulation (TMS) provides a noninvasive means to evaluate human neuronal circuits and to measure human regional cortical excitability. When applied over the motor cortex, TMS reliably elicits motor-evoked potentials (MEPs) in a contralateral target muscle group. Attributes of these MEPs, such as motor threshold, the slope and plateau of MEP recruitment curves, MEP amplitude, and signal integral may provide clinicians and researchers with an in vivo, real-time quantitative look into the neurophysiology of healthy and diseased states [1, 2].

TMS applied as a pair of pulses to the same cortical location separated by variable interstimulus intervals (ISIs) further allows

Alexander Rotenberg et al. (eds.), *Transcranial Magnetic Stimulation*, Neuromethods, vol. 89,
DOI 10.1007/978-1-4939-0879-0_6, © Springer Science+Business Media New York 2014

for the noninvasive assessment of excitatory and inhibitory corticocortical circuits [3–6]. Whereas single-pulse TMS can provide insight into the excitability and integrity of the corticospinal tract [1, 2, 7], paired-pulse TMS (ppTMS) can provide insight into the excitability and integrity of corticocortical connections. Paired-pulse stimulation is defined as the application of two sequential stimuli of variable intensities separated by preselected ISIs. These two sequential stimuli are commonly referred to as the initial conditioning stimulus (CS) and the subsequent test stimulus (TS). The provocation of inhibitory or excitatory cortical circuits has been shown to be a function of CS intensity and the ISI [4, 8–10]. TS intensity is more consistently held to a suprathreshold value and thus the importance of its variance is not fully understood.

Studies have demonstrated that ppTMS measurements provide an assessment specifically of intracortical processes. Di Lazzaro and colleagues [11–13] demonstrated that a subthreshold CS, within ppTMS protocols, failed to elicit an electromyographic response (e.g., MEP) and lacked the production of any descending epidural volleys recorded by cervical epidural electrodes. Accordingly, the authors concluded that ppTMS inhibition was likely the result of synaptic inhibition within the motor cortex. In addition, Kujirai and colleagues [4] demonstrated that conditioning stimuli capable of suppressing the motor response to the TS had no effect upon spinal H-reflexes or transcranial electrical anodal stimulation response, thought to evoke more distal activation along the corticospinal tract. Thus, ppTMS has long been accepted as a tool specifically for corticocortical circuit evaluation, whether intrahemispheric, interhemispheric, or interregional.

The effects of ppTMS have been demonstrated to be widespread within the human motor cortex. Studies have shown that measures of intracortical inhibition (ICI) and intracortical facilitation (ICF) are reproducible within multiple motor representations, including the rectus abdominis, biceps brachii, abductor pollicis brevis, quadriceps femoris, and the abductor hallucis, irrespective of the estimated strength of these different corticospinal trajectories [8]. Although the focus of this chapter will be on the applications of ppTMS to evaluate the activity of circuits involving the motor cortex, it is notable to mention that these techniques may also assess excitability in non-motor cortical regions. For example, Oliveri and colleagues [14] studied the effects of ppTMS on psychophysical measurements, namely a subject's perception of threshold tactile electrical stimuli. They showed that ppTMS to the parietal cortex could improve or worsen a healthy subject's performance on task discrimination depending upon the preselected ISI.

In ppTMS studies, one first identifies a subject's resting motor threshold (rMT) in the absence of spontaneous/volitional muscle activity. The intensities of the CS and TS are selected relative to the rMT and the paired-pulse paradigm is applied at preset ISIs.

Currently, with the pair of stimuli applied to the same cortical location, there are at least four corticocortical circuits one can activate. These include two inhibitory circuits: short-interval intracortical inhibition (SICI) and long-interval intracortical inhibition (LICI), as well as two facilitatory circuits: ICF and short-interval intracortical facilitation (SICF). Additionally, while the application of the TS often remains fixed to a given motor cortical region, the CS location may be varied to provoke inhibition or facilitation of motor cortical activity via circuits of more distal origins including the contralateral motor cortex, cerebellum, and peripheral nerves. These circuits correspond to pathways of interhemispheric inhibition (IHI), cerebellar inhibition (CBI), and short- (SAI) or long-latency afferent inhibition (LAI), respectively.

As a prelude to the discussion of individual paired-pulse protocols, a brief mention of the required equipment would be worthwhile. The equipment for paired-pulse studies is similar to many other TMS studies. However, two magnetic stimulators are required, often Magstim 200 stimulators (Magstim, Wales, UK), which provide monophasic pulses. For paired-pulse stimulation of the same cortical site, the stimulators must be connected through a BiStim Module (Magstim, Wales, UK) so that the dual output can be applied through a single coil of choice. The stimulator intensities are set on each TMS unit independently and the ISIs may be specified through the BiStim Module or other external hardware and/or software interfaces [1, 9, 15].

2 Paired-Pulse Protocols

Although ppTMS paradigms have been utilized for nearly two decades, our understanding of the mechanisms of the aforementioned circuits remains largely incomplete. Currently, SICI is believed to be the product of axonal refractoriness and low threshold $GABA_A$ receptor-mediated inhibition [5]. However, several discrete ISIs at which SICI is present appear to behave differently and might be the result of different mechanisms [6, 8, 9, 16–22]. LICI is believed to be the result of postsynaptic $GABA_B$ receptor activity as well as a function of cortical silent period, the period of suppressed voluntary muscle contraction following suprathreshold TMS [3, 22–27]. The mechanisms of ICF and SICF are less understood. Although it is largely believed that increased glutamatergic activity plays a role in these facilitatory circuits, pharmacologic studies have demonstrated variable findings, such as reduced ICF with the NMDA receptor blocker dextromethorphan but no change with other anti-glutamatergic drugs (e.g., gabapentin, memantine). Another study looked at the effects of ketamine and found a large yet insignificant increase in SICF thereby suggesting that SICF is mediated by a non-NMDA receptor-dependent glutamatergic process [9, 28–31].

Table 1
Intrahemispheric paired-pulse TMS protocols

	Circuit	ISI (ms)	CS	TS	Referenced studies
Inhibitory	SICI	1–6	50–90 % RMT <100 % AMT	TS 1 mV >100 % RMT 120 % RMT	[9, 18, 23, 25, 61, 62]
	LICI	50–200+	TS 1 mV 100–130 % RMT	TS 1 mV 120 % RMT	[3, 15, 22, 23, 25–27, 62]
Excitatory	ICF	8–30	90 + % RMT >80 % AMT	TS 0.5–1.5 mV	[4–6]
	SICF	1.0–1.5 2.3–3.0 4.1–5.0	>RMT TS 1 mV 100–130 % RMT	<RMT 90 % RMT	[9, 10, 30]

AMT active motor threshold—in the presence of isometric contraction (5–10 % of maximum voluntary muscle contraction), the minimum stimulator intensity required to evoke MEPs >100 μV peak-to-peak in 5 of 10 consecutive trials, *CS* conditioning stimulus, *ICF* intracortical facilitation, *ISI* interstimulus interval, *LICI* long-interval intracortical inhibition, *ms* millisecond, *mV* millivolt, *RMT* resting motor threshold—in the absence of volitional muscle activation, the minimum stimulator intensity required to evoke MEPs >50 μV peak-to-peak in 5 of 10 consecutive trials, *SICF* short-interval intracortical facilitation, *SICI* short-interval intracortical inhibition, *TS* test stimulus, *TS #.# mV* stimulator intensity required to elicit MEPs with peak-to-peak amplitudes of #.# mV [61, 62]

The individual paired-pulse protocols are rather similar and, accordingly, multiple measures can be easily implemented during any given study. SICI utilizes a subthreshold CS followed by a suprathreshold TS at ISIs of 1–6 ms. LICI uses a suprathreshold CS and TS at ISIs of 50–200 ms and longer [3, 22, 23, 25–27]. ICF uses the same CS and TS settings as SICI but is observed for ISIs of 8–30 ms [4–6]. Lastly, SICF, also known as facilitatory I-wave interaction, is evoked by a supra- or at-motor threshold CS followed by a subthreshold TS at ISIs similar to SICI but most robust at 1.1–1.5, 2.3–3.0, and 4.1–5.0 ms [9, 10, 30]. Refer to Table 1 for more details and reference studies.

Interhemispheric paired-pulse stimulation is most commonly achieved via application of the conditioning stimulus to the motor cortex ipsilateral to the target muscle group, e.g., contralateral to, and mirroring, the location of the test stimulus. The effects of interhemispheric ppTMS appear to be of cortical origin. IHI has been demonstrated for TMS-evoked MEPs but is unable to inhibit test responses evoked by anodal electrical stimulation, presumed to stimulate more distally along the corticospinal tract, bypassing these corticocortical circuits. Additionally, conditioning stimuli fail to effect ipsilateral H reflexes of relaxed target muscle groups [32]. These combined observations support the notion that measures of interhemispheric ppTMS are, at least in part, cortically derived. However, Gerloff and colleagues [33] demonstrated that the conditioning effects of IHI may also be observed on MEPs evoked by transcranial electrical stimulation at the level of the medullary pyramidal decussation (pdTES). This suggests that IHI measures likely derive from multiple sources, including isolated subcortical circuits. Interhemispheric ppTMS measures also allow for the assessment of excitatory circuits. Early interhemispheric facilitation, observed with shorter ISIs (4–5 ms), is believed to be mediated via transmission through the corpus callosum. This interhemispheric facilitation was reproduced using an anodal transcranial electrical conditioning stimulus, further validating the corticospinal origin of these facilitatory effects. It was further demonstrated that the selectivity between interhemispheric excitatory or inhibitory circuits was a function of the position, the direction of the induced current, and the magnitude of the conditioning stimulus [34].

As mentioned, interhemispheric ppTMS measures are dependent upon pulse locations. The strongest effects are observed when the conditioning stimulus is delivered to the mirrored cortical region ipsilateral to the target muscle group [34]. The directionality of the induced current, however, has been shown to have minimal effect on IHI measurements [35]. Although the mechanisms of interhemispheric ppTMS have not been fully elucidated, important relationships have been described. Chen and colleagues [35] postulated that IHI and the ipsilateral silent period are complementary but unique measures of ipsilateral inhibition. Furthermore, given that IHI at a shorter ISI (8 ms) is reduced during target muscle activation while IHI at a longer ISI (40 ms) is more closely related to the ipsilateral silent period, it appears IHI generated at varied ISIs are mechanistically different [35]. For specific elements of interhemispheric ppTMS protocols, refer to Table 2.

Test stimulation of the motor cortex can also be conditioned via input from more remote locations such as a peripheral nerve or

Table 2
Interhemispheric paired-pulse TMS protocols

	Circuit	ISI (ms)	CS	TS	Referenced studies
Ipsilateral motor cortex	I	12	130 % RMT	TS 1 mV	[18]
		6, 7, 12, 15	110–125 % RMT	TS 1.5 mV	[32]
		≥11	105–130 % AMT	TS 0.2–0.4 mV[a]	[34]
		8–50	45, 60, 75, 90 % MO	TS 1 mV	[35]
		6–70	102–159 % RMT	100–143 % RMT	[33]
	F	4–5	105–110 % AMT	TS 0.2–0.4 mV[a]	[34]

AMT active motor threshold—in the presence of isometric contraction (5–10 % of maximum voluntary muscle contraction), the minimum stimulator intensity required to evoke MEPs >100 μV peak-to-peak in 5 of 10 consecutive trials, *CS* conditioning stimulus, *F* facilitation, *I* inhibition, *ISI* interstimulus interval, *MO* machine output, *ms* millisecond, *mV* millivolt, *RMT* resting motor threshold—in the absence of volitional muscle activation, the minimum stimulator intensity required to evoke MEPs >50 μV peak-to-peak in 5 of 10 consecutive trials, *TS* test stimulus, *TS #.# mV* stimulator intensity required to elicit MEPs with peak-to-peak amplitudes of #.# mV

[a]Denotes measurement performed in active muscle as defined in the obtainment of AMT

within a cerebellar hemisphere. Peripheral afferent sensory stimulation is often applied to the median nerve or select digital cutaneous nerves. This form of stimulation, known as afferent inhibition, can be divided into short-latency (SAI) and long-latency (LAI) subclasses. As SAI and LAI afferent activations are different, they are likely mediated by different mechanisms. Whereas SAI typically activates the contralateral primary and secondary somatosensory cortices, LAI provides more widespread proximal activation, including the ipsilateral secondary somatosensory cortex and the contralateral posterior parietal cortex [36–39].

In afferent inhibition protocols, the most important parameters for predicting the effect of cutaneous nerve stimulation on the modulation of MEPs are the ISI and the intensity of the conditioning stimulus [40]. This form of paired-pulse stimulation is believed to be subcortical in nature, involving postsynaptic and presynaptic spinal inhibitory mechanisms, i.e., inhibition of spinal interneurons necessary for the baseline activation of spinal alpha-motor neurons. Researchers have hypothesized that this form of afferent inhibition is part of a protective inhibitory reflex providing distal inhibition with more proximal facilitation via A-delta fibers [40]. In addition, reduced motor excitability following peripheral stimulation (i.e., median nerve or digital stimulation) correlates with an event-related synchronization (ERS) of a 20-Hz rolandic rhythm arising predominantly from the motor cortex. This ERS is thought to represent an inactive, dormant cortical state promoting reduced excitability of the motor cortex [41–45]. Although gaps remain in our understanding of this afferent inhibition, several studies have alluded to potential neurochemical factors. For instance, when compared to age-matched healthy subjects, patients with Alzheimer's disease demonstrated reduced SAI and LAI. In addition, reductions in SAI have been shown to be inducible via dopaminergic medications [46].

CBI or cerebellar facilitation (CBF) involves stimulation with a double-cone coil used to inhibit or facilitate MEPs from the contralateral motor cortex with short ISIs in the range of 3–8 ms. Optimal positioning for stimulation is just rostral to the level of the foramen magnum ipsilateral to the target muscle group. The mechanism of CBI is thought to involve the activation of Purkinje cells in the cerebellar cortex which inhibits deep cerebellar nuclei (e.g., dentate nucleus) and reduces the excitatory output to the motor cortex [47, 48]. As many ataxic patients with an impaired cerebellar efferent system lack or display reduced CBI, it is thought that these efferent pathways play an important role CBI [48]. There is less known about the mechanism of CBF, however, there are theories citing the importance of the dentatothalamocortical pathway, which provides excitatory projections from the cerebellar nuclei to the motor cortex via the ventral thalamus [48]. For details on interregional paired-pulse paradigms involving distal afferent inputs and cerebellar stimulation, refer to Table 3.

Table 3
Interregional paired-pulse protocols

	Circuit	ISI (ms)	CS	TS	Target muscle group	Peripheral stimulation	References
Afferent inhibition	Short-latency (SAI)	19–21 20–40	~TTT 200–400 % PT	TS 0.5–2.0 mV TS 0.5–2.0 mV	APB, FDI APB, FDI	MN Digit II, digit III	[63, 65]
	Long-latency (LAI)	50, 100, 150, 200, 240	250 % PT	120 % RMT	FPB, ADM	MN	[62]
		40–140	5–100 % MTI 400 % PT 150 % PT	150–170 % RMT	TE	Digit II, digit V	[40]
		200–1,000 200–600	~TTT 200–800 % PT	TS 1 mV TS 1 mV	APB, FDI, ECR APB, FDI, ECR	MN Digit III	[41]
		200	~TTT	TS 0.2, 1.0, 4.0 mV	FDI	MN	[64]
Cerebellar stimulation	I	5–8	~PTT	TS 1 mV	FDI	Cerebellum	[47, 48]
	F	3–4	<PTT	TS 1 mV	FDI	Cerebellum	

ADM abductor digiti minimi, *APB* abductor pollicis brevis, *CS* conditioning stimulus, *ECR* extensor carpi radialis, *F* facilitation, *FDI* first dorsal interosseous, *FPB* flexor pollicis brevis, *I* inhibition, *ISI* interstimulus interval, *LAI* long-latency afferent inhibition, *MN* median nerve, stimulated with 0.2–0.5 ms square wave current pulses, *ms* millisecond, *MTI* maximum tolerable intensity for cutaneous stimulation, *mV* millivolt, *PT* perceptual threshold, *PTT* pyramidal tract threshold, *RMT* resting motor threshold—in the absence of volitional muscle activation, the minimum stimulator intensity required to evoke MEPs >50 μV peak-to-peak in 5 of 10 consecutive trials, *SAI* short-latency afferent inhibition, *TE* thenar eminence, *TS* test stimulus, *TS #.# mV* stimulator intensity required to elicit MEPs with peak-to-peak amplitudes of #.# mV, *TTT* thumb twitch threshold [63–65]

3 Paired-Pulse TMS and EEG: Evaluation Beyond the Motor Cortex

With the advancement of amplifier technologies, the combined application of TMS and EEG, providing continuous data acquisition, has been made possible. A considerable advantage of these combined technologies is that EEG allows for the evaluation of TMS-evoked potentials (TEPs) no longer confined to the motor cortex [49]. As such, one may investigate cortical inhibition or facilitation in non-motor cortical regions that may be more directly involved in certain neurologic or psychiatric disease states. Recently, Fitzgerald and colleagues investigated the ability of EEG to assess the effects of LICI applied to the motor cortex, the dorsolateral prefrontal cortex (DLPFC), and a region of the parietal lobe. They demonstrated that LICI, when applied to these cortical regions, could provoke cortical inhibition in the form of reduced mean cortical-evoked activity as measured via EEG. Further analysis demonstrated that significant inhibition of the mean cortical-evoked activity on EEG occurred at ISIs, similar to those used in prior TMS-EMG LICI studies. These findings support a similar $GABA_B$ receptor-mediated process occurring in these non-motor cortical regions [50–52]. As this growing area of TMS research might greatly broaden its clinical applications, a systematic comparison of TMS-EMG and TMS-EEG measures may be indicated to further define the technique of greatest clinical value.

4 Paired-Pulse Stimulation in Diseased States

Although the application of paired-pulse measures for the assessment of diseased states has yet to be formalized, there have been numerous intriguing findings that appear to be associated with varied neuropathologies. For example, SICI has shown reduction in Parkinson's disease, dystonia, Alzheimer's disease, and schizophrenia. Conversely, LICI has shown enhancement in Parkinson's disease and cerebellar degeneration. Additionally, the use of ppTMS techniques to assess neuronal plasticity (i.e., cortical disinhibition) in epilepsy and stroke has recently gained popularity and may hold immense translational value in enabling optimal treatment selection and more confident prognostic capacity [53–60]. Overall, given the ease of implementation, if the appropriate equipment is available, the incorporation of paired-pulse stimulation into any TMS study may greatly increase our understanding of these cortical processes at a minimal time and monetary cost. When feasible, these experimental modifications should be considered and, if conducted, our understanding of the underlying mechanisms of TMS and the neurophysiology of normal and diseased states may benefit greatly.

References

1. Kobayashi M, Pascual-Leone A (2003) Transcranial magnetic stimulation in neurology. Lancet Neurol 2:145–156
2. Ziemann U, Muellbacher W, Hallett M, Cohen LG (2001) Modulation of practice-dependent plasticity in human motor cortex. Brain 124:1171–1181
3. Valls-Sole J, Pascual-Leone A, Wassermann EM, Hallett M (1992) Human motor evoked responses to paired transcranial magnetic stimuli. Electroencephalogr Clin Neurophysiol 85:355–364
4. Kujirai T, Caramia MD, Rothwell JC, Day BL, Thompson PD, Ferbert A, Wroe S, Asselman P, Marsden CD (1993) Corticocortical inhibition in human motor cortex. J Physiol 471: 501–519
5. Ziemann U, Lonnecker S, Steinhoff BJ, Paulus W (1996) The effect of lorazepam on the motor cortical excitability in man. Exp Brain Res 109:127–135
6. Ziemann U, Rothwell JC, Ridding MC (1996) Interaction between intracortical inhibition and facilitation in human motor cortex. J Physiol 496(Pt 3):873–881
7. Rothwell JC (1997) Techniques and mechanisms of action of transcranial stimulation of the human motor cortex. J Neurosci Methods 74:113–122
8. Chen R, Tam A, Butefisch C, Corwell B, Ziemann U, Rothwell JC, Cohen LG (1998) Intracortical inhibition and facilitation in different representations of the human motor cortex. J Neurophysiol 80:2870–2881
9. Ilic TV, Meintzschel F, Cleff U, Ruge D, Kessler KR, Ziemann U (2002) Short-interval paired-pulse inhibition and facilitation of human motor cortex: the dimension of stimulus intensity. J Physiol 545:153–167
10. Peurala SH, Muller-Dahlhaus JF, Arai N, Ziemann U (2008) Interference of short-interval intracortical inhibition (SICI) and short-interval intracortical facilitation (SICF). Clin Neurophysiol 119:2291–2297
11. Di Lazzaro V, Oliviero A, Profice P, Saturno E, Pilato F, Insola A, Mazzone P, Tonali P, Rothwell JC (1998) Comparison of descending volleys evoked by transcranial magnetic and electric stimulation in conscious humans. Electroencephalogr Clin Neurophysiol 109: 397–401
12. Di Lazzaro V, Restuccia P, Oliviero A, Profice P, Ferrara L, Insola A, Mazzone P, Tonali P, Rothwell JC (1998) Effects of voluntary contraction on descending volleys evoked by transcranial stimulation in conscious humans. J Physiol 508:625–633
13. Di Lazzaro V, Oliviero A, Profice P, Insola A, Mazzone P, Tonali P, Rothwell JC (1999) Direct recordings of descending volleys after transcranial magnetic and electric motor cortex stimulation in conscious humans. Electroencephalogr Clin Neurophysiol Suppl 51:120–126
14. Oliveri M, Rossini PM, Filippi MM, Traversa R, Cicinelli P, Palmieri MG, Pasqualetti P, Caltagirone C (2000) Time-dependent activation of parieto-frontal networks for directing attention to tactile space A study with paired transcranial magnetic stimulation pulses in right-brain-damaged patients with extinction. Brain 123(9):1939–1947
15. Vahabzadeh-Hagh AM, Muller PA, Pascual-Leone A, Jensen FE, Rotenberg A (2011) Measures of cortical inhibition by paired-pulse transcranial magnetic stimulation in anesthetized rats. J Neurophysiol 105:615–624
16. Chen R (2004) Interactions between inhibitory and excitatory circuits in the human motor cortex. Exp Brain Res 154:1–10
17. Fisher RJ, Nakamura Y, Bestmann S, Rothwell JC, Bostock H (2002) Two phases of intracortical inhibition revealed by transcranial magnetic threshold tracking. Exp Brain Res 143:240–248
18. Florian J, Muller-Dahlhaus M, Liu Y, Ziemann U (2008) Inhibitory circuits and the nature of their interactions in the human motor cortex a pharmacological TMS study. J Physiol 586:495–514
19. Hanajima R, Furubayashi T, Iwata NK, Shiio Y, Okabe S, Kanazawa I, Ugawa Y (2003) Further evidence to support different mechanisms underlying intracortical inhibition of the motor cortex. Exp Brain Res 151:427–434
20. Kang Y, Kaneko T, Ohishi H, Endo K, Araki T (1994) Spatiotemporally differential inhibition of pyramidal cells in the cat motor cortex. J Neurophysiol 71:280–293
21. McCormick DA (1992) Neurotransmitter actions in the thalamus and cerebral cortex. J Clin Neurophysiol 9:212–223
22. Nakamura H, Kitagawa H, Kawaguchi Y, Tsuji H, Takano H, Nakatoh S (1995) Intracortical facilitation and inhibition after paired magnetic stimulation in humans under anesthesia. Neurosci Lett 199:155–157
23. Chu J, Gunraj C, Chen R (2008) Possible differences between the time courses of presynaptic and postsynaptic GABAB mediated

inhibition in the human motor cortex. Exp Brain Res 184:571–577

24. McCormick DA (1989) GABA as an inhibitory neurotransmitter in human cerebral cortex. J Neurophysiol 62:1018–1027
25. McDonnell MN, Orekhov Y, Ziemann U (2006) The role of GABA(B) receptors in intracortical inhibition in the human motor cortex. Exp Brain Res 173:86–93
26. Wassermann EM, Samii A, Mercuri B, Ikoma K, Oddo D, Grill SE, Hallett M (1996) Responses to paired transcranial magnetic stimuli in resting, active, and recently activated muscles. Exp Brain Res 109:158–163
27. Werhahn KJ, Kunesch E, Noachtar S, Benecke R, Classen J (1999) Differential effects on motorcortical inhibition induced by blockade of GABA uptake in humans. J Physiol 517(Pt 2):591–597
28. Cash RF, Benwell NM, Murray K, Mastaglia FL, Thickbroom GW (2009) Neuromodulation by paired-pulse TMS at an I-wave interval facilitates multiple I-waves. Exp Brain Res 193:1–7
29. Di Lazzaro V, Oliviero A, Profice P, Pennisi MA, Pilato F, Zito G, Dileone M, Nicoletti R, Pasqualetti P, Tonali PA (2003) Ketamine increases human motor cortex excitability to transcranial magnetic stimulation. J Physiol 547:485–496
30. Ziemann U, Chen R, Cohen LG, Hallett M (1998) Dextromethorphan decreases the excitability of the human motor cortex. Neurology 51:1320–1324
31. Ziemann U, Tergau F, Wischer S, Hildebrandt J, Paulus W (1998) Pharmacological control of facilitatory I-wave interaction in the human motor cortex. A paired transcranial magnetic stimulation study. Electroencephalogr Clin Neurophysiol 109:321–330
32. Ferbert A, Priori A, Rothwell JC, Day BL, Colebatch JG, Marsden CD (1992) Interhemispheric inhibition of the human motor cortex. J Physiol 453:525–546
33. Gerloff C, Corwell B, Chen R, Hallett M, Cohen LG (1998) The role of the human motor cortex in the control of complex and simple finger movement sequences. Brain 121(9):1695–1709
34. Hanajima R, Ugawa Y, Machii K, Mochizuki H, Terao Y, Enomoto H, Furubayashi T, Shiio Y, Uesugi H, Kanazawa I (2001) Interhemispheric facilitation of the hand motor area in humans. J Physiol 531:849–859
35. Chen R, Yung D, Li JY (2003) Organization of ipsilateral excitatory and inhibitory pathways in the human motor cortex. J Neurophysiol 89: 1256–1264
36. Allison T, McCarthy G, Wood CC (1992) The relationship between human long-latency somatosensory evoked potentials recorded from the cortical surface and from the scalp. Electroencephalogr Clin Neurophysiol 84: 301–314
37. Allison T, McCarthy G, Wood CC, Williamson PD, Spencer DD (1989) Human cortical potentials evoked by stimulation of the median nerve. II. Cytoarchitectonic areas generating long-latency activity. J Neurophysiol 62: 711–722
38. Forss N, Hari R, Salmelin R, Ahonen A, Hamalainen M, Kajola M, Knuutila J, Simola J (1994) Activation of the human posterior parietal cortex by median nerve stimulation. Exp Brain Res 99:309–315
39. Hari R, Reinikainen K, Kaukoranta E, Hamalainen M, Ilmoniemi R, Penttinen A, Salminen J, Teszner D (1984) Somatosensory evoked cerebral magnetic fields from SI and SII in man. Electroencephalogr Clin Neurophysiol 57:254–263
40. Kofler M, Fuhr P, Leis AA, Glocker FX, Kronenberg MF, Wissel J, Stetkarova I (2001) Modulation of upper extremity motor evoked potentials by cutaneous afferents in humans. Clin Neurophysiol 112:1053–1063
41. Chen R, Corwell B, Hallett M (1999) Modulation of motor cortex excitability by median nerve and digit stimulation. Exp Brain Res 129:77–86
42. Pfurtscheller G (1992) Event-related synchronization (ERS): an electrophysiological correlate of cortical areas at rest. Electroencephalogr Clin Neurophysiol 83:62–69
43. Pfurtscheller G, Stancak A Jr, Neuper C (1996) Post-movement beta synchronization. A correlate of an idling motor area? Electroencephalogr Clin Neurophysiol 98:281–293
44. Salenius S, Schnitzler A, Salmelin R, Jousmaki V, Hari R (1997) Modulation of human cortical rolandic rhythms during natural sensorimotor tasks. Neuroimage 5:221–228
45. Salmelin R, Hari R (1994) Spatiotemporal characteristics of sensorimotor neuromagnetic rhythms related to thumb movement. Neuroscience 60:537–550
46. Sailer A, Molnar GF, Paradiso G, Gunraj CA, Lang AE, Chen R (2003) Short and long latency afferent inhibition in Parkinson's disease. Brain 126:1883–1894
47. Ugawa Y, Uesaka Y, Terao Y, Hanajima R, Kanazawa I (1995) Magnetic stimulation over the cerebellum in humans. Ann Neurol 37:703–713
48. Iwata NK, Ugawa Y (2005) The effects of cerebellar stimulation on the motor cortical excit-

ability in neurological disorders: a review. Cerebellum 4:218–223
49. Miniussi C, Thut G (2010) Combining TMS and EEG offers new prospects in cognitive neuroscience
50. Fitzgerald PB, Daskalakis ZJ, Hoy K, Farzan F, Upton DJ, Cooper NR, Maller JJ (2008) Cortical inhibition in motor and non-motor regions: a combined TMS-EEG study. Clin EEG Neurosci 39:112–117
51. Fitzgerald PB, Maller JJ, Hoy K, Farzan F, Daskalakis ZJ (2009) GABA and cortical inhibition in motor and non-motor regions using combined TMS-EEG: a time analysis. Clin Neurophysiol 120:1706–1710
52. Rotenberg A (2010) Prospects for clinical applications of transcranial magnetic stimulation and real-time EEG in epilepsy. Brain Topogr 22:257–266
53. Blicher JU, Jakobsen J, Andersen G, Nielsen JF (2009) Cortical excitability in chronic stroke and modulation by training: a TMS study. Neurorehabil Neural Repair 23:486–493
54. Brodtmann A, Macdonell RA, Gilligan AK, Curatolo J, Berkovic SF (1999) Cortical excitability and recovery curve analysis in generalized epilepsy. Neurology 53:1347–1349
55. Butefisch CM, Netz J, Wessling M, Seitz RJ, Homberg V (2003) Remote changes in cortical excitability after stroke. Brain 126: 470–481
56. Butefisch CM, Wessling M, Netz J, Seitz RJ, Homberg V (2008) Relationship between interhemispheric inhibition and motor cortex excitability in subacute stroke patients. Neurorehabil Neural Repair 22:4–21
57. Cicinelli P, Pasqualetti P, Zaccagnini M, Traversa R, Oliveri M, Rossini PM (2003) Interhemispheric asymmetries of motor cortex excitability in the postacute stroke stage: a paired-pulse transcranial magnetic stimulation study. Stroke 34:2653–2658
58. Klimpe S, Behrang-Nia M, Bott MC, Werhahn KJ (2009) Recruitment of motor cortex inhibition differentiates between generalized and focal epilepsy. Epilepsy Res 84:210–216
59. Manganotti P, Acler M, Zanette GP, Smania N, Fiaschi A (2008) Motor cortical disinhibition during early and late recovery after stroke. Neurorehabil Neural Repair 22:396–403
60. Wittenberg GF, Bastings EP, Fowlkes AM, Morgan TM, Good DC, Pons TP (2007) Dynamic course of intracortical TMS paired-pulse responses during recovery of motor function after stroke. Neurorehabil Neural Repair 21:568–573
61. Ferreri F, Pasqualetti P, Maatta S, Ponzo D, Ferrarelli F, Tononi G, Mervaala E, Miniussi C, Rossini PM (2011) Human brain connectivity during single and paired pulse transcranial magnetic stimulation. Neuroimage 2010:1
62. Russmann H, Lamy JC, Shamim EA, Meunier S, Hallett M (2009) Associative plasticity in intracortical inhibitory circuits in human motor cortex. Clin Neurophysiol 120:1204–1212
63. Manganotti P, Zanette G, Bonato C, Tinazzi M, Polo A, Fiaschi A (1997) Crossed and direct effects of digital nerves stimulation on motor evoked potential: a study with magnetic brain stimulation. Electroencephalogr Clin Neurophysiol 105:280–289
64. Sailer A, Molnar GF, Cunic DI, Chen R (2002) Effects of peripheral sensory input on cortical inhibition in humans. J Physiol 544: 617–629
65. Tokimura H, Di Lazzaro V, Tokimura Y, Oliviero A, Profice P, Insola A, Mazzone P, Tonali P, Rothwell JC (2000) Short latency inhibition of human hand motor cortex by somatosensory input from the hand. J Physiol 523(Pt 2):503–513

Chapter 7

Repetitive Transcranial Magnetic Stimulation (rTMS) Protocols

Lindsay Oberman

Abstract

Repetitive TMS (rTMS) refers to a train of TMS pulses delivered over a single cortical region. Unlike single or paired pulses, rTMS is capable of inducing long-term effects—either facilitative or suppressive—that outlast the stimulation period. Parameters such as frequency and intertrain interval can influence the direction (facilitation versus suppression) of the effect. rTMS protocols have been developed both to study the brain and as a therapeutic intervention for neurological and psychological disorders. rTMS is considered to be generally safe in healthy individuals, however, adverse side effects have been reported in the literature. In 2009 a safety consensus paper was published which provided guidelines for parameters considered to be safe. The most severe side effect associated with rTMS is the risk of seizure induction, though this is considered rare, less than 0.1 % risk. This chapter aims to summarize the safety and utility of rTMS protocols in healthy and patient populations.

Key words rTMS, Safety, Parameters, Frequency, Facilitation, Suppression

1 What Is rTMS?

Repetitive TMS (rTMS) refers to a train of TMS pulses delivered at the same intensity over a single cortical region. The pulse per second frequency of rTMS trains typically ranges between 1 Hz (one pulse per second) and 50 Hz (50 pulses per second). 1 Hz paradigms are commonly applied continuously for a period of several minutes, while higher frequency paradigms are commonly applied in a "patterned" fashion: short high-frequency bursts interleaved by pauses of several 100 ms to several seconds (Fig. 1).

RTMS can be differentiated from single pulse and paired pulse paradigms not only by the frequency of stimulation, but also by its effect on the cortex. Unlike single or paired pulses which are generally thought to induce an effect that does not last beyond the duration of stimulation, rTMS is capable of inducing long-term effects—either facilitative or suppressive—that outlast the stimulation

Alexander Rotenberg et al. (eds.), *Transcranial Magnetic Stimulation*, Neuromethods, vol. 89,
DOI 10.1007/978-1-4939-0879-0_7, © Springer Science+Business Media New York 2014

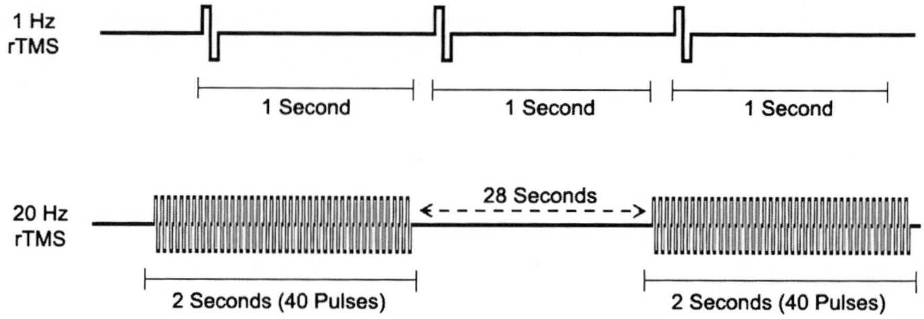

Fig. 1 Representative schematic of commonly applied protocols of rTMS. One Hz protocols are typically applied in a continuous fashion, while high-frequency protocols are typically applied in short bursts of stimulation with intertrain intervals ranging from several 100 ms to several seconds

period. The duration of this effect can last several minutes following a single rTMS session, or upwards of days to weeks following several consecutive rTMS sessions.

It should be noted that though the electrical field induced from the TMS machine is spatially localized to a small (1–2 cm) area of cortex, it does not follow that the impact of stimulation is confined to the region of cortex directly below the TMS coil. Numerous studies combining TMS with behavioral paradigms, TMS with EEG, TMS with MRI and fMRI, and TMS with PET have demonstrated the effects of TMS in brain regions distant from the site of stimulation. Thus, it cannot be assumed that a behavioral effect induced by TMS to a specific region is a direct result of activity in that region—it could result from any number of transsynaptic regions also affected by that stimulation.

2 Important Parameters: Interpulse Interval and Intertrain Interval

The direction of effect generated by rTMS, either facilitation or suppression, is largely dependent upon both the frequency of stimulation (interpulse interval) and the structure of the pulse train (presence and length of intertrain interval). For example, low-frequency (1 Hz) stimulation is generally applied continuously with no intertrain interval, while high-frequency (5–25 Hz) stimulation is most often applied for a short period of time (1–2 s) with a relatively long (20–30 s) intertrain interval during which no stimulation is applied. When evaluating these two categories, there is a general consensus that continuous low-frequency stimulation is suppressive while bursts of high-frequency stimulation are facilitatory. This is largely based on electrophysiological studies indicating that following 1 Hz stimulation to the primary motor cortex, the motor-evoked potential (MEP) recorded at a peripheral

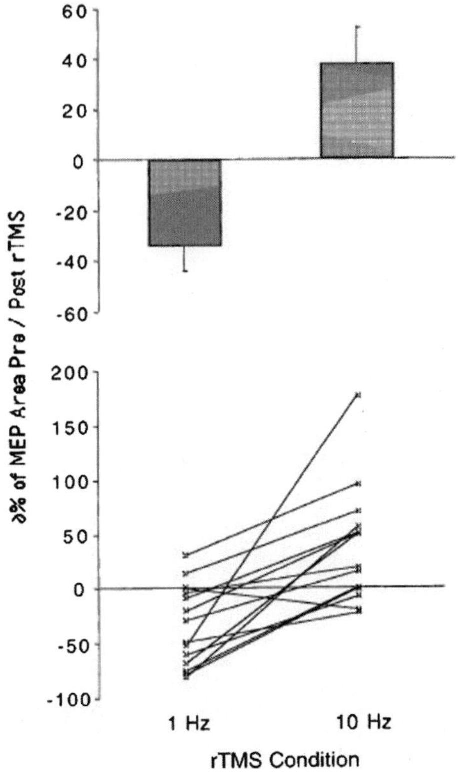

Fig. 2 One Hz stimulation typically results in a suppression of TMS-induced motor-evoked potentials (MEPs) in the peripheral nervous system as measured by EMG, while 10 Hz stimulation typically results in facilitation of MEPs (obtained with permission from [1])

muscle is suppressed for a period of time. Similarly, following high-frequency stimulation (10 Hz) the MEP is enhanced for a period of time (Fig. 2).

Based upon these limited protocols, it would be easy for one to make the generalization that high-frequency stimulation (with short interpulse intervals) is *always* facilitatory. Unfortunately, this is not the case. Several studies have demonstrated that the intertrain interval also plays a role in rTMS effect direction. The influence these intertrain intervals exert is likely a result of meta-plasticity (the ability of the synapse to undergo a secondary plastic change following the recent induction of plasticity). The impact of the intertrain interval is perhaps most clearly demonstrated by the contrasting effects of two types of Theta Burst Stimulation (TBS). TBS can be applied to the motor cortex in a continuous fashion with no intertrain interval. Known as continuous Theta Burst (cTBS), this stimulatory pattern generally serves to suppress neural activity. Conversely, the same frequency of stimulation can be

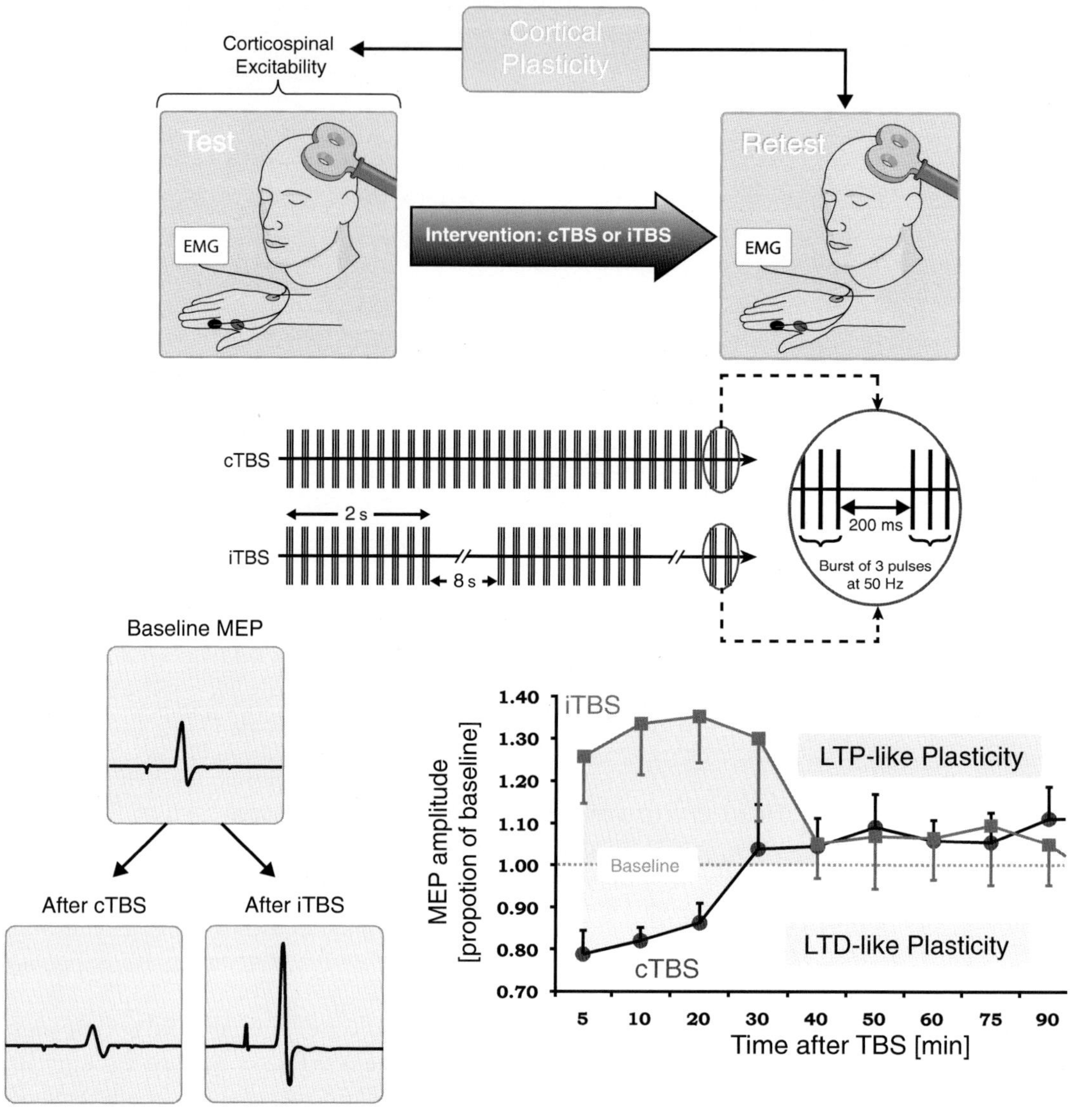

Fig. 3 Schematic summary of the theta burst stimulation (TBS) protocol. TBS involves applying bursts of high-frequency transcranial magnetic stimulation (3 pulses at 50 Hz) repeated at intervals of 200 ms. After TBS is applied to the motor cortex in an intermittent fashion (iTBS), single-pulse TMS-induced MEPs show increased amplitude for a period of 20–30 min, whereas continuous TBS (cTBS) leads to a suppression of the TMS-induced MEPs for approximately the same amount of time (obtained with permission from [2])

applied in 2 s bursts with an 8 s intertrain interval. Known as intermittent Theta Burst (iTBS), this stimulatory pattern generally serves to facilitate neural activity (Fig. 3). Both paradigms (described in detail later in the chapter) involve the same frequency of stimulation within the train, only differing in their interstimulus interval between trains.

3 What can rTMS Tell Us?

Because of its potential to online modulate brain activity in a given region and its capacity to induce long-term changes in cortical functioning, neuroscientists have utilized rTMS to study typical and atypical functioning of the brain, while clinicians have utilized this technology to modify cortical functioning in patients with various neurological conditions.

In Neuroscience, rTMS is often used to create "virtual lesions" or temporarily suppress functioning of a given region in order to establish its functional significance. TMS, in this way, can be used to complement and extend knowledge obtained from fMRI and PET studies which may implicate a given set of regions in a particular task. More specifically, TMS can selectively suppress activity in one of these regions allowing for an evaluation of the resulting change in behavior. Such studies have elucidated the role of primary visual cortex in visual imagery [3], primary motor cortex in mental rotation [4], and frontal cortex in random number generation [5] just to name a few.

Many rTMS protocols can be administered either "online" or "off-line" (Fig. 4). During online protocols, participants engage in

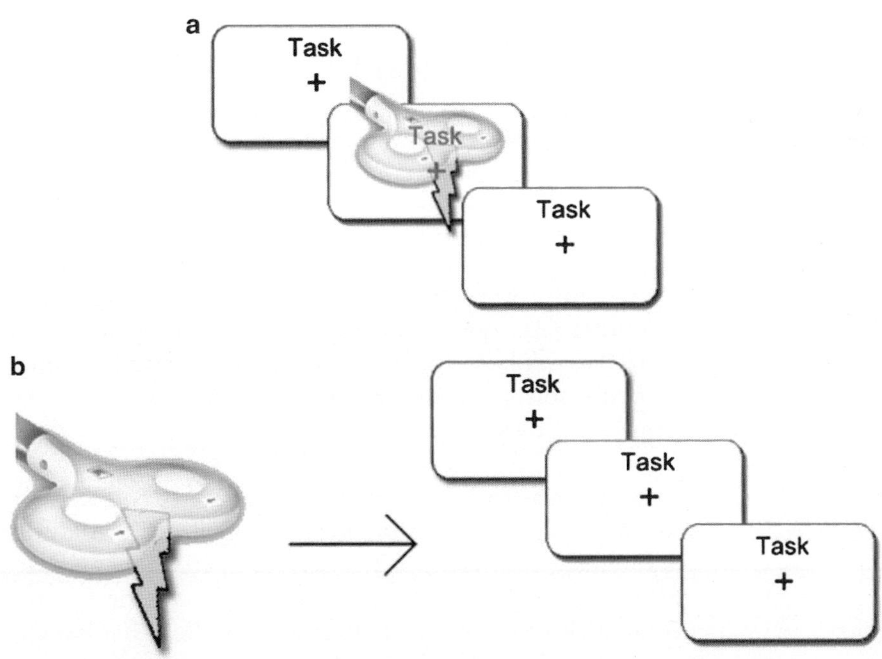

Fig. 4 Timeline of online (**a**) and off-line (**b**) protocols. Online protocols interleave short bursts of stimulation just prior to a task in order to induce a relatively short (few second) modulation in the functioning of the region of cortex where the stimulation is applied, while off-line protocols use longer trains of rTMS to induce a longer (several minute) modulation in the functioning of the region of interest

a task while short trains of high-frequency stimulation are applied. These short trains tend to create a sort of "noise" in the neural signal thereby temporarily impairing cortical processing and resultant task performance. This type of protocol makes it possible to study both the role of a given cortical region and at what stage of processing it plays said role. Off-line protocols take advantage of the potential of rTMS to induce long-lasting changes. In this type of procedure, participants undergo a full pattern of rTMS and then engage in a given task.

Neuroscientists interested in the mechanisms of cortical plasticity, or the ability of the cortex to change and adapt, often explore the local and distant cortical activity generated in response to excitatory or suppressive rTMS trains. This type of protocol can be done in healthy control participants to explore typical connectivity and circuit development as well as in patients to explore plastic reorganization or differences in capacity for plasticity following injury or disease.

Finally, and perhaps most frequently, rTMS has been utilized by clinicians as a novel therapeutic intervention to produce lasting functional improvements for a number of neurological conditions. For neurological conditions where the behavioral disability is due to too little or too much cortical activity, rTMS protocols can be applied to "normalize" such pathology and engender behavioral benefits. Such interventions have been explored in conditions ranging from depression, Parkinson's disease, and focal dystonia to chronic pain, intractable epilepsy, and stroke.

4 RTMS Protocols

Though all rTMS protocols can be administered using the same machine, it is important to realize effects will differ depending on the type of protocol used (as determined by frequency and inter-train interval), how it's presented (online versus off-line) and where it is applied. In this section, I will describe a few common protocols and some interesting discoveries that have been made using said protocols.

4.1 Online Suppressive Protocols

Though high-frequency protocols (as described above) typically lead to excitation of the given cortical region, during online paradigms, researchers often use high-frequency stimulation to induce a virtual lesion and transiently disrupt the functioning of a given cortical region. One interesting finding that has come out of this technique involves the role of primary visual cortex in a blind person's ability to read Braille.

Functional imaging studies have revealed that the primary visual cortex is active during Braille reading in early and congenitally blind participants [6]. However, it was not clear at the time of this study whether this neural activity was causally related to the tactile discrimination ability of the participants or an epiphenomenon. In order to answer this question, Cohen and colleagues [7, 8] asked blind and sighted participants to use their index finger to read Braille or embossed Roman letters while high-frequency rTMS was applied to primary visual cortex. They found that short trains of 10 Hz stimulation to primary visual cortex impaired the performance of congenitally and early blind patients, but had no effect on sighted or late-blind patients. This suggests the primary visual cortex of congenital and early-blind patients has undergone a functional plastic change that allows it to encode tactile information.

4.2 Off-Line Suppressive Protocols

The most common suppressive protocol is 1 Hz rTMS. In this protocol, a single TMS pulse is applied once per second for a period of 15–30 min resulting in a suppression of the targeted cortical region for a period lasting up to several minutes following the stimulation. The physiological mechanism leading to these lasting effects on cortical excitability is still unclear, however, recent studies have suggested that low-frequency rTMS induces a long-term depression (LTD)-like effect on cortical synapses. A recent study probed the role of GABAergic inhibition and *N*-methyl-D-aspartate (NMDA) receptors in 1 Hz rTMS suppression [9]. Patients medicated with either a positive modulator of the GABA-A receptor (lorazepam), an NMDA receptor antagonist (dextromethorphan), or a placebo were given suprathreshold 1 Hz rTMS for 15 min. While typical suppression was evident in the placebo group, those on either medication showed no lasting suppression in cortical excitability following stimulation. This study suggests that both GABA and NMDA receptors play a role in the off-line suppression of cortical excitability by 1 Hz rTMS.

Another off-line suppressive protocol that is fairly novel but growing in popularity is continuous (c) TBS. cTBS involves the application of 3 pulses of 50 Hz at 5 Hz for 20 s (300 total stimuli) or 40 s (600 stimuli). This protocol was originally introduced into the literature in 2005 after Huang and colleagues [10] demonstrated that this pattern induces suppression in motor cortical excitability (as measured by TMS-induced MEPs) for a period of approximately 30–40 min following less than a minute of stimulation.

TBS protocols were designed to mimic electrical stimulation protocols that were successful in inducing LTD in animal models. To evaluate whether, similar to 1 Hz stimulation, the LTD-like effects were a result of GABAergic mechanisms (like 1 Hz stimulation), Stagg and colleagues [11] performed a magnetic resonance spectroscopy (MRS) study during which cTBS was applied to the primary motor cortex. Afterwards, subjects were scanned using an MRS to measure local changes in cortical concentrations of GABA and glutamate. Researchers found that cTBS leads to a local increase in GABA and no significant change in glutamate concentrations.

4.3 Off-Line Facilitatory Protocols

Just as low frequency and cTBS protocols can lead to long-lasting suppression of cortical excitability, high frequency and iTBS protocols can lead to long-lasting facilitation of cortical excitability. High-frequency protocols (as described above) are most commonly applied in an intermittent fashion with trains of stimulation intermixed with periods of no stimulation. The most common application of such protocols is for the treatment of medication resistant depression. It was noted that patients with depression have hypoactivity in the left prefrontal cortex [12]. As such, many have explored the use of high-frequency TMS applied to the same region in an attempt to locally enhance cortical activity.

The exact protocol utilized differs between clinicians; however, the majority of recent studies have applied 10-Hz suprathreshold stimulation (110–120 % of resting motor threshold) for a period of 2–5 s separated by an intertrain interval of between 25 and 55 s. A recent meta-analysis [13] found that high-frequency rTMS applied to the left dorsal lateral prefrontal cortex is superior to sham stimulation and may be as effective as commercially available antidepressant medications.

iTBS protocols have been studied much less frequently, but these protocols may also hold promise for therapeutic interventions for a number of neurological conditions including depression [14], multiple sclerosis [15] and schizophrenia [16]. iTBS protocols, like cTBS, involves the application of 3 pulses of 50 Hz applied at 5 Hz. However, instead of applying these bursts continuously, a 10 s intertrain interval is applied every 2 s. This protocol lasts 190 s (still much shorter than most other high-frequency protocols) and off-line facilitatory effects tend to last approximately 30–40 min in healthy controls following a single application.

5 Safety of rTMS

rTMS is considered to be generally safe in healthy individuals using protocols that adhere to current safety guidelines. That being said, certain adverse side effects have been reported that deserve mention. The most severe side effect associated with rTMS is the risk of seizure induction. Since rTMS has been used in the clinic and research lab, a total of 16 incidents of seizure in previously healthy individuals have been reported. Though these cases should not be minimized, in light of the tens of thousands of individuals who have undergone rTMS, the risk is thought to be exceedingly small. Seizures are caused by abnormally highly synchronized neural discharges.

Table 1
Maximum safe duration (expressed in seconds) of single trains of rTMS

Frequency (Hz)	90 % rMT	100 % rMT	110 % rMT	120 % rMT	130 % rMT
1	>1,800	>1,800	>1,800	>360	>50
5	>10	>10	>10	>10	>10
10	>5	>5	>5	4.2	2.9
20	2.05	2.05	1.6	1.0	.055
25	1.28	1.28	0.84	0.4	0.24

Safety defined as the absence of seizure, spread of excitation or after discharge of EMG activity. Numbers proceeded by > are longest duration tested. Consensus has been reached for this table (obtained with permission from [17])

As rTMS protocols use increasingly higher frequency and shorter intertrain intervals, the risk of inducing seizures during or immediately following rTMS trains increase. It is also important to note that other factors such as concurrent seizure threshold reducing medication and large quantities of neurostimulants (such as caffeine and sleep deprivation) can also increase the risk of seizure induction. In 2009 a safety consensus paper was published which provided guidelines for parameters considered to be safe [17]; these are displayed in Table 1.

Other safety concerns include hearing impairment and headache, as well as cognitive and neuropsychological changes. The TMS coil produces an acoustic artifact that may exceed 140 dB [18]. Thus, the use of earplugs is highly recommended to avoid hearing loss or tinnitus. Headache is the most common side effect reported during TMS. It is unclear exactly what induces headaches during TMS, however trigeminal nerve stimulation and muscle contraction likely contribute to discomfort. Said discomfort has been reported in up to 40 % of subjects, depending upon the protocol, location, and duration of stimulation. In the majority of individuals, a common analgesic administered orally usually alleviates this pain. Typically, cognitive and neuropsychological changes are intentional outcomes of rTMS rather than unwanted side effects; however, it is important to monitor and report any unexpected or unexpectedly severe induced effects.

In general rTMS is considered safe when protocols that fit within the above safety guidelines are used and when patients are screened so as not be of greater risk for adverse events due to medication, comorbid neurological or psychological condition. Despite the relatively small risk of serious side effects, all rTMS protocols should be done under the oversight of a medically responsible physician familiar with TMS, TMS protocols, and the associated risks.

Also, as there is little safety data on the application of rTMS with pediatric populations, rTMS should only be used in these individuals when there is compelling clinical reasons (such as treatment of refractory epilepsy or specific psychiatric conditions).

In addition to safety considerations, ethical issues must also be considered. Though it is believed that rTMS does not carry with it any long-term negative side effects, it is unclear whether sustained rTMS treatments may lead to more permanent plastic changes in the cortex. Additionally, though it may be safe to apply rTMS to children, one must consider that the brain of a child is still developing and synapses are still quite plastic. Stimulation may lead to more drastic plastic changes in the developing cortex than comparable stimulation in an adult. Though it may be safe to apply rTMS to a neonate, one must consider what that stimulation may do to the very immature and plastic cortex that is developing.

6 Conclusions

rTMS is a unique technique capable of noninvasively inducing long-lasting effects in the brain. TMS is the only reliable tool of its kind and, thus, provides neuroscientists and clinicians with a means to study and modify brain activity in healthy individuals and those with neurological conditions. Depending on the exact parameters used and the location of stimulation, the effects of rTMS can range in both direction (facilitation or suppression) and duration. This chapter aimed to review the most commonly applied protocols and how they can be applied to study and in some cases treat the cortex. rTMS has tremendous potential for science and medicine, but must be applied with caution as there are both safety and ethical considerations that need to be addressed.

References

1. Maeda F, Keenan JP, Tormos JM, Topka H, Pascual-Leone A (2000) Modulation of corticospinal excitability by repetitive transcranial magnetic stimulation. Clin Neurophysiol 111(5):800–805
2. Oberman L, Eldaief M, Fecteau S, Ifert-Miller F, Tormos JM, Pascual-Leone A (2012) Abnormal modulation of corticospinal excitability in adults with Asperger's syndrome. Eur J Neurosci 36(6):2782–2788
3. Sparing R, Mottaghy FM, Ganis G, Thompson WL, Topper R, Kosslyn SM et al (2002) Visual cortex excitability increases during visual mental imagery—a TMS study in healthy human subjects. Brain Res 938(1–2):92–97
4. Ganis G, Keenan JP, Kosslyn SM, Pascual-Leone A (2000) Transcranial magnetic stimulation of primary motor cortex affects mental rotation. Cereb Cortex 10(2):175–180
5. Jahanshahi M, Profice P, Brown RG, Ridding MC, Dirnberger G, Rothwell JC (1998) The effects of transcranial magnetic stimulation over the dorsolateral prefrontal cortex on suppression of habitual counting during random number generation. Brain 121(Pt 8):1533–1544
6. Sadato N, Pascual-Leone A, Grafman J, Ibanez V, Deiber MP, Dold G et al (1996) Activation of the primary visual cortex by Braille reading in blind subjects. Nature 380(6574):526–528

7. Cohen LG, Celnik P, Pascual-Leone A, Corwell B, Falz L, Dambrosia J et al (1997) Functional relevance of cross-modal plasticity in blind humans. Nature 389(6647):180–183
8. Cohen LG, Weeks RA, Sadato N, Celnik P, Ishii K, Hallett M (1999) Period of susceptibility for cross-modal plasticity in the blind. Ann Neurol 45(4):451–460
9. Fitzgerald PB, Benitez J, Oxley T, Daskalakis JZ, de Castella AR, Kulkarni J (2005) A study of the effects of lorazepam and dextromethorphan on the response to cortical 1 Hz repetitive transcranial magnetic stimulation. Neuroreport 16(13):1525–1528
10. Huang YZ, Edwards MJ, Rounis E, Bhatia KP, Rothwell JC (2005) Theta burst stimulation of the human motor cortex. Neuron 45(2):201–206
11. Stagg CJ, Wylezinska M, Matthews PM, Johansen-Berg H, Jezzard P, Rothwell JC et al (2009) Neurochemical effects of theta burst stimulation as assessed by magnetic resonance spectroscopy. J Neurophysiol 101(6):2872–2877
12. Baxter LR Jr, Schwartz JM, Phelps ME, Mazziotta JC, Guze BH, Selin CE et al (1989) Reduction of prefrontal cortex glucose metabolism common to three types of depression. Arch Gen Psychiatry 46(3):243–250
13. Schutter DJ (2009) Antidepressant efficacy of high-frequency transcranial magnetic stimulation over the left dorsolateral prefrontal cortex in double-blind sham-controlled designs: a meta-analysis. Psychol Med 39(1):65–75
14. Holzer M, Padberg F (2010) Intermittent theta burst stimulation (iTBS) ameliorates therapy-resistant depression: a case series. Brain Stimul 3(3):181–183
15. Mori F, Codeca C, Kusayanagi H, Monteleone F, Boffa L, Rimano A et al (2010) Effects of intermittent theta burst stimulation on spasticity in patients with multiple sclerosis. Eur J Neurol 17(2):295–300
16. Demirtas-Tatlidede A, Freitas C, Cromer JR, Safar L, Ongur D, Stone WS et al (2010) Safety and proof of principle study of cerebellar vermal theta burst stimulation in refractory schizophrenia. Schizophr Res 124(1–3):91–100
17. Rossi S, Hallett M, Rossini PM, Pascual-Leone A (2009) Safety, ethical considerations, and application guidelines for the use of transcranial magnetic stimulation in clinical practice and research. Clin Neurophysiol 120(12):2008–2039
18. Counter SA, Borg E (1992) Sound perception induced by extracranial magnetic stimulation in deaf patients. Scand Audiol 21(4):239–243

Part III

Experimental Design

Chapter 8

Offline and Online "Virtual Lesion" Protocols

Shirley Fecteau and Mark Eldaief

Abstract

Transcranial Magnetic Stimulation (TMS) can be used to transiently disrupt the function of specified cortical targets. If this disruption is generated prior to the onset of a critical regional function, a suppression or arrest of the said function may result. This method of using TMS to disrupt functions has been dubbed the "virtual lesion" approach. In this chapter, we explore both the mechanistic and functional foundations of this novel technique. After briefly discussing the history and development of this paradigm, we outline practical applications of the "virtual lesion" in clinical, cognitive, and behavioral research. Finally, we offer practical considerations to ensure successful experimentation with and application of this technique.

Key words Transcranial magnetic stimulation, Virtual Lesion, Brain-behavior relationship, Online protocol, Offline protocol

1 Introduction

One of the central applications of transcranial magnetic stimulation (TMS) in cognitive and clinical neuroscience is the induction of the so-called "virtual lesions." Put simply, TMS is capable of non-invasively altering the activation, and, as a corollary, the function, of a targeted brain region. This process is referred to as creating a "virtual lesion" because the temporary and reversible interruption of the function of a circumscribed brain region can mimic the psychophysiological effects of a true cerebral lesion.

1.1 The Virtual Lesion

TMS can be used to transiently disrupt the function of specified cortical targets. If this disruption is generated prior to the onset of a critical regional function, a suppression or arrest of the said function may result. This method of using TMS to disrupt functions has been dubbed the "virtual lesion" approach.

Alexander Rotenberg et al. (eds.), *Transcranial Magnetic Stimulation*, Neuromethods, vol. 89,
DOI 10.1007/978-1-4939-0879-0_8, © Springer Science+Business Media New York 2014

1.2 What Changes in Cortical Physiology Occur as a Result of Virtual Lesions?

Suppression of cortical activity can be accomplished in one of two ways: by single pulse TMS or by low frequency *repetitive* TMS (rTMS). Physiologically, authors have attributed the inhibitory effect of TMS to several factors, including activation of GABAergic inhibitory interneurons [1] and another explanation concerns the introduction of "neural noise" into the neuronal cluster or network being stimulated [2, 3]. This "noise" is thought to be random and asynchronous to the ongoing and task-specific activity in the stimulated region thereby resulting in the observed behavioral dysfunction [4].

While virtual lesion studies typically seek to transiently disrupt the brain activity, it is also possible to facilitate or augment this activity. Several studies in humans and animals suggest that TMS is capable of either facilitating or suppressing cortical activity at the site of stimulation, depending on the type and frequency of stimulation [5–8]. While suppression is mediated by single pulse or low frequency TMS, facilitation is often accomplished with high frequency rTMS (e.g. at frequencies of 5 Hz and above). TMS can also enhance cortical excitation in an indirect manner. TMS inhibition to one hemisphere can release transcallosal inhibition to homologous areas in the contralateral hemisphere, thus engendering contralateral facilitation [9].

Nonetheless, an important caveat to the aforementioned studies is that the exact biophysical changes occurring at the site of TMS stimulation remain unclear. For instance, whereas most human subjects increase focal cortical activity with high frequency rTMS and decrease activity with low frequency rTMS, a minority of individuals shows the opposite pattern; and there exists a degree of inter-subject variability in response to each stimulation pattern [7]. Finally, the precise response of a cortical target to a given TMS regimen may be state-dependent—meaning the baseline activation pattern of the targeted neural region may differentially affect the response to stimulation [10].

2 The Utility of Virtual Lesions

2.1 Why Use the Virtual Lesion Approach?

Traditionally, virtual lesions have been used to correlate TMS-induced changes in cortical excitability to measured behavioral outcomes. Unlike other methodologies which study brain-behavior relationships, the virtual lesion approach has the advantage of allowing for inferential *causality* between the stimulation and the behavioral measure. Studies of patients with acquired lesions often use neuropsychological measures to track the effects of focal cortical damage. However, the associations drawn from these measures are complicated by the fact that lesions are typically not discrete and oftentimes extend beyond the region of interest to other areas of cortex which may generate secondary effects on the behavior being studied. For example, one might seek to investigate the role of a medial prefrontal lesion on emotional processing, but if the said lesion extends more laterally in the prefrontal cortex, it may encroach on regions involved with the cognitive control of emotion. Moreover, in classical lesion studies, observed behavioral changes may be the result of plasticity-based mechanisms meant to compensate for the cortical insult rather than being the result of the lesion, per se [4].

A main drawback to task-based activation paradigms (e.g. PET, fMRI, EEG, MEG, and NIRS), which seeks to deduce the neural instantiation of behavior by observing brain responses to task presentation, is that the deductions are inherently associative. That is, it is often difficult to discern whether an "activated" region is solely necessary for task execution or is necessary but not sufficient for task completion. In addition to such "false positives," many neuroimaging techniques may be prone to false negatives—i.e., a failure to associate cortical regions with their true functional significance [11]. Lastly, neuroimaging studies are incapable of directionally changing the brain activity in order to study brain-behavior relationships in a more sophisticated fashion. It should be mentioned, however, that effective connectivity analyses have attempted to address these limitations in fMRI by establishing temporal causality between activation in different brain regions.

TMS offers the ability to circumvent the inherently correlational nature of classical lesion studies and neuroimaging approaches. First, the effects of the TMS stimulation are known to temporally precede the behavioral change, thereby allowing a causative timeline to be created. Second, TMS can be conducted in normal subjects, essentially sidestepping the difficulty of assessing behavior in the cortically injured brain [12]. Third, TMS allows for the same anatomical "lesion" to be induced across several subjects and over repeated sessions, thereby bolstering the statistical power of the determined conclusions [13]. (Of note: cortical targeting reliability with TMS has been made possible with the advent of neuronavigation systems and optical tracking software). Fourth, the magnitude and location of the created virtual lesion can be controlled to some extent by modulating the parameters of the stimulation, such as the intensity, the frequency, and duration. Finally, TMS pulses can be delivered at different time points or via different interstimulus intervals (ISIs) thereby providing insight into the temporal pattern of information processing, a stimulation paradigm referred to as TMS *chronometry* [1, 12, 14]. TMS chronometry allows for inferences of *when* a brain region executes a behavior of interest in addition to *what* the brain region is responsible for said behavior.

2.2 Virtual Lesions to Study the Behavioral Ramifications of Network-Wide Changes Caused by TMS

It is becoming increasingly clear that many behaviors are instantiated not through the operation of single cortical regions, but through many regions connected on a network level. In addition to causing changes at the site of stimulation, TMS (particularly rTMS) can induce changes across cortical networks, thus allowing for the correlation of changes in network-wide activity to a given behavior. Local stimulation with TMS to an accessible network node can propagate, via trans-synaptic means, to distal inter-connected nodes with high spatial specificity [15–17]. Much of the evidence for this comes in the form of studies that have combined locally applied TMS with neuroimaging (e.g. PET or fMRI) or with EEG [18]. If one can establish that TMS is able to effect changes in cortical excitability that are relatively specific to a given network, then the directed changes in network activity can be correlated to changes in behavior. For example, Lou et al. [19] demonstrated that single pulse stimulation applied to different cortical regions of the brain's default network (the lateral parietal cortices and the medial prefrontal cortex), an interrupted subjects' ability to process self-specific adjectives. This finding served to corroborate the claim that the default network plays a role in self-referential processing. Future uses of the virtual lesion methodology will have to take into account what distributed effects are caused by focal stimulation, and future studies should endeavor to determine how TMS changes in network (not simply local) activity are translated to behavior.

2.3 The Utility of Virtual Lesions Beyond the Study of Brain-Behavior Relationships

The ability to reversibly alter cortical excitation in a predetermined neural region with TMS has implications for understanding multiple aspects of cortical organization and adaptability. For instance, TMS to specific brain regions can be used to chart connectivity across cortical networks. Paus [20] has argued that TMS is well suited to study network connectivity through a "perturb and measure" approach, in which a cortical area is locally disrupted with TMS and the resulting network effects are tracked via functional imaging. The network-wide effects of local stimulation can be measured eletrophysiologically (e.g. electromyographic recording of motor-evoked potentials or electroencephalographic recording of neural oscillations) to assess changes in the site contralateral to stimulation or distributed brain regions across a given network. Alternatively, the network effects of TMS can be studied by combining TMS with neuroimaging (e.g. fMRI and PET). With combined TMS and fMRI, it may be possible to gauge whether the application of TMS to a single network node increases or decreases the connectivity within the said network, shifts activation to secondary associated networks, or changes the connectivity between the stimulated network and neighboring networks. For a recent comprehensive review of the use of TMS to delineate network connectivity, the reader is directed to Hampson and Hoffman [21].

2.4 The Role of Virtual Lesions in Investigating Disease

The induction of transient increases or decreases in cortical excitability can also have therapeutic ramifications. The hypothesis underlying this approach is that the diseased brain will differentially respond to external perturbation. By comparing neural and behavioral responses to virtual lesions in healthy versus disease states, the mechanisms underlying clinical disorders can be better elucidated. Along the same lines as the discussion above, virtual lesions can elicit aberrant behavioral, electrophysiological, or neuroimaging responses in clinical populations. For example, Levit-Binnun and colleagues [22] showed that TMS to M1 disrupted cognitive sensorimotor networks in schizophrenic patients but not in normal individuals. Additionally, the same group demonstrated aberrations in cortical excitation in schizophrenia through a combined TMS-EEG paradigm [22]. Similarly, Ferrarelli et al. [23] showed that schizophrenics exhibited decreases in evoked EEG gamma oscillations in response to TMS. Moreover, Oberman et al. [24] recently demonstrated that Theta Burst Stimulation to M1 alters the amplitude of MEPs elicited in the contralateral hand for a significantly longer period in patients with Autism Spectrum Disorders as compared to healthy controls. In brief, using TMS to create virtual lesions can elucidate aspects of disease pathophysiology. Moreover, aberrant reactions to TMS perturbation may become biomarkers for diagnosing disease earlier in biologically or genetically susceptible populations.

3 Methodological Considerations

3.1 Online vs. Offline Protocols

Virtual lesion studies can be divided into "online" and "offline" protocols. An online virtual lesion protocol is characterized by the measurement of the consequences of TMS stimulation (e.g. changes in behavioral performance, changes in neurological status, changes in EEG recordings, changes in fMRI activation, etc.) *during* the course of stimulation. A well-known example of an online virtual lesion protocol, published 20 years ago, is the speech arrest induced in healthy subjects during rTMS [25]. In this case, subjects were asked to perform a speech production task, such as counting from one to ten, simultaneous to rTMS stimulation over Broca's area. Online protocols can also be performed using single pulse TMS. In this case, task stimuli and the stimulation pulses are interleaved such that each TMS pulse briefly interferes with cortical activity in a time-controlled manner. In contrast, "offline" virtual lesion protocols measure the effects of TMS before and after stimulation (or only after stimulation). Because in offline protocols, measurements are not done during stimulation, it is necessary to use a form of TMS engendered cortical effect that outlasts the duration of the stimulation.

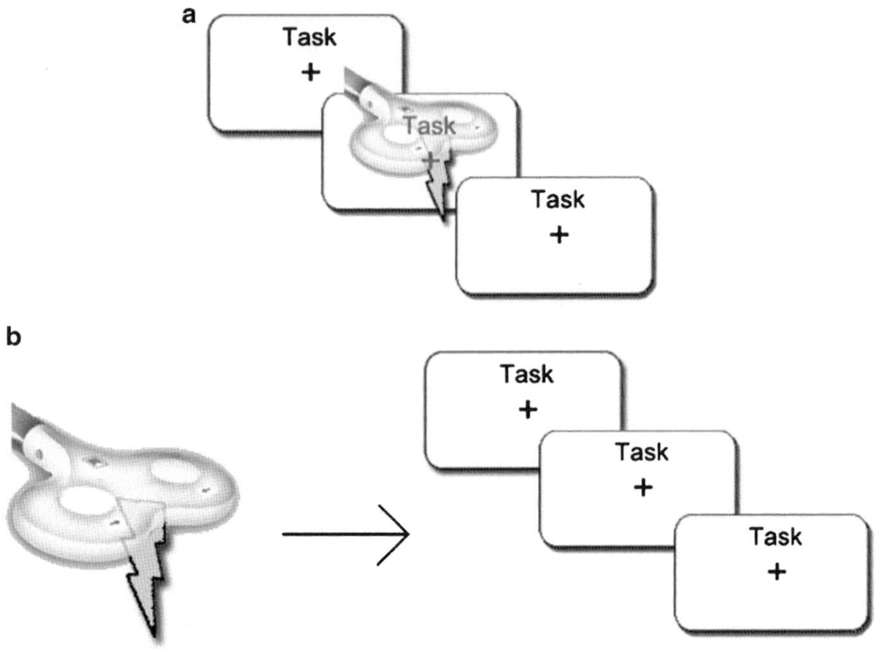

Fig. 1 Illustration of the differences between online and offline TMS protocols. In an online protocol (*upper panel*) TMS is administered during the measurement of interest (e.g. performance on a letter cancelation task, EEG activity or BOLD activation on fMRI). In an offline protocol (*lower panel*) TMS is administered either, before and after, or after TMS and the effects of the stimulation are assumed to outlast the period of stimulation properly

To achieve this, offline protocols utilize rTMS, as its effects have been estimated to last from 20 min [5, 6, 26] up to 1 h [27] post stimulation. Figure 1 schematically depicts the differences between online and offline protocols.

The choice of whether to use an online or offline protocol is dictated by several factors. Among these are whether the effects of TMS are known to be short-lived, as is the case with single pulse TMS, or if the effects of rTMS are known to be shorter in duration then the time needed to measure the outcome of interest. In this circumstance, an online protocol is preferable. Other considerations include whether it is logistically difficult to administer TMS during the outcome measure of interest, e.g. during certain combined TMS-fMRI paradigms. In this circumstance, an offline protocol may be preferable. Yet another consideration is the fact that if stimulation occurs during a behavioral task (as in an online protocol), neural populations being stimulated may already be in a state of activation due to task engagement [28]. This activation may affect and, accordingly, confound the measurement of interest.

3.2 Using a Control Condition

In order to accurately link changes in behaviors (e.g., cognitive, motor functions) or neural responses (e.g., BOLD signal, evoked responses) to the induced virtual lesion, it is necessary to collect measurements in the setting of a control stimulation condition. The use of a sham coil over the same targeted region is one control possibility, but has often been criticized owing to the fact that sham stimulation is qualitatively quite different from active stimulation (e.g., [29]). For instance, when targeting brain regions overlaid by large muscle groups, such as the pars triangularis of the inferior frontal gyrus, active stimulation will typically create noticeable twitches that are absent with sham stimulation. Although more recent sham coils deliver a clicking noise similar to sham stimulation, they still do not induce the same physical sensation as active coils. An alternative approach is to deliver active stimulation to a neural region which is not believed to be relevant to the studied behavioral function or which is anatomically and functionally disconnected from the network of regions in which evoked responses are expected to occur. For example, vertex stimulation is sometimes used as an active control condition. Nevertheless, identification of such a control condition is not always trivial and care must be taken to not inadvertently stimulate a region involved in the outcome being measured. For example, if one were to perform a study measuring the effects of TMS over the left inferior frontal gyrus during a picture naming task, it would be imperative, the control region was not involved with either visual processing or motor speech production.

3.3 Considerations Regarding the Behavioral Task Employed

Just as it is important to have a control stimulation, it is important to have a control condition for the behavioral task employed to track the effects of the induced virtual lesion. For example, imagine we are investigating the role of the dorsolateral prefrontal cortex (DLPFC) in working memory. To achieve this, we have designed an experiment in which TMS is used to disrupt the DLPFC while subjects are performing a n-back working memory task. Let us then imagine that we are using the subjects' response times measured via a task-relevant button press as our metric of behavioral performance. In this experiment, it may be prudent to measure the effects of DLPFC stimulation upon task-irrelevant button pressing, in order to ensure that reaction time measurements are not altered by the effects of TMS on motor performance.

The duration of the behavioral tasks employed is also important. This is especially true when considering offline TMS protocols. Although, as stated before, the effects of repetitive stimulation are known to outlast the period of stimulation, these effects are still time limited. Therefore, there is a restricted time window in which the effects of stimulation can be measured. As such, it is impractical to conduct an offline protocol in which stimulation is administered for 15 min and subsequent behavioral testing lasts for 90 min.

Proper experimental design should allow for enough time such that both the experimental and control tasks can be administered in a reasonable timeframe. As a corollary, if the effects of TMS are hypothesized to slow performance, additional time should be allotted to each behavioral task. A special circumstance may be, which, repeated measurements of some sort are taken before and after TMS. In this case, an investigator might seek to gauge the effects of TMS by assessing how long it takes for subjects to return to their pre-TMS, or baseline performance level after they have received a virtual lesion.

3.4 Double Dissociation Protocols

Finally, TMS virtual lesion protocols are often most powerful when they are able to illustrate a double dissociation. For instance, it may be possible to show that creating a virtual lesion in cortical region A impairs behavior X, but not behavior Y; while a virtual lesion to cortical region B impairs behavior Y, but not behavior X [30]. In a recent visual perception experiment, Romei et al. [31] demonstrated that short rhythmic bursts of TMS to the right parietal cortex at a theta frequency benefited processing of the global features of a visual stimulus, but not the local features of that stimulus. In contrast, short bursts of TMS to the right parietal cortex at a beta frequency improved processing of local features of a visual stimulus, but not the processing of its global features. This is a good example of virtual-lesion deduced double dissociation.

4 Conclusion

The virtual lesion approach allows for a causal timeline to be drawn between TMS-induced alterations in cortical excitation and a separately measured outcome—oftentimes performance on a behavioral task. By establishing such causality, virtual lesions can provide information about brain-behavior relationships which are not entirely possible to obtain via neuroimaging and classical lesion protocols. Moreover, in addition to being a useful tool in the study of brain-behavior relationships in healthy individuals, the virtual lesion approach can be used to assess the effects of focal cortical perturbation on functional interactions within and across neural networks (using combined TMS-fMRI and TMS-EEG paradigms) and to demonstrate differences in cortical reactivity between healthy and clinical populations. The virtual lesion protocol can be conducted in an online or offline fashion and can be modified to include variations in the TMS parameters. Finally, careful design of the virtual lesion experiment should include a controlled form of stimulation, a controlled behavioral experiment and, if possible, demonstrate a double dissociation. A summary of methodological considerations for TMS lesion studies is provided in Table 1.

Table 1
Summary of methodological considerations associated with virtual lesion protocols, including those related to how the virtual lesion is induced by TMS (left panel) and how this effect is measured

Related to TMS administration			Related to the outcome measure		
Suppression vs. excitation of cortical target	Low frequency or single pulse	High frequency	Anatomical locations of measurement	Region of stimulation	Across a cortical network
Control stimulation	Sham stimulation	Real stimulation to control site	Control task?	Yes	No
Temporal characteristics	Single delivery of TMS	Different temporal deliveries (TMS chronometry)	Timing of measurement	Online	Offline
Population stimulated	Healthy	Clinical	Type of outcome measured	Behavioral performance	Evoked neural response (e.g. fMRI, EEG)

References

1. Mottaghy FM, Gangitano M, Horkan C, Chen Y, Pascual-Leone A, Schlaug G (2003) Repetitive TMS temporarily alters brain diffusion. Neurology 60(9):1539–1541
2. Sack AT, Linden DE (2003) Combining transcranial magnetic stimulation and functional brain imaging in cognitive brain research: possibilities and limitations. Brain Res Rev 43(1):41–56
3. Harris JA, Clifford CW, Miniussi C (2008) The functional effect of transcranial magnetic stimulation: signal suppression or neural noise generation? J Cogn Neurosci 20(4):734–740
4. Walsh V, Cowey A (2000) Transcranial magnetic stimulation and cognitive neuroscience. Nat Rev Neurosci 1(1):73–79
5. Valero-Cabre A, Payne BR, Pascual-Leone A (2007) Opposite impact on 14C-2-deoxyglucose brain metabolism following patterns of high and low frequency repetitive transcranial magnetic stimulation in the posterior parietal cortex. Exp Brain Res 176(4): 603–615
6. Hilgetag CC, Theoret H, Pascual-Leone A (2001) Enhanced visual spatial attention ipsilateral to rTMS-induced 'virtual lesions' of human parietal cortex. Nat Neurosci 4(9): 953–957
7. Maeda F, Keenan JP, Tormos JM, Topka H, Pascual-Leone A (2000) Interindividual variability of the modulatory effects of repetitive transcranial magnetic stimulation on cortical excitability. Exp Brain Res 133:425–430
8. Speer AM, Kimbrell TA, Wassermann EM, D Repella J, Willis MW, Herscovitch P et al (2000) Opposite effects of high and low frequency rTMS on regional brain activity in depressed patients. Biol Psychiatry 48(12):1133–1141
9. Fecteau S, Pascual-Leone A, Theoret H (2006) Paradoxical facilitation of attention in healthy humans. Behav Neurol 17(3–4):159–162
10. Silvanto J, Muggleton NG, Cowey A, Walsh V (2007) Neural adaptation reveals state-dependent effects of transcranial magnetic stimulation. Eur J Neurosci 25:1874–1881
11. Robertson EM, Theoret H, Pascual-Leone A (2003) Studies in cognition: the problems solved and created by transcranial magnetic stimulation. J Cogn Neurosci 15(7):948–960
12. Pascual-Leone A, Walsh V, Rothwell J (2000) Transcranial magnetic stimulation in cognitive neuroscience—virtual lesion, chronometry, and functional connectivity. Curr Opin Neurobiol 10(2):232–237
13. Pascual-Leone A, Bartres-Faz D, Keenan JP (1999) Transcranial magnetic stimulation: studying the brain-behaviour relationship by induction of 'virtual lesions'. Philos Trans R Soc Lond B Biol Sci 354(1387):1229–1238

14. Hadipour-Niktarash A, Lee CK, Desmond JE, Shadmehr R (2007) Impairment of retention but not acquisition of a visuomotor skill through time-dependent disruption of primary motor cortex. J Neurosci 27(49):13413–13419
15. Bestmann S, Ruff CC, Blankenburg F, Weiskopf N, Driver J, Rothwell JC (2008) Mapping causal interregional influences with concurrent TMS-fMRI. Exp Brain Res 191(4):383–402
16. Ruff CC, Driver J, Bestmann S (2009) Combining TMS and fMRI: from 'virtual lesions' to functional-network accounts of cognition. Cortex 45(9):1043–1049
17. Siebner HR, Rothwell J (2003) Transcranial magnetic stimulation: new insights into representational cortical plasticity. Exp Brain Res 148(1):1–16
18. Thut G, Miniussi C (2009) New insights into rhythmic brain activity from TMS-EEG studies. Trends Cogn Sci 13(4):182–189
19. Lou HC, Luber B, Stanford A, Lisanby SH (2010) Self-specific processing in the default network: a single-pulse TMS study. Exp Brain Res 207(1–2):27–38
20. Paus T (2005) Inferring causality in brain images: a perturbation approach. Philos Trans R Soc Lond B Biol Sci 360(1457):1109–1114
21. Hampson M, Hoffman RE (2010) Transcranial magnetic stimulation and connectivity mapping: tools for studying the neural bases of brain disorders. Front Syst Neurosci 4:40
22. Levit-Binnun N, Litvak V, Pratt H, Moses E, Zaroor M, Peled A (2007) Differences in TMS-evoked responses between schizophrenia patients and healthy controls can be observed without a dedicated EEG system. Clin Neurophysiol 121(3):332–339
23. Ferrarelli F et al (2008) Reduced evoked gamma oscillations in the frontal cortex in schizophrenia patients: a TMS/EEG study. Am J Psychiatry 165(8):996–1005
24. Oberman L, Eldaief M, Fecteau S, Ifert-Miller F, Tormos JM, Pascual-Leone A (2012) Abnormal modulation of corticospinal excitability in adults with Asperger's syndrome. Eur J Neurosci 36(6):2782–2788
25. Pascual-Leone A, Gates JR, Dhuna A (1991) Induction of speech arrest and counting errors with rapid-rate transcranial magnetic stimulation. Neurology 41(5):697–702
26. Jin Y, Hilgetag C (2008) Perturbation of visuospatial attention by high-frequency offline rTMS. Exp Brain Res 189:121–128
27. Siebner HR, Hartwigsen G, Kassuba T, Rothwell JC (2009) How does transcranial magnetic stimulation modify neuronal activity in the brain? Implications for studies of cognition. Cortex 45(9):1035–1042
28. Silvanto J, Muggleton NG (2008) New light through old windows: moving beyond the "virtual lesion" approach to transcranial magnetic stimulation. Neuroimage 39:549–552
29. Borckardt JJ, Walker J, Branham RK, Rydin-Gray S, Hunter C, Beeson H, Reeves ST, Madan A, Sackeim H, George MS (2008) Development and evaluation of a portable sham transcranial magnetic stimulation system. Brain Stimul 1(1):52–59
30. Lu MK, Arai N, Tsai CH, Ziemann U (2012) Movement related cortical potentials of cued versus self-initiated movements: double dissociated modulation by dorsal premotor cortex versus supplementary motor area rTMS. Hum Brain Mapp 33:824–839
31. Romei V, Driver J, Schyns PG, Thut G (2011) Rhythmic TMS over parietal cortex links distinct brain frequencies to global versus local visual processing. Curr Biol 21(4):334–337

Chapter 9

State-Dependent Transcranial Magnetic Stimulation (TMS) Protocols

Juha Silvanto and Zaira Cattaneo

Abstract

In this chapter we describe a novel approach which enhances the functional resolution of transcranial magnetic stimulation (TMS) to a level that allows for differential stimulation of functionally distinct neuronal populations within a cortical area. It is based on the well-known principle of *state-dependency*: a phenomenon whereby the response of a system to an external stimulus is affected not only by the properties of that stimulus but also by the internal state of the system. With regard to TMS, the neural impact of an applied pulse is determined not only by the stimulation parameters but also by the initial activation state of the affected neurons; therefore, neurons within a cortical area will be differentially affected by TMS if their initial activation states at the time of stimulation are dissimilar. The basic idea in state-dependent TMS is to control this initial state/TMS interaction. By selectively increasing the susceptibility of a specific neuronal population via adaptation and priming, one can differentially stimulate this population from other neurons in the area. The main benefit of state-dependent TMS is that it allows TMS research to move beyond questions of "Is region X necessary for task Y" and investigate the functional neuronal properties within a targeted area. So far, this approach has been successfully used to investigate neuronal representations associated with a wide range of cognitive functions such as numerical cognition, action observation, and conceptual knowledge, and it may hold much promise for future research.

Key words Adaptation, Priming, Virtual lesion, Neural tuning, Cognition, Short-term memory

1 Introduction

In cognitive neuroscience, transcranial magnetic stimulation (TMS) is commonly used to investigate the necessity of cortical areas in perceptual and cognitive functions by the induction of the so-called virtual lesions. This term refers to behavioral impairments in perceptual/cognitive tasks which are believed to arise from TMS indiscriminately stimulating all neurons in a targeted area thereby interfering with the highly organized pattern of neuronal activity which underlies perception and behavior. While this approach has been very useful for determining the causal role of specific cortical areas in various functions, questions of cognitive neuroscience often go beyond *Is region X necessary for task Y*?

Alexander Rotenberg et al. (eds.), *Transcranial Magnetic Stimulation*, Neuromethods, vol. 89,
DOI 10.1007/978-1-4939-0879-0_9,

As each cortical area contains a wide range of functionally distinct neuronal populations, in order to link neural activity to behavior it is necessary to determine the selectivity of each of these populations. To accomplish this, a tool which can selectively stimulate a specific neural representation within the targeted area is required. Unfortunately, virtual lesion TMS lacks this level of specificity.

The objective of this chapter is to describe a novel approach which enhances the functional resolution of TMS to a level that allows for differential stimulation of functionally distinct neuronal populations within a cortical area. This methodological refinement comes not from modifying the physical mode of action, but rather by exploiting factors which modulate neural susceptibility to TMS. It is based on the well-known principle of *state-dependency*: a phenomenon whereby the response of a system to an external stimulus is affected not only by the properties of that stimulus but also by the internal state of the system. With regard to TMS, the neural impact of an applied pulse is determined not only by the stimulation parameters but also by the initial activation state of the affected neurons (see [1], for a review). As such, neurons within a cortical area will be differentially affected by TMS if their initial activation states at the time of stimulation are dissimilar. The basic idea in state-dependent TMS is to control this initial state/TMS interaction. By selectively increasing the susceptibility of a specific neuronal population via adaptation and priming, one can differentially stimulate this population from other neurons in the area.

This chapter will proceed as follows: We will first describe the factors which determine the level of specificity required from techniques used in cognitive neuroscience. This will be followed by an evaluation of how the standard "virtual lesion" approach meets and fails to meet these factors. We will then discuss how the principle of state-dependency can be used to overcome some of these limitations and we will provide a detailed overview of how to utilize the state-dependent TMS approach. Finally, we will provide examples of previous research demonstrating the potential of state-dependent TMS in cognitive neuroscience.

2 TMS in Cognitive Neuroscience

2.1 Factors Determining the Level of Specificity Required from Tools in Cognitive Neuroscience

Cognitive neuroscience, as any field of science, operates under certain principles and assumptions that guide empirical work and place specific demands upon research techniques. One such principle is functional specialization—the idea that each cortical area is specialized to perform a specific cognitive or perceptual function. There are numerous examples of functional specialization in the visual cortex. For example, the area V5/MT is believed to be specialized to process motion information. Likewise, the region V4 is believed to be specialized to process color perception [2].

In order to investigate functional specialization within a specific brain area, one must select a technique that can selectively assess neuronal function independent of neighboring neural regions. A second principle is temporal: the involvement of a cortical area can (and likely does) change during a perceptual/cognitive process. A good example of this can be seen in the primary visual cortex, where early responses reflect local details of a stimulus and later responses (after approximately 100 ms of response onset) reflect figure–ground segregation—a more global process [3]. Temporal resolution in the millisecond range is thus a desirable feature when one is selecting a technique for cognitive neuroscience research. A third principle arises from the fact that each cortical area contains a functionally heterogeneous set of neuronal populations. For example, the early visual cortex contains some neuronal populations tuned to motion direction and others tuned to wavelength. Heterogeneity is found also in areas which show functional specialization. For example, although all V5/MT neurons are selective for visual motion, there are differences in the direction of motion and motion speed to which they are specialized [4]. In a similar vein, the lateral occipital cortex is selectively activated by the presentation of unique objects, but different types of objects activate distinct neuronal populations within this area (e.g., [5, 6]). Therefore, in order to link neuronal activity to behavior, one must select a technique which can selectively assess the function of a specific neuronal population independently of other neurons in the same cortical area. A similar idea is that of tuning width, which refers to the *range* of stimuli that drives a given neuron. For instance, a neuron selective for a particular object might be activated whenever an exemplar of that object is presented, regardless of the size, orientation, and/or color of said object. Such invariance to low-level stimulus features is believed to be characteristic of high-level sensory areas (e.g., [6, 7]). In contrast, neurons in the early visual areas respond strongly to differences in low-level attributes, such as wavelength, contrast, and orientation. The assessment of tuning width and invariance requires a technique that can measure the function of a neuronal representation in isolation of other neurons.

2.2 How Does TMS Fare as a Technique in Cognitive Neuroscience?

The usefulness of any technique in cognitive neuroscience is determined by its ability to investigate factors such as those described above. How is TMS equipped to deal with these? The traditional virtual lesion approach has been very useful in answering questions dealing with the first aforementioned principle: the functional specialization of cortical areas. The spatial resolution of virtual lesions is believed to be in the range of 1–2 cm (see [8] for a detailed discussion). This allows it to differentially target neighboring cortical areas selectively. The excellent temporal resolution of TMS also allows investigation of the second aforementioned principle: neural chronometry. This feature of

TMS has been exploited since the earliest pioneering studies of Amassian et al. [9] and has continued to provide novel insights into how the role of cortical areas in cognitive and perceptual functions evolves over time (see e.g., [8, 10]).

However, the virtual lesion approach fares less well with questions addressing neuronal selectivity. As mentioned above, each cortical area contains functionally distinct neuronal populations. One aim of cognitive neuroscience is to determine the functional characteristics of these populations and describe how they contribute to cognition and perception. This requires a tool which can isolate each neuronal population such that it can be investigated independently from other neurons in the area. Unfortunately, the virtual lesion approach is not well suited to this task. While its resolution may be sufficient for *between-region* comparisons, it is too crude a method for assessing *within-region* differences in functional properties. This is because virtual lesion TMS is believed to randomly affect all neuronal populations within the stimulated area. This seriously limits the range of conclusions that can be drawn from TMS studies and hinders the overall utility of TMS in cognitive neuroscience: the finding that region X plays a causal role in function Y says nothing about the computations taking place within region X, or of the neuronal populations contributing to the cognitive/perceptual function under investigation.

To provide an example of this limitation, let us assume that a cortical area plays a causal role in two distinct functions, of which a virtual TMS lesion generated in this region disrupts both. This begs the question: is the same neuronal population in the targeted area responsible for both of the disrupted functions, or are both functions computed by different neuronal representations within the stimulated area? Such questions are not uncommon in cognitive neuroscience and an example can be seen in the recent literature on the neural basis of numerical cognition. There has been much discussion about the invariance of neural representations selective for numerical magnitude. The question is whether the same neuronal population encodes a given magnitude regardless of the format in which it is presented (e.g., as a digit "5" or a word "FIVE"). A standard virtual lesion experiment cannot address this issue, as a disruptive effect on the detection of both types of numbers would not reveal whether they are encoded by the same neuronal representation. The only way to address this issue is to determine the width of neural tuning. Unfortunately, this cannot be achieved with virtual lesion TMS as it does not offer a sufficient level of specificity.

In summary, virtual lesion TMS is useful in assessing functional specialization of cortical areas, as well as the time course of their involvement in cognitive/perceptual functions. However, by restricting its scope to questions of functional necessity, it forces one to deal with each cortical area as though it were a homogenous region. As a consequence, TMS has had very little to say about the functional properties of the various neuronal populations within a targeted area and it has made a very limited contribution to the literature on neuronal selectivity. This is a serious limitation, as there is limited value to assessing causality without being able to investigate the underlying mechanisms.

3 Forms of State-Dependency

3.1 State-Dependency in Psychology and Cognitive Neuroscience

The challenge is to improve the spatial resolution of TMS to a level which enables selective targeting of specific neuronal populations within a cortical area. At first consideration, this might seem impossible; the limitations imposed by the physical principles underlying TMS are likely to prevent its mode of action to ever be refined to such an extent. However, methodological refinements can be made in other ways, of which another widely used tool in cognitive neuroscience, functional magnetic resonance imaging (fMRI), is a fine example. In the 1990s, most fMRI studies adopted an approach similar to the virtual lesion, addressing questions of "Is region X activated by task Y?" More detailed questions were out of the scope of what fMRI could reveal. The spatial resolution of standard fMRI is a few millimeters, and, although superior to virtual lesion TMS, this resolution is insufficient to address questions of neural tuning and selectivity of specific neuronal populations.

A phenomenon which greatly enhanced the functional resolution of fMRI, and which offers a solution also for TMS, is state-dependency. As briefly discussed in the introduction, state-dependency refers to the phenomenon that a system's response to an external stimulus is affected not only by the stimulus properties but also by the internal state of the system. State-dependency has a long history in psychology, where paradigms such as visual adaptation and priming have been used since the nineteenth century to investigate the organization of perceptual systems. Visual adaptation (dubbed *the psychologist's microelectrode*) in particular has been widely used as the resulting visual aftereffects can reveal response properties of neural mechanisms that are activated by adapting stimuli. In visual adaptation, prolonged exposure to sensory stimulation can bias the perception of subsequently presented stimuli away from the adapting stimulus [11–13]. For example, if an observer is exposed to a rightward moving stimulus, subsequent presentation of that stimulus is likely to appear stationary or even move in the opposite direction. This is due to an imbalance in the motion system during which the neuronal representations encoding the adapted motion become less excitable than the representations encoding other motion directions. As motion perception is determined (to put it in a simplistic fashion) by the relative activation levels of neuronal representations with different directional tunings, reduced excitability of any one representation will bias perception towards the direction encoded by the unaffected representations. Mirroring this perceptual/behavioral effect, electrophysiological and functional neuroimaging studies consistently find reduced neuronal responses when a stimulus is repeatedly presented (see [14, 15] for explanations).

In psychology, this phenomenon is often exploited to investigate the selectivity of perceptual representations through the determination of the range of test stimuli for which such reduced sensitivity is observed. For example, if visual adaptation to an object viewed from one angle reduces the observer's sensitivity to that object viewed from other angles, this indicates that the neuronal representation which gives rise to the perception of that stimulus is invariant to the viewpoint. When combined with neuroimaging techniques, adaptation paradigms enable one to ask *where* in the brain such representations exist [14]. This combination enabled fMRI studies to move beyond questions of "Is region X activated by task Y?" to questions concerning neuronal selectivity and, thus, significantly increased its functional resolution.

The underlying logic of adaptation paradigms is that the initial state of a specific component of the perceptual system is changed by the prolonged presentation of a stimulus. This will result in reduced sensitivity to subsequent presentations said stimulus. The critical issue for the present discussion is that state-dependent phenomena offer a tool for selectively affecting the sensitivity of specific neuronal representations as adaptation does not affect the sensitivity of the whole perceptual/cognitive system, but only of those representations which are tuned to the adapting stimulus. The major implication for TMS is simple: if the impact of TMS varies as a function of the initial activation state, it should be possible to selectively stimulate a specific neuronal representation by manipulating its initial state prior to application of TMS. The empirical evidence in support of this possibility will be discussed in the next session.

3.2 State-Dependency of TMS Effects

This principle of state-dependency applies to TMS as it does to any other external stimulus: the behavioral and neural impact of TMS is modulated by factors such as attention, arousal, and illness (see [16] for review). There have also been systematic attempts to manipulate neural activation state to modulate the impact of TMS. Siebner et al. [17] showed that "facilitatory preconditioning" with anodal transcranial direct current stimulation (tDCS) (which increases neuronal excitability) caused a subsequent period of offline 1 Hz rTMS to reduce corticospinal excitability below baseline levels for 20 min. Conversely, "inhibitory preconditioning" with cathodal tDCS (which reduces neuronal excitability) resulted in 1 Hz rTMS *increasing* corticospinal excitability. Thus, changing the initial state of the motor cortex by a period of DC polarization reversed the conditioning effects of 1 Hz rTMS. Similar effects have been observed with online TMS as well. In a study by Silvanto et al. [18], when TMS was applied over the motion-selective region V5/MT during a simple motion detection task, subjects' motion detection ability was impaired. Similarly, suppression of V5/MT activity using offline 1 Hz rTMS disrupted performance in a subsequent motion detection task. However, paradoxically, online V5/MT TMS facilitated motion detection if V5/MT had been suppressed by offline 1 Hz rTMS prior to the motion detection task. These results demonstrate that online TMS can have an unexpected facilitatory effect on behavior when the targeted neural population is in a suppressed state.

In these two studies, preconditioning was used to uniformly change the initial state of *all* neurons in the targeted area. However, the specificity of state-dependency required for present purposes is substantially higher: the initial state manipulation must affect only a specific neuronal population rather. This specificity was attempted by Silvanto et al. [19], who used visual adaptation to bring wavelength-selective neurons in the early visual cortex to different initial activation states. The early visual cortex (V1/V2) contains different types of wavelength-selective neuronal populations, some of which are driven by the wavelength corresponding to the color *green* and suppressed by wavelengths corresponding to the color *red* (other neurons show the opposite behavior (e.g., [20])). The impact of the neural activation state on the effect of TMS was assessed by TMS-induced visual percepts (phosphenes) as well as by a psychophysical task. As predicted, after adaptation to a color stimulus, the appearance of phosphenes induced from the early visual cortex was considerably affected: namely, the phosphenes took on the color qualities of the adapting stimulus (see Fig. 1). Phosphene color was affected even after the visual afterimage induced by the adapter had faded, indicating that this effect is not merely an interaction between the visual afterimage and TMS. In the psychophysical task in which TMS was applied at an intensity below the phosphene threshold, TMS similarly facilitated the perception of the adapted attributes. The critical aspect of these results is that TMS *differentially* affected adapted neurons relative to other neurons in the targeted area. Furthermore, as neurons encoding the adapted attribute were made less active/excitable by adaptation, the finding that phosphenes took on the color of the adapting stimulus implies that TMS perceptually/behaviorally facilitates the less active/excitable neural populations relative to the more active neural populations. Similar findings have been obtained in motion perception after adaptation to simple translational motion, with TMS facilitating the detection of the adapted direction and impairing the detection of the opposite direction [21, 22].

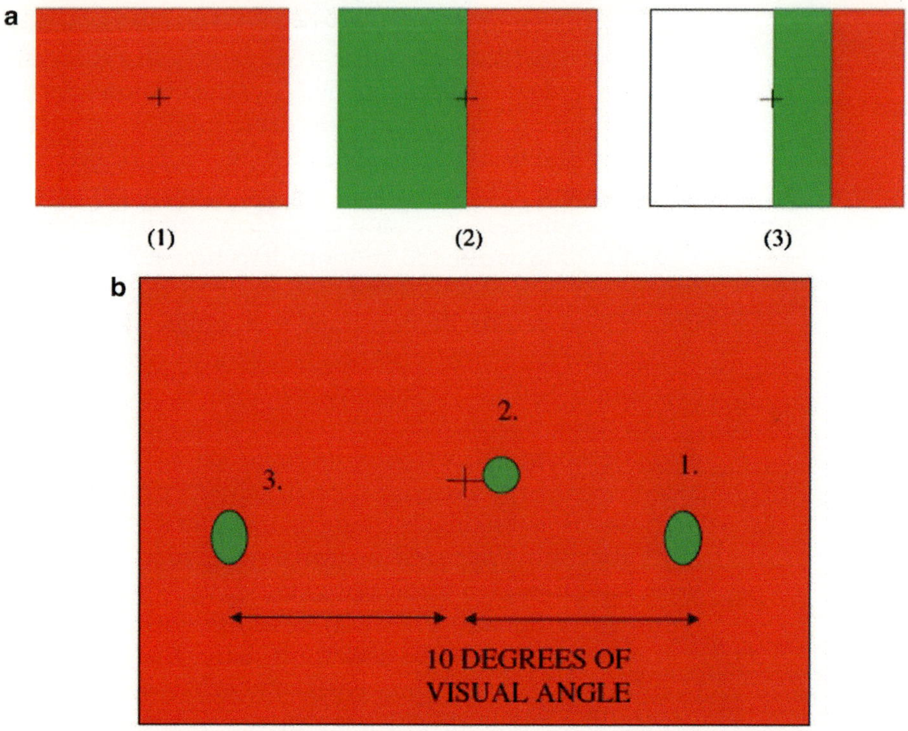

Fig. 1 (**a**) Adaptation displays in the study of Silvanto et al. [19]. Subjects adapted to single-color or dual-color displays. The single-color stimuli completely filled the CRT displays. For the dual-color displays, the stimulus filled (2) the whole display or (3) either its *left* or *right* side. (**b**) Schematic illustrations of subjects' phosphenes. This figure illustrates subjects' percept after adapting to a green single-color display. While the visual afterimage was *red*, the phosphenes induced from the occipital cortex appeared as vivid green blobs within the afterimage. Colored phosphenes were consistently induced from the lateral posterior occipital site (drawings 1 and 3 illustrate phosphenes induced from the *left* and *right* lateral occipital sites, respectively) and the medial occipital site (drawing 2)

The major implication of the modulatory impact of adaptation on TMS is that neurons within a cortical area will be differentially affected by TMS if they are in different activation states at the time of stimulation. Therefore, by systematically bringing functionally distinct neuronal populations within a cortical area into different initial activation states, one can presumably differentiate between said populations using TMS. This can allow researchers to address questions of neuronal selectivity.

3.3 How Does State-Dependent TMS Reveal Neuronal Selectivity?

As discussed above, the aim of state-dependent TMS is to differentially stimulate a specific neuronal population relative to other neurons in that cortical area. This is made possible by the selective modulation of a population's initial state and making use of the fact that online TMS facilitates attributes encoded by neurons that are less active/excitable at the time of stimulation.

Investigations of tuning properties of neuronal representations can be carried out with the following logic: if the impact of TMS is affected by the initial state manipulation, this indicates that neurons in the targeted area were sensitive to said manipulation. For example, if, after adaptation, TMS facilitates the detection of the adapted attribute, this indicates that the targeted area contains a neuronal population affected by (and thus tuned to) the adapting stimulus. By contrast, if no TMS-induced facilitation is found after adaptation, this indicates that adaptation did *not* modulate neural activity in that region; this in turn implies that neurons in the stimulated region are not tuned to the adapted attribute.

The width of neuronal tuning (i.e., the range of stimuli that drives a given neuronal representation) can be investigated by determining the range of test stimuli which are affected by the initial state manipulation. For example, if TMS facilitates the detection of stimulus B after adaptation to stimulus A, this implies that the neuronal representation affected by the adapter A also encodes the target B. In other words, we can conclude that the targeted area contains a neuronal representation which is driven by stimuli A *and* B. To provide a more concrete example: let us say that our objective is to determine the tuning width of motion-selective neuronal populations in the visual area V5/MT. This can be investigated by adapting observers to a stimulus moving in one direction and then, at the test phase, presenting test stimuli which deviate from the adapted direction (for example, in steps of 1°). The largest deviation at which TMS facilitates motion detection indicates the width of neuronal tuning. If TMS facilitates the detection of a motion stimulus which deviates up to 10° from the adapted direction, this tells us that the tuning width of motion-selective neuronal populations is approximately 10°. Further examples of using state-dependent TMS to determine tuning width are provided in Sect. 9.6 of this chapter.

3.4 TMS–Adaptation Paradigm

The initial activation state of specific neural populations may be modulated using well-known psychophysical paradigms, such as priming and adaptation. As discussed above, in adaptation, prolonged exposure to a sensory stimulation reduces sensitivity of the observer for test stimuli which have similar properties as the adapting stimulus. Importantly, adaptation effects are found not only with low-level sensory stimuli (such as motion direction or color) but also at higher levels of representation (such as semantics). One example of this is semantic satiation, in which repeated exposure to exemplars of a semantic category impairs subsequent processing of other exemplars belonging to the same category (e.g., [23–25]).

Figure 2 illustrates the TMS–adaptation protocol in the context of motion processing. In this study, participants were adapted to either leftward or rightward translational motion stimuli for 60 s [21].

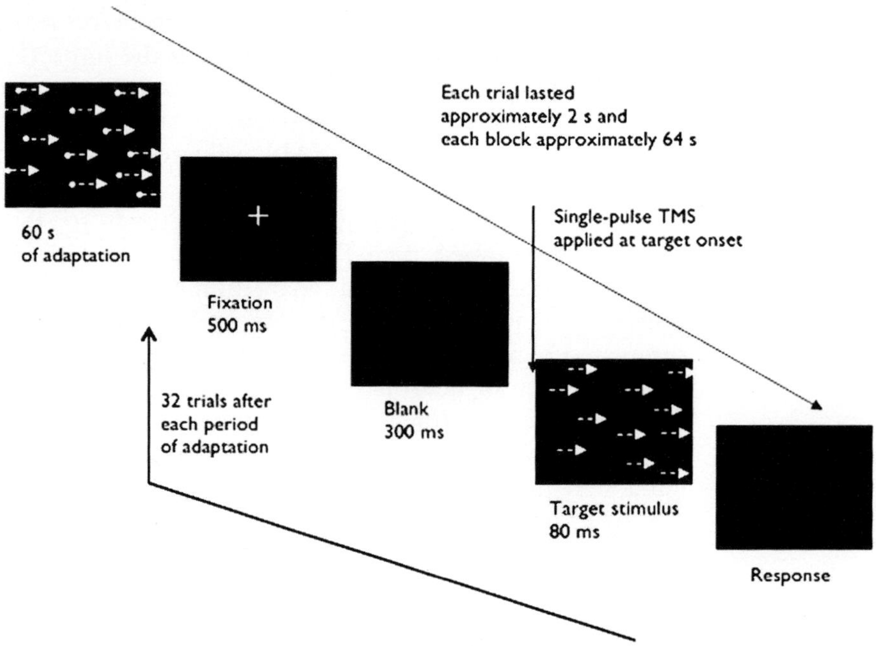

Fig. 2 Timeline of an experimental block in the study by Cattaneo and Silvanto [21]. An adaptation period (to either leftward or rightward motion) of 60 s was followed by 32 experimental trials (16 trials of leftward and rightward motion). Participants were asked to indicate the direction of the motion stimulus. A single pulse of TMS was applied over the left V5/MT+ complex at stimulus onset on each experimental trial

The adaptation phase was followed by 32 trials consisting of motion stimuli moving either in the same direction as the adapter (i.e., congruent trials) or in the opposite direction (i.e., incongruent trials). Participants were asked to indicate the direction of the motion stimulus on each trial. In the TMS condition, single-pulse TMS was applied over the motion-selective cortical area V5/MT+ at target onset. Without TMS, participants were generally worse in detecting the adapted direction (i.e., on congruent trials). This demonstrates the adaptation was behaviorally effective in biasing perception *away* from the adapting stimulus. Critically, TMS applied over left V5/MT+ *facilitated* the detection of the adapted direction and impaired the detection of the opposite direction [21, 22]. In other words, TMS reversed the activity imbalance induced by adaptation, improving the detection accuracy of the adapted attribute.

There are a number of factors that need to be considered when employing TMS–adaptation protocols. Perhaps the most important is the duration of the initial state change induced by adaptation. The neural impact of adaptation peaks immediately following the adaptation period and quickly decays thereafter. The rate of this decay is an important variable to consider, as it determines whether selective facilitatory TMS effects can be obtained; the lack of TMS-induced facilitation could be due to the neural activation state having returned to baseline. This problem is compounded by the possibility that TMS itself may accelerate the decay of neuronal adaptation [26].

Cattaneo and Silvanto [22] investigated the decay of adaptation by presenting post-adaptation blocks of 24 trials and comparing the impact of TMS throughout the block. Facilitation of the adapted attribute was found for both the first and second halves of the post-adaptation block. This suggests the decay of adaptation in single-pulse paradigms is not so fast as to render the TMS–adaptation paradigm impractical. Nevertheless, it is clear that activation state imbalance does decay and the post-adaptation test block should not be too long. Current evidence suggests that a safe number of trials in the post-adaptation block, particularly when studying high-level cognition functions, is approximately 10 (e.g., [27, 28]).

Another aspect to consider is the strength of the adaptation effect at the behavioral level. When studying low-level perceptual functions, adaptation typically leads to lower detection accuracy of the adapted attribute. However, when studying higher level processes, a behavioral effect of adaptation may not be readily apparent at the baseline condition. For example, in TMS–adaptation studies on numerical cognition and action observation [27, 28], adaptation effects were not obtained. This lack of a behavioral manifestation per se does not reflect the lack of adaptation at the neural level; TMS-induced facilitation of the adapted attribute can still be obtained even if the adaptation induces no behavioral effect at baseline (see, for instance, [27, 28] for examples). Similar effects have been observed after adaptation to low-level perceptual attributes: after adaptation to colored stimulus, phosphenes induced from the early visual cortex take on the color of the adapting stimulus even after the perceptual aftereffect (the visual afterimage in the complementary color) induced by the adapter has faded [19].

3.5 TMS–Priming

Together with adaptation, priming is another psychophysical technique that is effective in modulating the state of specific components of perceptual and semantic systems. In contrast to adaptation, the effect of priming is manifested after a single brief exposure to a stimulus, and it is reflected at the behavioral level by higher accuracy or faster response to the primed attribute (see [29–32]). At the neural level, priming is associated with a decreased hemodynamic response for primed versus unprimed stimuli, although priming-related response increases have also been found. A detailed discussion of the neural mechanisms mediating priming can be found in Grill-Spector et al. [11]. The important aspect for the present discussion is that, like adaptation, priming offers a tool for selectively changing the activation state of a specific neuronal representation. It is thus useful for state-dependent TMS.

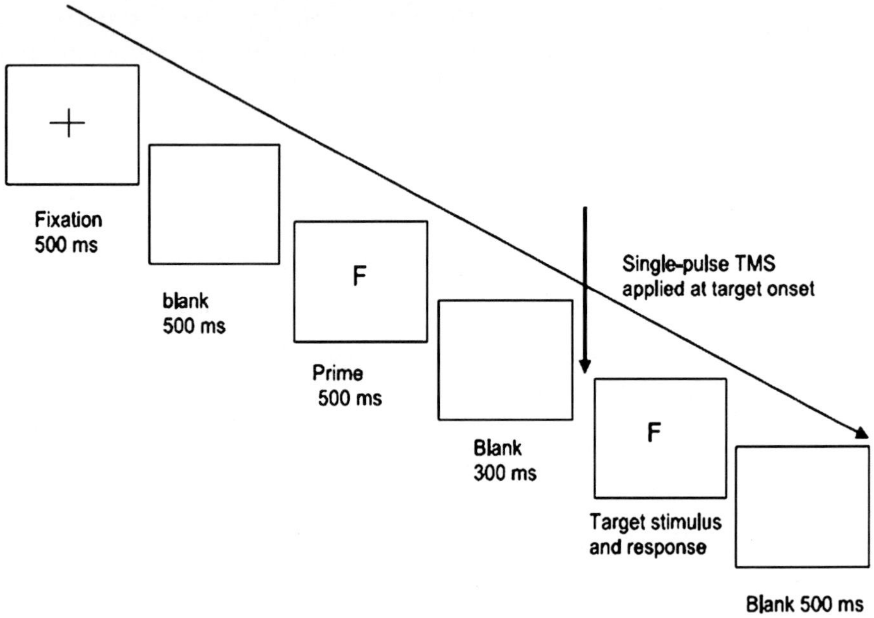

Fig. 3 Timeline of an experiment trial in the TMS–priming protocol used by Cattaneo et al. [33]. In each experimental trial, subjects were primed to one of four letters (V, A, E, or F), and one of these letters was also presented as the target stimulus. Subjects were asked to indicate whether the target letter was a vowel or a consonant. A single pulse of TMS was applied at stimulus onset in each experimental trial over the left posterior parietal cortex (location P3 according to the 10–20 EEG system)

Figure 3 shows a TMS–priming protocol that was used to investigate letter selectivity in the left posterior parietal cortex (PPC) [33]. In that study, participants were first presented with one of four different letters (V, A, E, or F) that worked as primes for 500 ms. Afterwards, one of the same four letters was presented as target. Participants had to decide whether the presented letter was a consonant or a vowel. A single pulse of TMS was applied at stimulus onset over the left PPC (corresponding to the electrode location P3, according to the international 10–20 EEG system) on each experimental trial. In the baseline condition (during which TMS was not applied), participants were faster in responding when the target and the prime letters were identical, reflecting the effect of priming. Critically, single-pulse TMS over left PPC reversed the effect of priming: participants' responses became slower for primed letters compared to unprimed ones. This effect was due to TMS facilitating the detection of the unprimed letters while not affecting the detection of primed letters. This TMS/priming interaction and selective effect on nonprimed letters indicate that the targeted region contains neurons that are tuned to the specific primed features.

The precise nature of the interaction between priming and TMS is not fully clear; there are reports of both facilitation of unprimed trials [34] and impairment of primed trials [35, 36]. The situation is thus different from TMS–adaptation in which facilitation of the adapted attribute is consistently found. Differences in susceptibility between cortical areas to priming effects [37] and the specific features being primed (e.g., perceptual versus conceptual—see [38]) may account for these inconsistencies. While more research is needed to settle the issue, it is clear that priming does interact with TMS effects. TMS–priming may thus become a useful paradigm, complementing TMS–adaptation.

4 Mechanisms Underlying State-Dependent TMS: Stochastic Resonance

That TMS parameters, which normally impair behavior, can have *facilitatory* effects when the target attribute has been adapted or suppressed may seem puzzling. Furthermore, that TMS behaviorally/perceptually facilitates attributes encoded by an adapted/suppressed neuronal population may seem counterintuitive. These phenomena can be accounted for by stochastic resonance [39], a phenomenon occurring in man-made or natural systems that contain a measurement threshold. In such systems, information transfer is enhanced by the injection of low levels of noise that effectively lower the response threshold. There is increasing evidence that stochastic resonance plays a role in sensory signal detection in the human brain [40–43]. The idea of stochastic resonance is shown in Fig. 4. In the image at the top of the figure, the signal is weak and below the sensory threshold for detection rendering the image very difficult to identify. In the images in the middle and at the bottom, random luminance noise has been added to each pixel. While the high level of noise drowns out the signal in the lower image, the middle image is actually enhanced, becoming the clearest of the three. This clarity is due to the low level noise pushing the intensity of some signal pixels above threshold. Thus, the addition of low-level noise can effectively enhance the detection of a weak signal.

A recent study [44] investigated whether TMS displays the characteristics of stochastic resonance. It was reasoned that if online TMS introduced neuronal noise the effects of stimulation would be characteristic of stochastic resonance: at low intensity TMS would facilitate the behavioral detection of weak signals. Conversely, at high intensity TMS would lead to disruption of behavioral performance. The results showed that TMS effects are, indeed, characteristic of stochastic resonance. While high intensity TMS *disrupted* performance for discriminating the direction of clearly distinguishable high coherence random dot stimuli, low intensity TMS *facilitated* the discrimination of low coherence motion signals (relative to the no TMS condition—see Fig. 5).

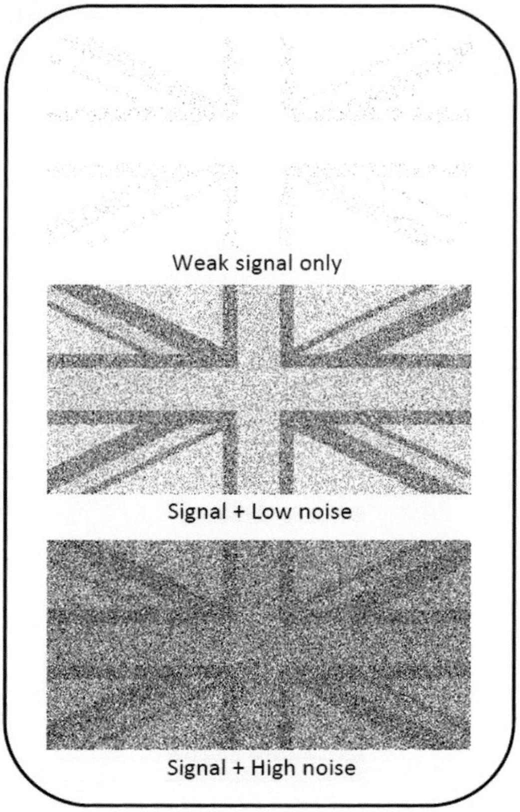

Fig. 4 The principle of stochastic resonance. *Top*: when the signal (picture of a flag) is weak it is difficult to recognize the image as only few pixels pass the threshold. *Bottom*: adding high levels of noise to the image drowns out the signal and recognition remains difficult. *Middle*: recognition can be best when an optimal amount of noise is added. This is because the noise pushes some signal pixels above the threshold

Stochastic resonance may provide a useful general framework for TMS effects, as it helps explain why TMS can induce both behavioral disruptions and facilitations. For example, it may explain findings that TMS facilitates the detection of stimuli detected by neuronal populations that have been subject to neuronal adaptation [19, 28, 33, 34], even though the same TMS parameters in the absence of adaptation impair performance. In the context of stochastic resonance effects, there are two relevant differences that result from adaptation and the subsequent reduction in neuronal excitability. First, presentation of the target stimulus induces a lower level of activation and, thus, the ability to detect the target stimulus is reduced (akin to making the stimulus more difficult). Second, TMS is less effective in activating the neurons, as the neurons

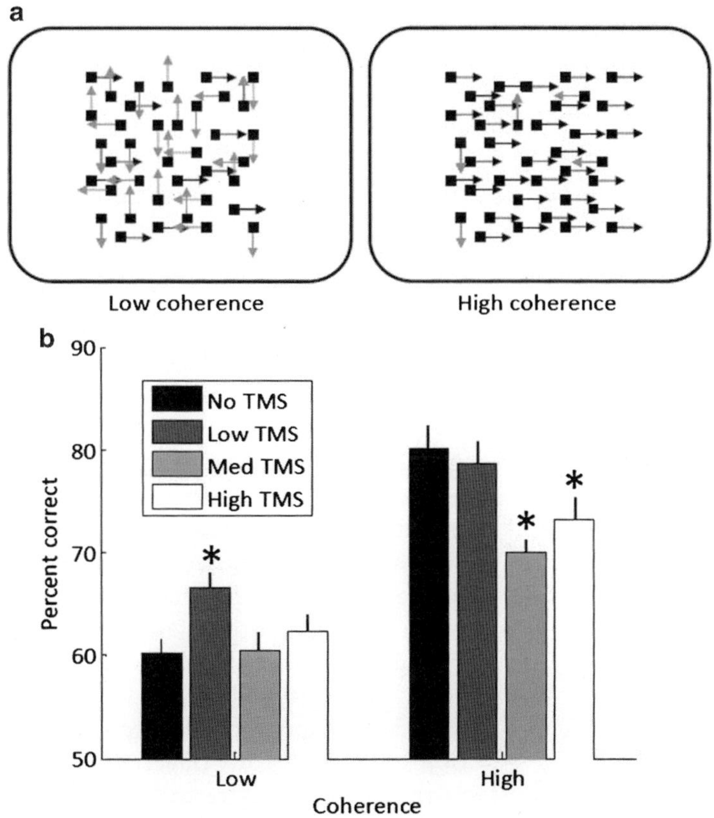

Fig. 5 The results of the study of Schwarzkopf et al. [44]. (**a**) On each trial participants were instructed to discriminate the direction of motion of a random dot stimulus where a proportion of dots moved either left or right (*black arrows*), while the direction of remaining dots was random (*grey arrows*). We tested two levels of coherence affording 60 % and 85 % correct (*left* and *right panel*), respectively. TMS was delivered simultaneously with stimulus onset at one of three intensities or not delivered at all. (**b**) Percent correct discrimination plotted for the two coherence levels and the four levels of TMS intensity. The *asterisks* above a bar indicate that performance in this condition was significantly (two-tailed *t*-test $p < 0.05$) different from trials without TMS (*black bars*)

are less susceptible after adaptation (akin to reducing the TMS intensity). Therefore, the facilitatory effect of TMS when applied after adaptation may be due to adaptation having a similar effect to that caused by reducing the TMS intensity and increasing the difficulty of the target stimulus—thus inducing conditions in which stochastic resonance is observed. The same explanation may apply to the facilitatory effect of TMS when the targeted region is in a suppressed state [18]. While this question requires further study [45, 46], stochastic resonance may provide an explanation for facilitatory online TMS effects across a wide range of paradigms in.

5 Examples of State-Dependent Studies

State-dependent TMS, by allowing the causal role and tuning properties of functionally distinct neural representations within a cortical area to be isolated, has been successfully used across a wide range of topics. These include rotation invariance in the visual cortex, representation of numerical magnitude in the parietal cortex, and neural selectivity to actions and semantic categories in the motor cortex. What follows is a summary of these studies. Common to all of them is that they have utilized TMS to provide information on neuronal selectivity not possible to discern using the standard virtual lesion paradigm.

5.1 Rotation Invariance in Extrastriate Cortex

A hallmark of object recognition is the integration of individual features into a unified gestalt independent of variations in orientation, size, and position of said object. Behavioral priming and adaptation effects indicate the existence of neural representations immune to these factors as priming and adaptation occur even when the orientation, size, or position of an object changes but the configural relationship between features is held constant. Silvanto et al. [35] used TMS in combination with priming to study invariant shape recognition (i.e., the capacity to recognize an object independently of variations in orientation, size, and position of the object) in extrastriate visual cortex. Specifically, the aim was to investigate whether there are rotation-invariant and shape-selective neural representations in the lateral occipital region (LO) and the occipital face area (OFA) (e.g., [5, 47]) that play a casual role in shape processing. Visual priming was used to manipulate the initial state of neurons selective for two-dimensional shapes: on each trial, one of four mirror-symmetric shapes was presented as the prime in an upright position, followed—after a brief delay—by the target stimulus (i.e., either the primed shape or one of the other shapes) embedded in a cluttered background of distracters and rotated either 20° or 70° compared to the prime stimulus. Participants' task was to judge the direction of rotation of the target shape (either left or right). On experimental trials, three pulses of TMS were applied over LO or OFA during the delay between the prime and the target. In the no TMS condition, as expected, discrimination accuracy was higher during trials in which the prime shape and the target shape were the same (relative to trials during which they were different). This demonstrates a rotation-invariant priming effect. However, when TMS was applied over the LO cortex, the behavioral benefit of shape priming was reduced for targets rotated 20°. Additionally, when TMS was applied over the OFA region, the behavioral benefit of shape priming was reduced for targets rotated 70°. These findings suggest that LO cortex plays a causal role in rotation-invariant representations for smaller rotations whereas OFA is causally involved in rotation-invariant representation of larger rotations. These findings indicate that shape invariance may be accomplished through stages, such that smaller degrees of rotational invariance are associated with neural populations different from those associated with larger degrees of rotational invariance.

The lack of any effect of OFA TMS at the 20° rotation highlights a critical difference between TMS and neuroimaging techniques: namely the lack of an effect does not mean that neurons in this region do not respond to stimuli with rotations of this magnitude (as any neuronal representation invariant to a rotation of 70° is very likely to also respond to stimuli rotated by a smaller angle). However, the issue here is whether such neural responses are *causally* required for accurate detection or discrimination. Indeed, a benefit of state-dependent TMS is that it combines the fine functional specificity of priming paradigms with such an assessment of causality. These results therefore show that shape-selective neuronal populations in the OFA are not necessary for the detection of shapes at small rotations (although they are likely to respond to such stimuli), possibly because such representations exist in other areas (such as the LO cortex).

5.2 Emotion Recognition

TMS in combination with priming has also been used to investigate emotion recognition. In particular, Mattavelli et al. (under revision) carried out an experiment to assess the role of the medial prefrontal cortex (mPFC) and the right somatosensory cortex (rSC) in recognizing happy and angry facial expressions. Although there is evidence that both the mPFC and the SC are involved in emotion processing and facial expression recognition (e.g., [48–51]), the question of unique circuits for specific emotions remains unanswered. On each trial, participants were primed to an emotional word (related to either happiness or anger) or to a neutral stimulus (consisting of a series of symbols). Participants were then asked to categorize faces presented during the test phase as being happy or angry. Single-pulse TMS was applied over the mPFC, the rSC, and Vertex (control site) after prime offset. At the behavioral level, the effect of priming was manifested by participants categorizing faster during trials in which the prime and the test stimulus contained the same emotion (relative to trials during which the prime was neutral). Critically, TMS over mPFC reversed this priming effect. This suggests that the mPFC contains distinct neural representations for different emotions. Conversely, TMS over rSC increased overall response latencies, regardless of the priming history. This indicates that the somatosensory cortex is involved in expression recognition but not in emotional differentiation.

5.3 Numerical Representation in the Intraparietal Sulcus: Abstract or Notation Dependent

In the literature concerning numerical cognition, there has been much debate over the abstract nature of the neuronal representation of numerical magnitude. Because numbers can appear in many different formats—for example, as digits ("7") or words ("SEVEN")—it is of great theoretical interest whether there are neuronal representations which encode magnitude regardless of format (see for example [52, 53]). This issue cannot be investigated with standard virtual lesion TMS. If TMS applied over a given area impaired the processing of a numerical magnitude presented in differing formats, it would be unclear whether all those formats were encoded by the same neuronal population in the targeted area, or whether the region contained a large number of distinct populations, each encoding a different format. This question is, however, ideally suited for state-dependent TMS.

Cohen-Kadosh et al. [28] used TMS–adaptation to examine whether the human intraparietal sulcus (IPS) contains neuronal populations which encode numerical magnitude in abstract fashion, or whether its representations are format dependent. If neuronal representations are abstract, then adapting to a numerical magnitude presented in one format would modulate the impact of TMS when that magnitude is presented in a different format. Such a cross-adaptation TMS effect would occur if the same neuronal representation was activated in both cases and the initial state of that representation had been changed by adaptation. In contrast, if numerical magnitudes presented in different formats activated distinct neuronal populations, the impact of TMS would not be affected by adaptation (as the initial state of the neuronal representation activated by the target would not be affected by adaptation).

Participants were adapted to either a digit ("7") or a verbal number ("SEVEN"). The adaptation period consisted of stimuli presented at different locations on a computer monitor in different fonts and sizes (in order to avoid adaptation to purely visual features). Adaptation was followed by a physical same–different task on pairs of digits or number words, with TMS applied on each trial after target onset. The results revealed a behavioral TMS-induced facilitation dependent upon the numerical distance between the adaptation stimulus and the target. Facilitation was largest when the adapter and target shared the same magnitude and it diminished as the numerical distance between the two increased. This result is consistent with electrophysiological evidence of the tuning of magnitude-selective neurons in the IPS. Critically, the TMS-induced facilitation was notation specific: it was only found when the adapting stimulus and the target shared the same notation. These results, thus, suggest that neuronal representations of numerical magnitude are notation dependent.

5.4 Abstract Encoding of Actions

Another example of the use of state-dependent TMS in higher level cognitive processes is a study carried out by Cattaneo and colleagues [27], who examined neuronal representation of action at different levels of cortical hierarchy ([54]; see [55] for a review). Specifically, they examined whether neuronal representations in regions such as ventral premotor cortex encode actions independently of the effector performing the action. This was accomplished by adapting participants to movies showing either a hand or a foot performing different actions on different objects. The adaptation phase was followed by the presentation of static pictures displaying a different act performed by a different effector, the same act performed by a different effector, a different act performed by the same effector, or the same act performed by the same effector. Participants were asked to indicate whether the depicted motor act was the same or different as the one in the adapting movie. Critically, TMS applied over the left and right ventral premotor cortex and the left supramarginal gyrus facilitated the detection of adapted actions *regardless* of the effector used. This cross-adaptation effect, in which action adaptation with one effector modulated the impact of TMS on action observation involving another effector, suggests that neuronal populations in these cortical areas encode action in an abstract manner. Conversely, when TMS was applied over the superior temporal sulcus, responses were facilitated only when both the action and effector matched the adapter. This implies that neurons in this region can recognize motor behavior, but they do not generalize its semantic valence across all possible physical variants. Thus, the representation of action in this region is non-abstract.

This study is another example of state-dependent TMS providing information on neuronal selectivity that could not be obtained with standard virtual lesion TMS (as the disruptive effect of virtual lesion TMS on action observation would provide no information on whether the neuronal representation underlying the behavioral effect was abstract).

5.5 Conceptual Knowledge

Functional segregation by semantic category has been observed in a number of fMRI studies. For example, the medial aspect of the fusiform gyrus has been shown to differentially respond to both pictures and words of tool-based stimuli, whereas the lateral aspect of the fusiform gyri differentially responds to animal-based stimuli [56, 57]. In the lateral temporal cortex, items corresponding to biological categories such as faces and animals differentially activated the superior temporal sulcus [56]. In contrast, pictures of tools activated more inferior regions centered on the left middle temporal gyrus [58, 59].

The ventral premotor cortex (PMv) is known to be part of the neural network mediating the representation of tool/graspable objects (e.g., [59]). However, it is not well established whether tool-selective neuronal representations in the PMv play a causal role in the encoding of the meaning of "tool" words independent of an associated motor action. To investigate this issue, Cattaneo et al. [34] used TMS in combination with priming. A category name (either "Tool" or "Animal") was used as a prime to modulate the initial activation state of the PMv prior to application of TMS and the presentation of the target stimulus. Following the prime, a word was presented and participants were asked to categorize it as being an animal or a tool. When the target word was an exemplar of the "Tool" category, the effects of TMS applied over PMv (but not PM dorsal) interacted with priming history by facilitating reaction times on incongruent trials (i.e., trials in which the target did not belong to the primed category) while not affecting congruent trials (i.e., trials in which the target word was an exemplar of the primed category). TMS applied over PMv had no behavioral effect when the target stimulus was an exemplar of the "Animal" category, regardless of whether the target word was congruent or incongruent with the prime. These findings indicated that the left PMv contains neuronal representations associated with graspable objects which play a causal role in semantic processing.

6 State-Dependent TMS Studies on Visual Imagery and Short-Term Memory

In the TMS–adaptation and TMS–priming paradigms, the aim of initial state manipulation is to determine if a neuronal population can be differentially activated—the direction of this state change is not typically of issue. However, there are research questions where the precise nature of this activation change is of interest. We might for example wish to examine how attention and arousal modulate neuronal activity, and to assess the extent to which they engage similar neuronal mechanisms.

One line of research where questions of this kind are important involves the neural basis of visual imagery and short-term memory (STM). STM refers to the encoding and recall of a previously experienced visual percept, whereas visual imagery refers to the internal visual representation generated after a verbal description of the target stimulus [60, 61]. The behavioral similarity of these two processes begs the question: are their neural implementations similar? In a recent study, Cattaneo et al. [62] used state-dependent TMS to investigate this issue. The experiment consisted of an imagery condition and a memory condition. In the former, participants were shown a digital time and were asked to imagine clock hands associated with this time. In the latter, clock hands were presented and participants were required to maintain that image in memory. At test (after a 2-s imagery/retention period), participants were asked to judge whether a target dot appeared inside or outside the imagined/remembered clock hands. In the first experiment, a single pulse of TMS was applied over the early visual cortex at the end of the retention/imagery phase. Previous evidence has shown that imagery increases the excitability of neurons encoding the mental image [63]. Accordingly, if both imagery and STM enhance early visual cortex excitability, TMS should preferentially activate neurons encoding the mental image. This would result in enhanced performance during both tasks. Such facilitation was indeed found, demonstrating that both imagery and STM are associated with increased neuronal excitability in the early visual cortex.

In a second experiment, using the same paradigm, TMS was applied over the early visual cortex *at the beginning* of the 2-s imagery/retention period. Interestingly, TMS impaired performance in the memory condition, whereas no effect was found in the imagery condition. The lack of TMS effect in the imagery condition demonstrates that imagery modulates visual cortical excitability only at a later stage in the conscious maintenance of the generated image [64], with the actual transformation of the digital time into a mental image likely being subserved by a frontoparietal network (e.g., [65, 66]). The disruptive effect of TMS in the STM condition indicates that, at the beginning of the delay period, the early visual cortex is still primarily responding to the visual presentation of the memory item. In other words, the visual cortical activity is visually driven rather memory driven. As the delay period progresses, there is a shift to memory-driven activation associated with an increase in neuronal excitability. Taken together, these two experiments demonstrate how state-dependent TMS can be used to trace neuronal activation state associated with cognitive functions and how these states change over time.

7 Summary and Future Directions

The objective of this chapter was to describe a novel TMS approach which combines the fine functional specificity of adaptation and priming paradigms with an equally effective assessment of causality: state-dependent TMS. The main benefit of state-dependent TMS is that it allows TMS research to move beyond questions of "Is region X necessary for task Y" and investigate the functional neuronal properties within a targeted area. So far, this approach has been successfully used to investigate neuronal representations associated with a wide range of cognitive functions such as numerical cognition, action observation, and conceptual knowledge. As such, this technique holds much promise for future research. Studies assessing the neural correlates of state-dependent TMS effects by using EEG and fMRI are required to shed light on its mechanisms.

References

1. Silvanto J, Pascual-Leone A (2008) State-dependency of transcranial magnetic stimulation. Brain Topogr 21:1–10
2. Zeki S, Watson JD, Lueck CJ, Friston KJ, Kennard C, Frackowiak RS (1991) A direct demonstration of functional specialization in human visual cortex. J Neurosci 11:641–649
3. Lamme VA, Supèr H, Landman R, Roelfsema PR, Spekreijse H (2000) The role of primary visual cortex (V1) in visual awareness. Vision Res 40:1507–1521
4. Maunsell JH, Van Essen DC (1983) Functional properties of neurons in middle temporal visual area of the macaque monkey. I. Selectivity for stimulus direction, speed, and orientation. J Neurophysiol 49:1127–1147
5. Malach R, Reppas J, Benson R, Kwong K, Jiang H, Kennedy W, Ledden P, Brady T, Rosen B, Tootell R (1995) Object-related activity revealed by functional magnetic resonance imaging in human occipital cortex. Proc Natl Acad Sci U S A 92:8135–8139
6. Grill-Spector K, Kourtzi Z, Kanwisher N (2001) The lateral occipital complex and its role in object recognition. Vision Res 41(10–11):1409–1422
7. Rolls ET (2000) Functions of the primate temporal lobe cortical visual areas in invariant visual object and face recognition. Neuron 27:205–218
8. Walsh V, Pascual-Leone A (2003) Transcranial magnetic stimulation: a neurochronometrics of mind. MIT Press, Cambridge
9. Amassian VE, Cracco RQ, Maccabee PJ, Cracco JB, Rudell A, Eberle L (1989) Suppression of visual perception by magnetic coil stimulation of human occipital cortex. Electroencephalogr Clin Neurophysiol 74:458–462
10. Silvanto J, Cowey A, Lavie N, Walsh V (2005) Striate cortex V1 activity gates awareness of motion. Nat Neurosci 8:143–144
11. Grill-Spector K, Henson R, Martin A (2006) Repetition and the brain: neural models of stimulus-specific effects. Trends Cogn Sci 10:14–23
12. Gibson J, Radner M (1937) Adaptation, after-effect and contrast in the perception of tilted lines. J Exp Psychol 12:453–467
13. Mather G, Verstraten F, Anstis S (1998) The motion aftereffect: a modern perspective. Oxford University Press, Oxford
14. Grill-Spector K, Kushnir T, Edelman S, Avidan G, Itzchak Y, Malach R (1999) Differential processing of objects under various viewing conditions in the human lateral occipital complex. Neuron 24:187–203
15. Grill-Spector K, Malach R (2001) fMR-adaptation: a tool for studying the functional properties of human cortical neurons. Acta Psychol 107:293–321
16. Silvanto J, Muggleton NG (2008) New light through old windows: moving beyond the "virtual lesion" approach to transcranial magnetic stimulation. Neuroimage 39:549–552
17. Siebner HR, Lang N, Rizzo V, Nitsche MA, Paulus W, Lemon RN et al (2004) Preconditioning of low-frequency repetitive transcranial magnetic stimulation with transcranial direct current stimulation: evidence for homeostatic plasticity in the human motor cortex. J Neurosci 24:3379–3385
18. Silvanto J, Cattaneo Z, Battelli L, Pascual-Leone A (2008) Baseline cortical excitability determines whether TMS disrupts or facilitates behavior. J Neurophysiol 99:2725–2730
19. Silvanto J, Muggleton NG, Cowey A, Walsh V (2007) Neural adaptation reveals state-dependent effects of transcranial magnetic stimulation. Eur J Neurosci 25:1874–1881
20. Engel SA (2005) Adaptation of oriented and unoriented color-selective neurons in human visual areas. Neuron 45:613–623
21. Cattaneo Z, Silvanto J (2008) Investigating visual motion perception using the TMS-adaptation paradigm. Neuroreport 19: 1423–1427
22. Cattaneo Z, Silvanto J (2008) Time course of the state-dependent effect of transcranial magnetic stimulation motion in the TMS-adaptation paradigm. Neurosci Lett 443: 82–85
23. Cattaneo Z, Devlin JT, Vecchi T, Silvanto J (2009) Dissociable neural representations of grammatical gender in Broca's area investigated by the combination of satiation and TMS. Neuroimage 47:700–704
24. Smith L, Klein R (1990) Evidence for semantic satiation: repeating a category slows subsequent semantic processing. J Exp Psychol Learn Mem Cogn 16:852–861
25. Smith LC (1984) Semantic satiation affects category membership decision time but not lexical priming. Mem Cogn 12:483–488
26. Théoret H, Kobayashi M, Ganis G, Di Capua P, Pascual-Leone A (2002) Repetitive transcranial magnetic stimulation of human area MT/V5 disrupts storage of the motion aftereffect. Neuropsychologia 40:2280–2287
27. Cattaneo Z, Devlin JT, Salvini F, Vecchi T, Silvanto J (2010) The causal role of category-specific neuronal representations in the left ventral premotor cortex (PMv) in semantic processing. Neuroimage 49:2728–2734

28. Cohen Kadosh R, Muggleton N, Silvanto J, Walsh V (2010) Double dissociation of format-dependent and number-specific neurons in human parietal cortex. Cereb Cortex 209: 2166–2171
29. Maljkovic V, Nakayama K (1994) Priming of pop-out: I. Role of features. Mem Cognit 22:657–672
30. Maljkovic V, Nakayama K (1996) Priming of pop-out: II. The role of position. Percept Psychophys 58:977–991
31. Campana G, Cowey A, Walsh V (2002) Priming of motion direction and area V5/MT: a test of perceptual memory. Cereb Cortex 12:663–669
32. Magnussen S, Greenlee MW (1999) The psychophysics of perceptual memory. Psychol Res 62:81–92
33. Cattaneo Z, Rota F, Walsh V, Vecchi T, Silvanto J (2009) TMS-adaptation reveals abstract letter selectivity in the left posterior parietal cortex (PPC). Cereb Cortex 19:2321–2325
34. Cattaneo L, Sandrini M, Schwarzbach J (2010) State-dependent TMS reveals a hierarchical representation of observed acts in the temporal, parietal, and premotor cortices. Cereb Cortex 209:2252–2258
35. Silvanto J, Schwarzkopf DS, Gilaie-Dotan S, Rees G (2010) Differing causal roles for lateral occipital cortex and occipital face area in invariant shape recognition. Eur J Neurosci 32:165–171
36. Mattavelli G, Cattaneo Z, Papagno C (2011) Transcranial magnetic stimulation of medial prefrontal cortex modulates face expressions processing in a priming task. Neuropsychologia 49(5):992–998
37. Pavan A, Campana G, Maniglia M, Casco C (2010) The role of high-level visual areas in short- and longer-lasting forms of neural plasticity. Neuropsychologia 48:3069–3079
38. Voss JL, Schendan HE, Paller KA (2010) Finding meaning in novel geometric shapes influences electrophysiological correlates of repetition and dissociates perceptual and conceptual priming. Neuroimage 49:2879–2889
39. Miniussi C, Ruzzoli M, Walsh V (2010) The mechanism of transcranial magnetic stimulation in cognition. Cortex 46(1):128–130
40. Kitajo K, Nozaki D, Ward LM, Yamamoto Y (2003) Behavioral stochastic resonance within the human brain. Phys Rev Lett 90:218103
41. Lugo E, Doti R, Faubert J (2008) Ubiquitous crossmodal Stochastic Resonance in humans: auditory noise facilitates tactile, visual and proprioceptive sensations. PLoS ONE 3:e2860
42. Sasaki H, Sakane S, Ishida T, Todorokihara M, Kitamura T, Aoki R (2008) Suprathreshold stochastic resonance in visual signal detection. Behav Brain Res 193:152–155
43. Simonotto E, Riani M, Seife C, Roberts M, Twitty J, Moss F (1997) Visual perception of stochastic resonance. Phys Rev Lett 78:1186
44. Schwarzkopf DS, Silvanto J, Rees G (2011) Stochastic resonance effects reveal the neural mechanisms of transcranial magnetic stimulation. J Neurosci 31(9):3143–3147
45. Ruzzoli M, Marzi CA, Miniussi C (2010) The neural mechanisms of the effects of transcranial magnetic stimulation on perception. J Neurophysiol 103(6):2982–2989
46. Ruzzoli M, Abrahamyan A, Clifford CW, Marzi CA, Miniussi C, Harris JA (2011) The effect of TMS on visual motion sensitivity: an increase in neural noise or a decrease in signal strength? J Neurophysiol (1):138–143
47. Gilaie-Dotan S, Nir Y, Malach R (2008) Regionally-specific adaptation dynamics in human object areas. Neuroimage 39: 1926–1937
48. Adolphs R, Damasio H, Tranel D, Cooper G, Damasio AR (2000) A role for somatosensory cortices in the visual recognition of emotion as revealed by three-dimensional lesion mapping. J Neurosci 20:2683–2690
49. Heberlein AS, Padon AA, Gillihan SJ, Farah MJ, Fellows LK (2008) Ventromedial frontal lobe plays a critical role in facial emotion recognition. J Cogn Neurosci 20:721–733
50. Mah LWY, Arnold MC, Grafman J (2005) Deficits in social knowledge following damage to ventromedial prefrontal cortex. J Neuropsychiatry Clin Neurosci 17:66–74
51. Pitcher D, Garrido L, Walsh V, Duchaine C (2008) Transcranial Magnetic Stimulation disrupts the perception and embodiment of facial expression. J Neurosci 28:8929–8933
52. Piazza M, Pinel P, Le Bihan D, Dehaene S (2007) A magnitude code common to numerosities and number symbols in human intraparietal cortex. Neuron 53:293–305
53. Dehaene S, Dehaene-Lambertz G, Cohen L (1998) Abstract representations of numbers in the animal and human brain. Trends Neurosci 21:355–361
54. Grafton ST, Hamilton AF (2007) Evidence for a distributed hierarchy of action representation in the brain. Hum Motor Sci 26:590–616
55. Cattaneo L, Rizzolatti G (2009) The mirror neuron system. Arch Neurol 66:557–560
56. Chao LL, Haxby JV, Martin A (1999) Attribute-based neural substrates in temporal cortex for perceiving and knowing about objects. Nat Neurosci 2:913–919
57. Devlin JT, Rushworth MF, Matthews PM (2005) Category-related activation for written

words in the posterior fusiform is task specific. Neuropsychologia 43:69–74

58. Damasio H, Grabowski TJ, Tranel D, Hichwa RD, Damasio AR (1996) A neural basis for lexical retrieval. Nature 380:499–505
59. Martin A, Wiggs CL, Ungerleider LG, Haxby JV (1996) Neural correlates of category specific knowledge. Nature 379:649–652
60. Baddeley A, Salame P (1986) The unattended speech effect: perception or memory? J Exp Psychol Learn Mem Cogn 12:525–529
61. Cornoldi C, Vecchi T (2003) Visuo-spatial working memory and individual differences. Psychology Press, Hove
62. Cattaneo Z, Vecchi T, Pascual-Leone A, Silvanto J (2009) Contrasting early visual cortical activation states causally involved in visual imagery and short-term memory. Eur J Neurosci 30(7):1393–1400
63. Sparing R, Mottaghy FM, Ganis G, Thompson WL, Töpper R, Kosslyn SM, Pascual-Leone A (2002) Visual cortex excitability increases during visual mental imagery—a TMS study in healthy human subjects. Brain Res 938: s92–97
64. Kosslyn SM, Ganis G, Thompson WL (2001) Neural foundations of imagery. Nat Rev Neurosci 2:635–642
65. Sack AT, Jacobs C, De Martino F, Staeren N, Goebel R, Formisano E (2008) Dynamic premotor-to-parietal interactions during spatial imagery. J Neurosci 28:8417–8429
66. Sack AT, Sperling JM, Prvulovic D, Formisano E, Goebel R, Di Salle F, Dierks T, Linden DE (2002) Tracking the mind's image in the brain II: transcranial magnetic stimulation reveals parietal asymmetry in visuospatial imagery. Neuron 35:195–204

Part IV

Multimodal Considerations

Chapter 10

Combination of Transcranial Magnetic Stimulation (TMS) with Functional Magnetic Resonance Imaging

Joan A. Camprodon and Mark A. Halko

Abstract

Transcranial magnetic stimulation (TMS) is a noninvasive brain stimulation and neuromodulation technique widely used in systems neurophysiology, cognitive neuroscience, and neuropsychiatry. Unlike other neuroimaging modalities, it is interventional and therefore able to establish causal relationships between brain activity and behavior. Despite the many advantages of independent TMS studies to demonstrate causal influence of brain areas on cognitive tasks, these studies assume TMS modulation of cortex underneath the coil but do not obtain any neurobiological measures. Multimodal approaches that combine TMS with neuroimaging, though technically challenging, are able to circumvent this problem. In this chapter, we will discuss the challenges, solutions, and applications of the combination of TMS with functional magnetic resonance imaging.

Key words TMS, fMRI, Neuromodulation, Neuroimaging, Systems neuroscience

1 Multimodal Transcranial Magnetic Stimulation

Transcranial magnetic stimulation (TMS) is a noninvasive brain stimulation and neuromodulation technique widely used in systems neurophysiology, cognitive neuroscience, and neuropsychiatry. Multiple stimulation paradigms are available allowing for great experimental flexibility and the capacity to provide unique solutions to neuroscientific problems. One unique characteristic of neuromodulation techniques, particularly when compared to standard brain imaging modalities, is their capacity to establish causal relationships between brain activity and cognitive function. In a standard TMS design, we utilize an intervention that affects brain activity (independent variable) and we measure consequential changes in cognition or behavior (dependent variable). Neuroimaging techniques, on the contrary, are not interventional but observational: they utilize a cognitive task (independent variable) while measuring simultaneous parallel changes in brain activity (now a dependent variable) with the aim of establishing a correlation

Alexander Rotenberg et al. (eds.), *Transcranial Magnetic Stimulation*, Neuromethods, vol. 89,
DOI 10.1007/978-1-4939-0879-0_10,

between the two. In this paradigm, we observe behavior and brain activity but do not interfere in the system therefore limiting our explanatory power to the level of correlations. A limitation of this approach is the inability to differentiate between patterns of brain activity that are truly causally involved in the computations necessary for the proposed cognitive task and mere collateral epiphenomena [1].

Despite the popularity of neuromodulation techniques in general, and of TMS in particular, our knowledge of their mechanisms of action is still incomplete. A major limitation has been the difficulty of obtaining neurophysiological measures of the effects of stimulation beyond peripheral measures of motor activity with motor-evoked potentials. The classic experimental paradigm utilizes modulation of neuronal activity in a given cortical area followed by the measurement of associated cognitive or behavioral changes, but without a direct measure of the effects induced by the stimulation at the neurobiological level. The nature of this design allows for the establishment of a causal relationship between neural activity and cognition. However, because we do not experimentally measure the neurobiological effects of the intervention, we do not know precisely what physiological changes are causally related to the cognitive and behavioral effects we observe, even if we make assumptions extrapolated from animal studies or in silico models. Neuroimaging techniques offer the advantage of measuring physiological variables with varying degrees of temporal and spatial resolution. The combination of TMS with neuroimaging techniques (multimodal TMS) is a promising methodological innovation that allows, among other things, for the adding of causal resolution to imaging studies and the introduction of neurobiological outcome measures to neurostimulation paradigms. Hence, multimodal TMS solves some of the major limitations these techniques have separately. This advance offers novel solutions to neuroscientific problems with the potential to bring new answers to difficult-to-address questions, particularly in humans and in vivo. Today, several imaging techniques can be combined with TMS. The question is how to combine them and why. It remains crucial that, despite the growing availability of multimodal approaches, the technical decision follows the experimental question.

Neuroimaging techniques can be combined with TMS by using scans to plan, guide, and document the location of stimulation. Prior to the introduction of these tools, the target of stimulation was selected in relation to the primary motor or visual cortex, easily localizable using the motor or phosphene threshold method. An alternative approach was the use of the international 10–20 EEG reference system. The development and popularization of stereotactic systems offers greater precision in localization, as it allows for target selection on the basis of an individual scan and dynamic stimulation guidance throughout the session [2]. In addition, stereotactic guidance allows for the determination and recording of accurate coil position relative to the cortex and orientation of the magnetic field in 3D at all times during stimulation. Perhaps the most interesting aspect of this application is the ability to guide TMS using functional magnetic resonance imaging (fMRI). Gross structural anatomy can serve as a good reference for some studies, but its correlation with functional divisions is not always accurate and is subject to great variability. Similarly, probabilistic maps using coordinate systems, such as the Talairach, based on previous functional studies can be useful in certain designs [3, 4], but they may present a degree of variability and uncertainty unacceptable to certain experimental question. The use of fMRI as a reference to locate the area of stimulation allows for the utilization of individualized activation maps and, therefore, can increase the certainty that we are affecting the functional systems of interest in each individual subject. Sack and colleagues [5] compared the precision of TMS when guided by individual fMRI, structural MRI, Talairach coordinates, and the 10–20 EEG system in a parietal task. They noted that all systems were potentially valid, although the accuracy was greater and the variability smaller with fMRI. They calculated the statistical power required for each method and concluded that, for the same behavioral paradigm, 5 subjects were needed to obtain statistically significant results when the stimulation was guided by fMRI, 9 subjects with structural MRI, 13 with Talairach coordinates, and 47 with the 10–20 EEG system. Using fMRI to guide TMS is also a valuable approach to investigate hypotheses about the causal relevance and physiological chronometry of specific brain areas. Based on an activation map that correlates brain activity with a cognitive task, one could use TMS to modulate a specific area of the map and determine whether it is causally related to the task or, alternatively, it is a mere epiphenomenon. This procedure can also be extended to clinical populations, where fMRI-guided TMS can be used either to validate stimulation sites [6] or perhaps to improve the reliability of targeting individualized locations [7].

The focus of this chapter will be the use of neuroimaging, in particular fMRI, to measure the biological effects of TMS. Two main approaches to measuring the neurophysiological changes induced by stimulation exist: the sequential or "offline" approach, and the simultaneous or "online" approach. Both offer different and complementary measures.

2 Sequential, or "Offline," TMS-fMRI

Despite the many advantages of independent TMS studies to demonstrate causal influence of brain areas on cognitive tasks, these types of studies assume TMS modulation of cortex underneath the coil rather than measure the effect of TMS on the brain. In offline stimulation paradigms, repetitive stimulation is assumed to change local cortical excitability. These paradigms are sometimes termed "virtual lesions" if the cortical excitability change disrupts task performance. These effects can last for 30 min or longer, depending on the parameters of stimulation. When these repetitive stimulation paradigms are applied over the course of multiple days, usually in the context of clinical therapeutic protocols, longer effects may be obtained [8, 9]. Although local effects of TMS are assumed, it is expected that TMS also affects larger brain networks. Put simply: if local excitability changes, then so too must the dynamics between the target site and its afferent and efferent connections. In some cases, changes in local excitability may not cause distal effects; in other cases local excitability changes may lead to profound reorganization of a cortical network. Thus, imaging can be used to investigate compensatory responses across the entire brain to local excitability changes from TMS.

The networks of the human brain are driven by predetermined connectivity patterns, and, as a result of this white-matter connectivity, regions of the brain which are anatomically connected show coherence within their activity pattern [10]. The mechanism of action of TMS is to induce changes in the firing rate of neurons by externally driving their activity. These neurons are not unconnected, but rather, exist in larger scale networks.

The earliest evidence of the brain's compensatory reaction to rTMS was indicated by Fox et al. [11] where low frequency (1 Hz) stimulation was delivered to the left primary motor cortex (M1). Under these parameters, it is expected that cortical excitability at the local site of stimulation would be reduced. However, using PET, Fox et al. found a local increase in cerebral blood flow which persisted for 20 min following stimulation.

To investigate circuit-wide effects of stimulation, O'Shea et al. [12] conducted a combined TMS-fMRI study to investigate the network response to 1 Hz rTMS to the motor system. This rTMS suppression of left premotor cortex resulted in an increase in activity in right premotor cortex. O'Shea et al. argue that this response is in line with a theory of recovery from stroke wherein the unaffected hemisphere recovers some of the function that is impaired within the affected hemisphere.

Additional investigations have shown that rTMS effects the functional relationship of nodes within a network. Lee et al. [13] stimulated primary motor cortex (M1) with 30 min of 1 Hz rTMS and investigated the response within the brain using PET. TMS increased cortical activity within the motor cortex and motor network areas such as supplementary motor cortex (SMA) and premotor cortex (PMd). The critical observation was that the magnitude of activity under the site of rTMS was correlated to the magnitude of activity within PMd and SMA, which was not observed under sham stimulation. Using fMRI, Plager et al. [14] demonstrated an inverse relationship between M1 and primary somatosensory cortex (S1) following 5 Hz rTMS to S1. Increased activation within S1 following rTMS was paired with decreased activation following rTMS within M1. Subjects with higher S1 and lower M1 responses showed improved somatosensory discrimination task performance.

Network-wide responses to rTMS are also found in patient groups where existing network dynamics can be altered via rTMS. In patients with stroke, contralesional TMS to M1 increases functional coupling between SMA and ipsilesional M1, which suggests that improvements in motor function due to rTMS treatment may be the result of a change in network dynamics within the entire affected network [15].

TMS changes neuronal excitability locally and globally throughout distributed network systems. Accordingly, it would be expected that not only cortical activity levels change throughout a network but also related neurotransmitter response profiles change. PET offers the ability to create radioligands which bind to specific neuroreceptors to identify changes in neurotransmitter responses. [11C]Raclopride, one such radioligand, provides an inverse measure of dopamine response within the striatum. Strafella et al. [16] demonstrated an increase of dopamine response in the left caudate following 10 Hz rTMS to left dorsolateral prefrontal cortex. Because [11C]raclopride is limited to the striatum, a followup study was performed using [11C]FLB 457, which has a higher signal-to-noise ratio and can, therefore, be used to image dopamine responses in the cortex. Cho et al. [17] found that 10 Hz rTMS to left, but not right, dorsolateral prefrontal cortex increased dopamine release in the subgenual cingulate cortex, pregenual cingulate, and medial orbital frontal cortex. These areas, especially subgenual cingulate cortex [18], are critically affected in the major depressive disorder, which can be treated with TMS to dorsolateral prefrontal cortex.

The central theme of the “virtual lesion” paradigm is to interfere with one node of a brain network to impair task performance. Because the brain is largely interconnected, the assumption of an isolated node changing response properties alone without additional distributed network changes simply doesn’t hold. Thus, imaging should be combined with cognitive virtual-lesion experiments to identify how brain networks change and compensate in the context of a virtual lesion. Two studies have utilized offline designs to examine the brain network responses to virtual lesions in cognitive tasks. Andoh and Paus [19] used an offline design where 10 Hz rTMS was applied to Wernike’s area either on the left or on the right hemisphere. rTMS to left Wernike’s area improved task performance whereas rTMS to the right impaired task performance. In both cases, rTMS increased functional activation contralateral to the site of stimulation. Thus, the network compensatory action revealed behaviorally different responses despite similarities in the observed network response properties. Mevorach et al. [20] examined the contribution of intraparietal sulcus (IPS) to visual salience. In their study they applied 1 Hz rTMS to left and right IPS areas, which respond to spatial salience selection. Movorach et al. found a saliency circuit between primary visual cortex and IPS, and that during modulation of IPS with TMS, V1 correspondingly modulated in an interaction with the task. In other words, they find that IPS activity is linked to V1 activity and this link modulates along with the behavioral salience task. A behavioral dissociation between salience conditions was found between left and right IPS stimulation. Furthermore, this was matched in V1 activation for the respective conditions. Thus, Movorach et al. demonstrate that a circuit between IPS and V1 is responsible for enhancing relative salience of features of the visual environment and that this circuit is modulated with rTMS.

Neuroimaging can also be used to assess cortical changes that result from chronic stimulation protocols. There are two outcome measures that can be assessed following stimulation: structural measures, which assess the integrity of cortical or subcortical structures, and functional measures, which assess changes in cortical processing under specific circumstances. As of yet, very few publications have discussed structural changes following TMS. Initial investigations of T1-weighted and T2-weighted anatomical images revealed no gross structural changes following an rTMS protocol for depression (10 daily left dorsolateral prefrontal cortex stimulation sessions at either 5 or 20 Hz) as assessed using qualitative measures and quantitative volume measures of prefrontal cortex [21]. One may expect the cortical changes induced by stimulation to not be gross anatomical, but rather to be small regional changes within structures underneath the stimulation site. To examine this, May et al. [22] used voxel-based morphometry to look at 5 days of 1 Hz rTMS to left superior temporal gyrus (left primary auditory cortex). Following 5 days of stimulation, an increase in local gray matter volume was found at the site of stimulation. Although these results are far from conclusive, additional studies should be undertaken to compare the effects of TMS over extended periods. If permanent plastic changes are intended, as in therapeutic protocols, it is likely that the cortical substrate will reflect these changes. Additionally, activity-dependent changes may be required. Put simply, stimulation alone may not sufficiently change gray- or white-matter structure, but, when coupled with a functional task, TMS may promote structural changes. Accordingly, the results of May et al. must be interpreted with great caution—the consensus within the literature suggests that chronic stimulation results in functional, but not structural, changes [6].

3 Simultaneous, or "Online," TMS-fMRI

3.1 Methodological Principles

TMS has been combined simultaneously with EEG [23–30], SPECT [31], and PET [4, 11, 32–38]. In this section, we will focus on the simultaneous combination of TMS with fMRI. This is the combination that offers both the most attractive promises and the most difficult technological challenges. Compared with PET, this combination offers greater spatial and temporal resolution and independence from radioactive ligands, which allows almost unlimited capacity of repeated measures. PET remains a preferable combination when the biological measures of interest are neurotransmitter dynamics. Compared to EEG, this technique also offers greater spatial resolution and the ability to obtain whole brain measures, including direct measures of subcortical structures. New advances in EEG signal analysis, such as the inverse solution algorithms, use mathematical approaches to infer topographic information including subcortical sources, but these models are still far from the resolution and accuracy of fMRI. The advantage of TMS-EEG remains its temporal resolution, far superior to any radiological technique.

The technological challenges posed by the simultaneous combination of TMS and fMRI are important. Many of the complexities are specific to each lab setting, MRI center, and experimental design and often require individualized solutions. However, we will review the major and more universal methodological problems and propose some solutions. At the end of the chapter, we will describe in more detail the methodological steps we used for one of our experiments, but we also invite the readers to explore the literature (see references above) and compare some of the proposed solutions to determine which are most applicable to their needs.

Three main types of artifacts must be addressed: (1) static artifacts, (2) dynamic artifacts, and (3) sensory or nonspecific artifacts. First, we must consider the *static artifacts* created by the mere presence of metallic objects inside the scanner. Metallic objects present a significant safety risk due to the potential for projectile effects, which can harm individuals (research subjects and staff) and damage the scanner. They can also generate significant noise and distortion of the T1 or blood oxygen level-dependent (BOLD) signal thereby impeding adequate measurement of brain structure and function. The solution requires either removing all ferromagnetic components of the TMS coil or moving them to a position outside the influence of the scanner's magnetic field. The wiring of most commercial coils is copper, which is not ferromagnetic, but other components, such as the security buttons many coils have in the handle, pose a safety risk and can generate large artifacts [39]. Many researchers have opted to place this safety

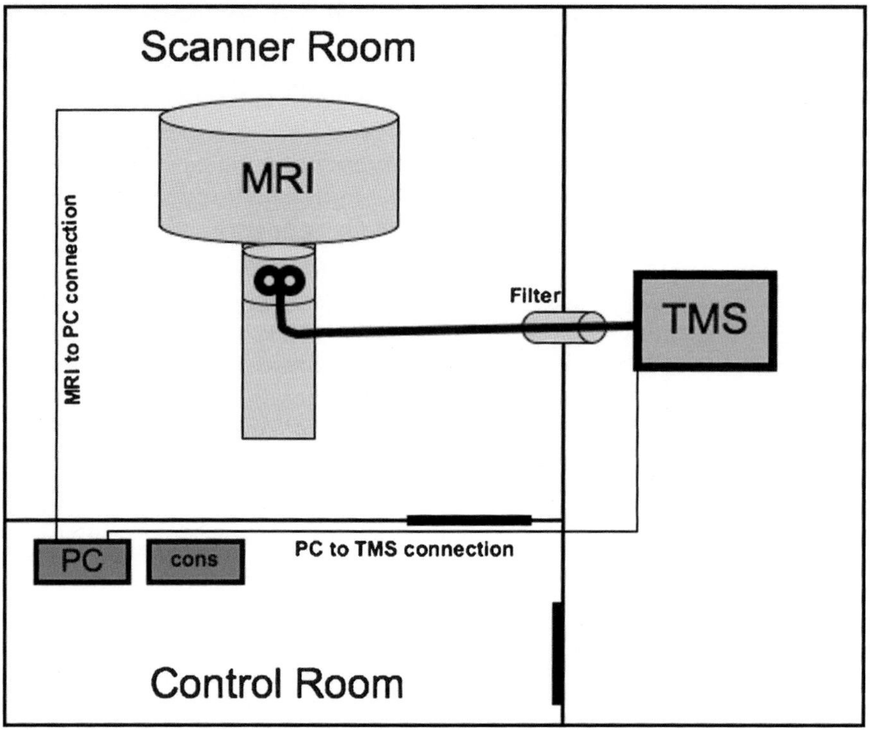

Fig. 1 Simultaneous TMS-fMRI set-up schematic

mechanism in a specially designed box located next to the stimulator rather than on the coil handle. In addition, coils are designed with long cables, usually at least 8 m, which connect to a standard unmodified TMS stimulator. The stimulator, which is heavy and ferromagnetic, should be located outside the scanner room in an adjoining space and the cable connecting to the coil should enter the scanner room via designed openings with appropriate electromagnetic noise filters. Filters may also be required to avoid not only electromagnetic noise from entering the scanner room but also backward interference to the TMS stimulator system. See Fig. 1 for a schematic of a simultaneous TMS-fMRI setting. Today, several companies sell MRI-compatible coils and systems, although most solutions must be ultimately individualized to each particular setting. This process requires a series of changes to the MRI suite, but these tend to be minimal and do not affect other operations of the scanner.

The second problem is the *dynamic artifact* resulting from the interference of the TMS magnetic pulse with the magnetic field of the scanner. The space inside the MRI coil is under the influence of a strong magnetic field (typically 1.5–3 T) with a small homogeneous variation in strength with respect to position (magnetic field gradient). It is essential for the image reconstruction process that the field be constant and homogeneous with respect to space. The

TMS magnetic pulse can have a strength of 1.5–2.5 T, which disrupts this homogeneity by introducing a large amount of noise that interferes with the image reconstruction process. Bestmann et al. [39] measured the intensity and duration of the magnetic field disruption caused by a TMS pulse in a 2 T MRI scanner at 2, 4, and 6 cm from the TMS coil. They reported that the intensity and duration of distortion were inversely proportional to the distance to the TMS coil. Accordingly, the maximum distortion occurred 2 cm from the coil and it took 70–100 ms for the MRI magnetic field to return to its undistorted baseline. The 2 cm distance is particularly relevant because this is the approximate gap between the surface of the human skull, where the TMS coil is applied, and the superficial layers of the cerebral cortex. Because the intensity of the TMS magnetic pulse decays with the square of the distance, neurons in these superficial cortical layers are the only ones TMS can directly depolarize and modulate. Beyond this distance the magnetic field generated by TMS is too weak to induce neuronal action potentials. In summary, the experiment from Bestmann et al. shows that, during the 70–100 ms after the TMS pulse, the distortion of the MRI magnetic field is significant and measurements are unreliable due to excessive noise. After this time, the magnetic field returns to baseline and it becomes possible to measure and reconstruct images on the basis of the NMR signal again. Different groups have proposed different solutions to this problem, all requiring finely synchronizing the TMS pulses with the measurement of BOLD signal. This requires an external computer connected to the scanner and the TMS stimulator with a customized program to coordinate the MRI measurements and the TMS stimulation. The computer must receive, usually via TTL pulses from the scanner, the precise time for the beginning of each volume (TR) or each slice. With this information, and taking into account the desired experimental design and MR pulse sequence parameters (TR, number of slices per volume, time between slices, etc.), the program will trigger a TMS pulse via another TTL pulse sent to the stimulator. If the experiment requires the presentation of stimuli (visual, auditory, etc.) and the recording of behavioral responses, the same computer and program will usually coordinate these as well. Despite these difficult parameters, several groups have successfully combined TMS with fMRI using single pulse, triple pulses, and trains of low and high frequency rTMS.

The third problem, common to all combinations of TMS with neuroimaging regardless of modality, is the *sensory* or *nonspecific artifact*. TMS selectively modulates the activity of targeted populations of neurons and their circuits. When we combine TMS multimodally, we want to measure these specific changes. But the experience of participating in a TMS experiment also activates a number of circuits nonspecifically. The noise of each TMS pulse, caused by the acute tension of the coil wiring when the current passes through it, will naturally activate the auditory cortex.

Similarly, the tapping sensation on the scalp will activate specific somatotopic regions of the somatosensory system, and if the tapping is painful it will also activate nociceptive areas. In addition, subjects may experience anxiety, fear, or relaxation with increased introspection, all of which will activate specific circuits. The activation of these regions is nonspecific and independent from the target location, its connectivity, and the TMS parameters (although more intensity may induce more noise and a stronger tap). These circuits will appear as "activated" in imaging maps thereby making it difficult to determine if their activation is due to the direct effects of TMS on the stimulation target or to nonspecific sensory phenomena. No major technological solutions are needed to deal with this form of artifact. Instead, investigators should develop appropriate experimental designs and analysis strategies that control for these artifacts and subtract them from their maps in order to avoid mistakenly attributing said activations to the electromagnetic modulation of a cortical area.

3.2 Applications

In this section, we will review some applications of simultaneous TMS-fMRI, which are numerous and promising. This combination offers the unprecedented ability to study functional connectivity causally in humans and in vivo, to explore mechanisms of neuroplasticity at the circuit level, and to examine the neurobiological substrates of TMS "virtual lesions." In addition, it offers the opportunity to address clinical questions, ranging from pathophysiology to biomarker development to therapeutic mechanisms of action, by creating models of human disease or studying patients directly.

The simultaneous combination of TMS and fMRI is particularly useful in studies of connectivity, offering new possibilities for describing the structure and function of neural networks. Neuroimaging studies of functional connectivity correlate fluctuations of activity between two or more areas over time, determine patterns of interaction, and reveal functional networks [23, 40, 41]. However, these studies do not provide causal information about the interregional interactions; they are limited to mere correlations. This is similar to the problem posed by the correlation of a pattern of activity with a cognitive task, but in this case applied to the co-activation of populations of neurons. In recent years, there has been growth in the popularity of mathematical models that offer an approximation of causality, such as Dynamic Causal Modelling [42] or Granger Connectivity [43]. Nevertheless, these models only offer a probabilistic approximation of causality, not an empirical measurement of causal interactions. The simultaneous combination of TMS with fMRI offers such a direct empirical measurement of causal interactions. We intervene in a circuit by stimulating one of its nodes (a point source) and then measure the changes in activation of cortical or subcortical areas that are directly (monosynaptically) or indirectly (polysynaptically) connected to it.

Initial studies focused on methodological questions applied to the connectivity of the motor system, since this is a well-known circuit in human and nonhuman primates, allows for a clear testable hypothesis, is easy to localize with TMS, and has easily measurable behavioral outcomes [44–51]. Subsequent studies examined the connectivity of the dorsolateral prefrontal cortex [52, 53], FEF [54], and attentional nodes in the parietal cortex [55–58].

In addition to providing measures of connectivity, the combination of TMS and fMRI offers the possibility of studying mechanisms of plasticity in unique ways. Brain networks are dynamic and patterns of connectivity changing. These plastic properties are essential to understand the computational strategies of neural circuits, particularly as related to cognition, the pathophysiology of neuropsychiatric disorders, and the mechanisms of action of treatments. TMS-fMRI designs have been used to study the plastic properties of motor, visuospatial, and somatosensory circuits revealing how the functional use of connections changes depending on the characteristics of external stimuli (i.e., stimulus- or task-dependent plasticity) [54, 56] or internal states (state-dependent plasticity) [55, 59, 60]. These mechanisms are crucial for understanding how the brain adapts to the needs of the environment.

This multimodal combination can also be used to understand the mechanism of action of TMS, and reveal the plastic adaptation strategies inherent to neural circuits in response to the pulse of TMS [55]. This is of crucial importance for understanding the mechanisms of action of the technique. Classically, TMS paradigms understood that the cognitive changes observed after the application of TMS were the result of the activation or inhibition of a selected cortical area. Multimodal TMS studies, among other approaches, have shown that TMS does not only modulate the target area directly but also cause indirect changes in areas connected within a functional circuit, including subcortical and contralateral regions. In addition to the local and distal effects of stimulation, we now know there is a third process to take into account: intrinsic compensatory plastic adaptation to the artificial effects of TMS. Circuits respond to the influence of the pulse in an attempt to maintain homeostasis by reactively changing patterns of internal connectivity, activating and inhibiting local and distal regions [55]. It is important to highlight that the overall cognitive, behavioral, or clinical effects of TMS are the compound result of the local effects on the target region, the distal modulation of interconnected areas within a given network, and the overall compensatory adaptation to the stimulation. These findings clarify the mechanisms of action of TMS, but also force us to reconsider the interpretation of previous experiments using classic "unimodal" TMS designs (i.e., limited to behavioral measures).

The ability to stimulate inside the scanner allows for the investigation of the neurobiological mechanisms of TMS "virtual lesions." We can perform a classic TMS cognitive paradigm by which we stimulate a target region and measure behavioral changes (reaction times and/or performance accuracy) in addition to changes in brain activity, thereby providing a mechanistic understanding of the physiological processes that sustain the observed psychometric changes [58]. This approach moves TMS studies beyond the "black box" model by which we stimulate a brain region but have no access to the changes we induce.

Finally, the simultaneous combination of TMS and fMRI offers significant promises for the clinical neurosciences. One can study patients [52, 61] or use the virtual lesion approach to create models of human disease [55] and examine pathophysiological mechanisms at the systems level. The same approach can be used to study not only mechanisms of disease but also mechanisms of recovery and rehabilitation [55, 61] or the therapeutic mechanisms of action of neuromodulation clinical protocols [21]. Finally, TMS-fMRI can be used in the development of neuropsychiatric biomarkers.

In summary, we reviewed different application for which the multimodal combination of TMS and fMRI has been used. It allows for the study of basic mechanisms, such as neural connectivity, in humans and in vivo with causal explanatory power. It is also a great tool to study network properties, such as neuroplasticity (stimulus- or state-dependent) or the intrinsic active adaptation and circuit-wide changes in response to the artificial influence of TMS. It overcomes the limitations of classic TMS virtual lesion designs by allowing us to understand not only how TMS can change cognition and behavior but also what the neural mechanisms are that sustain such change. Finally, it offers an opportunity to address translational clinical questions and study mechanisms of disease, recovery, and treatment.

4 Method: Offline rTMS Design for Assessing Acute Plastic Resting State fMRI Functional Connectivity Changes: rTMS to Parietal Node of Default Network

Subjects: Most fMRI designs assess changes across groups. Common fMRI designs assessing similarities within groups will use 15–25 subjects. For comparing across populations (e.g., patient groups compared to normal controls), one should use similar sizes for each group, 15–25 subjects per group. Subjects should be prescreened for TMS and MRI contraindications, including but not limited to claustrophobia and a history of epilepsy. Female subjects should be screened for pregnancy. All subjects should be free of neurological or psychiatric conditions.

fMRI: Three visits are necessary for this study. One baseline session, which will be used to determine the site of stimulation, and two TMS visits to modify cortical dynamics. Optionally, a non-stimulation/sham stimulation visit may be added. All resting state imaging should follow standard procedures for resting state imaging [62–64]. BOLD images should be acquired. A 3 T MRI

Fig. 2 Effective TMS placement near MRI room

capable of functional imaging should be used. For the present protocol, 40 slices are acquired axially (TR = 2,500 ms, TE = 28 ms, flip angle = 90°, voxel size = 3 × 3 × 3 mm). Each run of BOLD should be at least 6 min, resulting in 144 volumes per run. At least three 6 min runs should be collected for each resting state measurement. Subjects should fixate a small white dot on a black screen for the duration of the scan. For alignment purposes, an anatomical T1 should be obtained for each fMRI session, using the MPRAGE sequence (TR = 6.787 ms, TE = 3.13 ms, flip angle = 80°, voxel size = 1.0 × 1.0 × 1.2 mm).

TMS: The TMS equipment should be placed as close as possible to the MRI. Figure 2 shows the placement of the portable rTMS lab next to the MRI suite. In some cases, this may be a hallway or storage room. Take special care to note that the TMS equipment can disrupt electronic devices, and should not be performed in rooms which house MRI electronics such as image reconstruction computers. A portable TMS laboratory can be comprising the TMS machine, cooling equipment (if air-cooled), a frameless stereotaxy system (a laptop and a camera), and a motor-evoked potential recording system (either a computer with electrodes or integrated with other devices).

Prior to rTMS, the resting motor-evoked potential threshold will be used to determine the level of stimulation. rTMS should be applied within current safety guidelines [65]. The current method utilizes high (20 Hz) and low frequency (1 Hz) stimulation, although other repetitive stimulation paradigms, such as theta burst stimulation, can be used as well. For both stimulations, 1,800 pulses will be delivered. At low frequency, all pulses will be delivered continuously; at high frequency, 2 s trains of 20 Hz stimulation are followed by 28 s of inter-train interval. All stimulation is applied at 110 % of resting motor threshold (note, this intensity should be adjusted accordingly for theta burst paradigms).

TMS-fMRI: For TMS-fMRI visits, two resting state studies will be performed. First, prior to TMS, a resting state data set (three 6-min runs, as outlined above) should be collected, as well as a high resolution anatomical for registration purposes. Typically, the anatomical is collected first and followed by the resting state runs. After this initial fMRI session, the subject receives rTMS, and is immediately placed back in the MRI for a post-TMS-fMRI scan. During this acquisition, care should be taken to move quickly from TMS to the scanner. To quickly acquire fMRI data following stimulation, the MRI bed should be left in the highest position (if it lowers to assist subjects entering the scanner) and all of the MRI equipment, such as headphones or earplugs, should be ready for immediate use. As the effects of TMS are transient, it is best to acquire fMRI images first, before the second anatomical sequence.

Resting state data analysis: Custom software and routines from major data analysis packages (FSL, SPM, 4dfp, freesurfer) were used to analyze resting state data [66], but "off-the-shelf" analysis packages are available for resting state analysis (see http://www.nitrc.org for software packages which do independent component analysis or resting state cross-correlation). Briefly, preprocessing of fMRI data involves spatial normalization to the MNI 152 template, slice timing correction, motion artifact removal, spatial smoothing (8 mm Gaussian), and temporal smoothing. A second stage of pre-processing serves to remove nuisance variables (time courses from global mean, motion, white matter, and CSF) with regression. After preprocessing, a region of interest is placed in left inferior parietal lobe corresponding to the site of stimulation. For every voxel, a correlation coefficient was computed between the region of interest time course and the voxel. These coefficients are converted to z-scores and used for subsequent testing. Figure 3 shows the comparison of changes due to low vs. high frequency stimulation to the left inferior parietal cortex node of the default network.

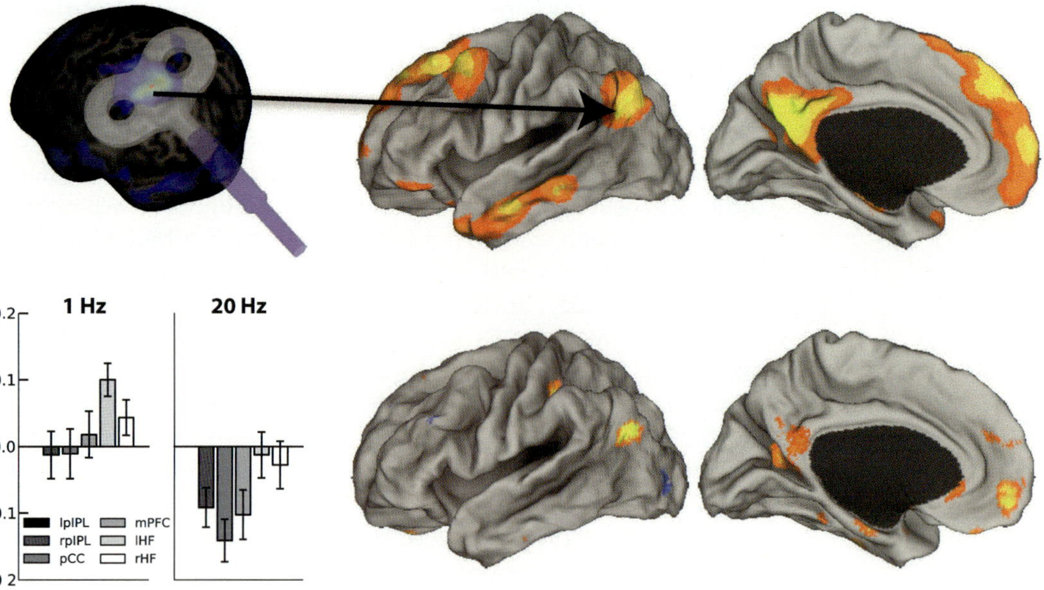

Fig. 3 Low- vs. high frequency rTMS effects on the left inferior parietal cortex of the default network

References

1. Sack AT (2006) Transcranial magnetic stimulation, causal structure-function mapping and networks of functional relevance. Curr Opin Neurobiol 16(5):593–599
2. Gugino LD, Romero JR, Aglio L, Titone D, Ramirez M, Pascual-Leone A et al (2001) Transcranial magnetic stimulation coregistered with MRI: a comparison of a guided versus blind stimulation technique and its effect on evoked compound muscle action potentials. Clin Neurophysiol 112(10):1781–1792
3. Paus T (1996) Location and function of the human frontal eye-field: a selective review. Neuropsychologia 34(6):475–483
4. Paus T, Jech R, Thompson CJ, Comeau R, Peters T, Evans AC (1997) Transcranial magnetic stimulation during positron emission tomography: a new method for studying connectivity of the human cerebral cortex. J Neurosci 17(9):3178–3184
5. Sack AT, Cohen Kadosh R, Schuhmann T, Moerel M, Walsh V, Goebel R (2009) Optimizing functional accuracy of TMS in cognitive studies: a comparison of methods. J Cogn Neurosci 21(2):207–221
6. Bystritsky A, Kaplan JT, Feusner JD, Kerwin LE, Wadekar M, Burock M, Wu AD, Iacoboni M (2008) A preliminary study of fMRI-guided rTMS in the treatment of generalized anxiety disorder. J Clin Psychiatry 69(7):1092–1098
7. Fitzgerald PB, Hoy K, McQueen S, Maller JJ, Herring S, Segrave R, Bailey M, Been G, Kulkarni J, Daskalakis ZJ (2009) A randomized trial of rTMS targeted with MRI based neuronavigation in treatment-resistant depression. Neuropsychopharmacology 34(5):1255–1262
8. George MS, Lisanby SH, Avery D, McDonald WM, Durkalski V, Pavlicova M et al (2010) Daily left prefrontal transcranial magnetic stimulation therapy for major depressive disorder: a sham-controlled randomized trial. Arch Gen Psychiatry 67(5):507–516
9. O'Reardon JP, Solvason HB, Janicak PG, Sampson S, Isenberg KE, Nahas Z et al (2007) Efficacy and safety of transcranial magnetic stimulation in the acute treatment of major depression: a multisite randomized controlled trial. Biol Psychiatry 62(11):1208–1216
10. Greicius MD, Supekar K, Menon V, Dougherty RF (2009) Resting-state functional connectivity reflects structural connectivity in the default mode network. Cereb Cortex 19(1):72–78
11. Fox P, Ingham R, George MS, Mayberg H, Ingham J, Roby J et al (1997) Imaging human intra-cerebral connectivity by PET during TMS. Neuroreport 8(12):2787–2791
12. O'Shea J, Sebastian C, Boorman ED, Johansen-Berg H, Rushworth MF (2007) Functional specificity of human premotor-motor cortical interactions during action selection. Eur J Neurosci 26(7):2085–2095
13. Lee L, Siebner HR, Rowe JB, Rizzo V, Rothwell JC, Frackowiak RS, Friston KJ (2003) Acute remapping within the motor system induced by low-frequency repetitive transcranial magnetic stimulation. J Neurosci 23(12):5308–5318
14. Pleger B, Blankenburg F, Bestmann S, Ruff CC, Wiech K, Stephan KE, Friston KJ, Dolan RJ (2006) Repetitive transcranial magnetic stimulation-induced changes in sensorimotor coupling parallel improvements of somatosensation in humans. J Neurosci 26(7):1945–1952
15. Grefkes C, Nowak DA, Wang LE, Dafotakis M, Eickhoff SB, Fink GR (2010) Modulating cortical connectivity in stroke patients by rTMS assessed with fMRI and dynamic causal modeling. Neuroimage 50(1):233–242
16. Strafella AP, Paus T, Barrett J, Dagher A (2001) Repetitive transcranial magnetic stimulation of the human prefrontal cortex induces dopamine release in the caudate nucleus. J Neurosci 21(15):RC157
17. Cho SS, Strafella AP (2009) rTMS of the left dorsolateral prefrontal cortex modulates dopamine release in the ipsilateral anterior cingulate cortex and orbitofrontal cortex. PLoS One 4(8):e6725
18. Drevets WC, Savitz J, Trimble M (2008) The subgenual anterior cingulate cortex in mood disorders. CNS Spectr 13(8):663–681
19. Andoh J, Paus T (2011) Combining functional neuroimaging with off-line brain stimulation: modulation of task-related activity in language areas. J Cogn Neurosci 23(2):349–361
20. Mevorach C, Hodsoll J, Allen H, Shalev L, Humphreys G (2010) Ignoring the elephant in the room: a neural circuit to downregulate salience. J Neurosci 30(17):6072–6079
21. Nahas Z, Lomarev M, Roberts DR, Shastri A, Lorberbaum JP, Teneback C et al (2001) Unilateral left prefrontal transcranial magnetic stimulation (TMS) produces intensity-dependent bilateral effects as measured by interleaved BOLD fMRI. Biol Psychiatry 50(9):712–720
22. May A, Hajak G, Ganssbauer S, Steffens T, Langguth B, Kleinjung T, Eichhammer P

(2007) Structural brain alterations following 5 days of intervention: dynamic aspects of neuroplasticity. Cereb Cortex 17(1):205–210

23. Ilmoniemi R, Aronen H (2000) Cortical excitability and connectivity reflected in fMRI, MEG, EEG and TMS. In: Moneen C, Bandettini P (eds) Functional MRI. Springer, New York, pp 453–463
24. Ilmoniemi RJ, Virtanen J, Ruohonen J, Karhu J, Aronen HJ, Naatanen R et al (1997) Neuronal responses to magnetic stimulation reveal cortical reactivity and connectivity. Neuroreport 8(16):3537–3540
25. Izumi S, Takase M, Arita M, Masakado Y, Kimura A, Chino N (1997) Transcranial magnetic stimulation-induced changes in EEG and responses recorded from the scalp of healthy humans. Electroencephalogr Clin Neurophysiol 103(2):319–322
26. Paus T, Sipila PK, Strafella AP (2001) Synchronization of neuronal activity in the human primary motor cortex by transcranial magnetic stimulation: an EEG study. J Neurophysiol 86(4):1983–1990
27. Thut G, Northoff G, Ives JR, Kamitani Y, Pfennig A, Kampmann F et al (2003) Effects of single-pulse transcranial magnetic stimulation (TMS) on functional brain activity: a combined event-related TMS and evoked potential study. Clin Neurophysiol 114(11):2071–2080
28. Thut G, Pascual-Leone A (2010) Integrating TMS with EEG: how and what for? Brain Topogr 22(4):215–218
29. Thut G, Theoret H, Pfennig A, Ives J, Kampmann F, Northoff G et al (2003) Differential effects of low-frequency rTMS at the occipital pole on visual-induced alpha desynchronization and visual-evoked potentials. Neuroimage 18(2):334–347
30. Virtanen J, Ruohonen J, Naatanen R, Ilmoniemi RJ (1999) Instrumentation for the measurement of electric brain responses to transcranial magnetic stimulation. Med Biol Eng Comput 37(3):322–326
31. Catafau AM, Perez V, Gironell A, Martin JC, Kulisevsky J, Estorch M et al (2001) SPECT mapping of cerebral activity changes induced by repetitive transcranial magnetic stimulation in depressed patients. A pilot study. Psychiatry Res 106(3):151–160
32. Mottaghy FM, Pascual-Leone A, Kemna LJ, Topper R, Herzog H, Muller-Gartner HW et al (2003) Modulation of a brain-behavior relationship in verbal working memory by rTMS. Brain Res Cogn Brain Res 15(3):241–249
33. Paus T, Jech R, Thompson CJ, Comeau R, Peters T, Evans AC (1998) Dose-dependent reduction of cerebral blood flow during rapid-rate transcranial magnetic stimulation of the human sensorimotor cortex. J Neurophysiol 79(2):1102–1107
34. Siebner HR, Peller M, Willoch F, Minoshima S, Boecker H, Auer C et al (2000) Lasting cortical activation after repetitive TMS of the motor cortex: a glucose metabolic study. Neurology 54(4):956–963
35. Siebner HR, Takano B, Peinemann A, Schwaiger M, Conrad B, Drzezga A (2001) Continuous transcranial magnetic stimulation during positron emission tomography: a suitable tool for imaging regional excitability of the human cortex. Neuroimage 14(4):883–890
36. Siebner HR, Willoch F, Peller M, Auer C, Boecker H, Conrad B et al (1998) Imaging brain activation induced by long trains of repetitive transcranial magnetic stimulation. Neuroreport 9(5):943–948
37. Speer AM, Willis MW, Herscovitch P, Daube-Witherspoon M, Shelton JR, Benson BE et al (2003) Intensity-dependent regional cerebral blood flow during 1-Hz repetitive transcranial magnetic stimulation (rTMS) in healthy volunteers studied with H215O positron emission tomography: I. Effects of primary motor cortex rTMS. Biol Psychiatry 54(8):818–825
38. Speer AM, Willis MW, Herscovitch P, Daube-Witherspoon M, Shelton JR, Benson BE et al (2003) Intensity-dependent regional cerebral blood flow during 1-Hz repetitive transcranial magnetic stimulation (rTMS) in healthy volunteers studied with H215O positron emission tomography: II. Effects of prefrontal cortex rTMS. Biol Psychiatry 54(8):826–832
39. Bestmann S, Baudewig J, Frahm J (2003) On the synchronization of transcranial magnetic stimulation and functional echo-planar imaging. J Magn Reson Imaging 17(3):309–316
40. Friston K (1994) Functional and effective connectivity in neuroimaging: a synthesis. Hum Brain Mapp 2(1–2):56–78
41. McIntosh AR, Nyberg L, Bookstein FL, Tulving E (1997) Differential functional connectivity of prefrontal and medial temporal cortices during episodic memory retrieval. Hum Brain Mapp 5(4):323–327
42. Friston KJ, Harrison L, Penny W (2003) Dynamic causal modelling. Neuroimage 19(4): 1273–1302
43. Roebroeck A, Formisano E, Goebel R (2005) Mapping directed influence over the brain using Granger causality and fMRI. Neuroimage 25(1):230–242
44. Baudewig J, Siebner HR, Bestmann S, Tergau F, Tings T, Paulus W et al (2001) Functional

MRI of cortical activations induced by transcranial magnetic stimulation (TMS). Neuroreport 12(16):3543–3548
45. Bestmann S, Baudewig J, Siebner HR, Rothwell JC, Frahm J (2003) Subthreshold high-frequency TMS of human primary motor cortex modulates interconnected frontal motor areas as detected by interleaved fMRI-TMS. Neuroimage 20(3):1685–1696
46. Bestmann S, Baudewig J, Siebner HR, Rothwell JC, Frahm J (2004) Functional MRI of the immediate impact of transcranial magnetic stimulation on cortical and subcortical motor circuits. Eur J Neurosci 19(7): 1950–1962
47. Bohning DE, Shastri A, McConnell KA, Nahas Z, Lorberbaum JP, Roberts DR et al (1999) A combined TMS/fMRI study of intensity-dependent TMS over motor cortex. Biol Psychiatry 45(4):385–394
48. Bohning DE, Shastri A, Nahas Z, Lorberbaum JP, Andersen SW, Dannels WR et al (1998) Echoplanar BOLD fMRI of brain activation induced by concurrent transcranial magnetic stimulation. Invest Radiol 33(6):336–340
49. Bohning DE, Shastri A, McGavin L, McConnell KA, Nahas Z, Lorberbaum JP et al (2000) Motor cortex brain activity induced by 1-Hz transcranial magnetic stimulation is similar in location and level to that for volitional movement. Invest Radiol 35(11):676–683
50. Bohning DE, Shastri A, Wassermann EM, Ziemann U, Lorberbaum JP, Nahas Z et al (2000) BOLD-f MRI response to single-pulse transcranial magnetic stimulation (TMS). J Magn Reson Imaging 11(6):569–574
51. Denslow S, Lomarev M, George MS, Bohning DE (2005) Cortical and subcortical brain effects of transcranial magnetic stimulation (TMS)-induced movement: an interleaved TMS/functional magnetic resonance imaging study. Biol Psychiatry 57(7):752–760
52. Li X, Nahas Z, Kozel FA, Anderson B, Bohning DE, George MS (2004) Acute left prefrontal transcranial magnetic stimulation in depressed patients is associated with immediately increased activity in prefrontal cortical as well as subcortical regions. Biol Psychiatry 55(9):882–890
53. Nahas Z, DeBrux C, Chandler V, Lorberbaum JP, Speer AM, Molloy MA, Liberatos C, Risch SC, George MS (2000) Lack of significant changes on magnetic resonance scans before and after 2 weeks of daily left prefrontal repetitive transcranial magnetic stimulation for depression. J ECT 16(4):380–390
54. Ruff CC, Blankenburg F, Bjoertomt O, Bestmann S, Freeman E, Haynes JD et al (2006) Concurrent TMS-fMRI and psychophysics reveal frontal influences on human retinotopic visual cortex. Curr Biol 16(15):1479–1488
55. Camprodon JA, Sack AT, Pascual-Leone A (2010) Developing strategies for therapeutic neuromodulation: simultaneous Transcranial Magnetic Stimulation and functional Magnetic Resonance Imaging to study the mechanisms of Hemispatial Neglect recovery. In: 49th Annual Meeting of the American College of Neuropsychopharmacology (ACNP), Miami, FL
56. Ruff CC, Bestmann S, Blankenburg F, Bjoertomt O, Josephs O, Weiskopf N et al (2008) Distinct causal influences of parietal versus frontal areas on human visual cortex: evidence from concurrent TMS-fMRI. Cereb Cortex 18(4):817–827
57. Ruff CC, Blankenburg F, Bjoertomt O, Bestmann S, Weiskopf N, Driver J (2009) Hemispheric differences in frontal and parietal influences on human occipital cortex: direct confirmation with concurrent TMS-fMRI. J Cogn Neurosci 21(6):1146–1161
58. Sack AT, Kohler A, Bestmann S, Linden DE, Dechent P, Goebel R et al (2007) Imaging the brain activity changes underlying impaired visuospatial judgments: simultaneous FMRI, TMS, and behavioral studies. Cereb Cortex 17(12):2841–2852
59. Bestmann S, Ruff CC, Blankenburg F, Weiskopf N, Driver J, Rothwell JC (2008) Mapping causal interregional influences with concurrent TMS-fMRI. Exp Brain Res 191(4):383–402
60. Blankenburg F, Ruff CC, Bestmann S, Bjoertomt O, Josephs O, Deichmann R et al (2010) Studying the role of human parietal cortex in visuospatial attention with concurrent TMS-fMRI. Cereb Cortex 20(11):2702–2711
61. Bestmann S, Swayne O, Blankenburg F, Ruff CC, Teo J, Weiskopf N et al (2010) The role of contralesional dorsal premotor cortex after stroke as studied with concurrent TMS-fMRI. J Neurosci 30(36):11926–11937
62. Damoiseaux JS, Rombouts SA, Barkhof F, Scheltens P, Stam CJ, Smith SM, Beckmann CF (2006) Consistent resting-state networks across healthy subjects. Proc Natl Acad Sci U S A 103(37):13848–13853
63. Fox MD, Snyder AZ, Vincent JL, Corbetta M, Van Essen DC, Raichle ME (2005) The human brain is intrinsically organized into dynamic, anticorrelated functional networks. Proc Natl Acad Sci U S A 102(27):9673–9678

64. Greicius MD, Krasnow B, Reiss AL, Menon V (2003) Functional connectivity in the resting brain: a network analysis of the default mode hypothesis. Proc Natl Acad Sci U S A 100(1):253–258
65. Rossi S, Hallett M, Rossini PM, Pascual-Leone A (2009) Safety, ethical considerations, and application guidelines for the use of transcranial magnetic stimulation in clinical practice and research. Clin Neurophysiol 120(12): 2008–2039
66. Halko MA, Eldaief MC, Horvath JC, Pascual-Leone A (2010) Combining transcranial magnetic stimulation and FMRI to examine the default mode network. J Vis Exp 46:2271

Chapter 11

Electroencephalography During Transcranial Magnetic Stimulation: Current Modus Operandi

Marine Vernet and Gregor Thut

Abstract

Transcranial magnetic stimulation (TMS) is a widely used research and clinical device that has the potential to modulate and interact with brain activity. However, its mechanisms of action, whether used to explore brain functions in healthy participants or to induce meaningful therapeutic effects in patients, are still not fully understood. One method allowing bridging the gap between TMS administration and its behavioral consequences is the simultaneous recording of brain activity with electroencephalography (EEG). Unfortunately, the acquisition and interpretation of EEG data during TMS is still not straightforward because of the contamination of the EEG by artifacts, despite the introduction of several TMS-compatible EEG systems. This chapter is providing a step-by-step guide to online TMS-EEG experimentation, from the selection of appropriate material, to conducting the experiment and later data analysis. We first review the multiple possible sources of EEG contaminations related to a TMS discharge (of electromagnetic, mechanical, and physiological origin). We then examine various methods that have been proposed in the literature to minimize these artifacts or isolate them from genuine neuronal responses to TMS. We finally survey insights in cognitive and clinical neurosciences that have been gained from the TMS-EEG combination between its introduction in 1997 (first TMS-compatible EEG systems) up to 2011.

Key words TMS-EEG coupling, Artifacts, EEG data acquisition, EEG data processing, Cognitive neuroscience, Clinical neuroscience

1 Introduction

Transcranial magnetic stimulation (TMS) was originally developed as an alternative to the often unpleasant transcranial electric stimulation via surface electrodes attached to the skin [1]. Today, TMS is widely used as a research device in both cognitive and clinical neuroscience and has recently been implemented in neurotherapeutics [2–6]. However, in order to take advantage of this method's full potential for brain research and therapeutic control, it is essential to understand its mechanisms of action.

Alexander Rotenberg et al. (eds.), *Transcranial Magnetic Stimulation*, Neuromethods, vol. 89,
DOI 10.1007/978-1-4939-0879-0_11, © Springer Science+Business Media New York 2014

Although much can be inferred indirectly by the observation of TMS effects on behavior, direct insight into TMS action can only be obtained through the recording of neuronal signals during and after TMS. For instance, TMS not only has different effects depending on the stimulation parameters, but those effects might differ as a function of regions of the brain [7]. Moreover, TMS effects are not limited to the targeted area but reflect the ability of the entire brain to cope with the local change [8]. Ultimately, understanding how TMS acts upon the brain will help to more accurately interpret the induced behavioral consequences, e.g., in terms of network effects [9–11], or their dependence on pre-TMS brain states (e.g., [12]), etc. This will aid, in turn, in the overall understanding of the brain-behavior relationship and may well lead to the development of new therapeutic TMS protocols.

One method that can be used to explore neuronal signals and network dynamics in real-time during TMS is electroencephalography (EEG). Unfortunately, due to the TMS-induced magnetic field contaminating the brain signals, EEG recordings during TMS are not straightforward. The first measures of TMS-evoked brain responses using EEG were performed by Cracco et al. [13], who used one scalp electrode to register responses to TMS over a cortical area contralateral to the stimulation site. Eight years later, a TMS-compatible EEG recording system was introduced, allowing measurement of TMS brain responses over multiple scalp locations, including at the stimulation site [14]. In the last decade, further methodological improvements have been made and recording systems designed.

In this chapter, we provide a step-by-step guide to the methods of TMS-EEG. We first examine possible artifacts in the EEG recordings resulting from TMS (Sect. 2). We define artifacts as any contamination of the brain signal of interest. Consequently, TMS artifacts can be not only of electromagnetic nature (Sect. 2.1.1) but also of mechanical (Sect. 2.1.2) and physiological origin (Sect. 2.1.3). We will emphasize the difficulty in artifact avoidance (Sect. 2.2) and illustrate the methods that have been applied to minimize these artifacts or isolate them from the brain signals of interest (Sect. 3). We proceed by examining practical issues to consider before the TMS-EEG experiment (Sect. 3.1), during experimentation (Sects. 3.2–3.3), and after recording (Sect. 3.4). A summary of the main points can be found in Table 1. Finally, we attempt to provide a survey of recent insights in the fields of cognitive and clinical neurosciences stemming from the TMS-EEG combination (see Sects. 4 and 5).

Table 1
Step-by-step guide to TMS-EEG

Points to consider	Methods proposed	Key references	Used by
Electrodes	Plastic pellet electrodes coated with a thin layer of silver epoxy (Ag/AgCl surface)	[27]	
Amplifiers	Sample-and-hold	[14, 27]	[19, 23, 26, 37]
	Limited slew-rate	[20, 30]	[16]
	MR-EEG system	[46]	[29]
	DC-amplifiers with wide dynamic range	[42]	[31, 43, 44]
	DC-amplifiers with adjustable operational range	[17, 47]	
Synchronization between TMS discharge and EEG recordings	Synchronization box	[20, 30]	
Electrode and skin preparation	Low impedance	[17]	
	Mini-puncturing technique	[18]	
Wire and coil placement	Procedure of wire rearrangement	[48]	
Sound masking	Noise adjusted on the TMS click	[50]	[51, 52]
	Layer of foam between the coil and scalp to attenuate bone conduction	[50]	[51, 52]
Control conditions	Sham stimulation	[55]	
	Changing coil orientation	[42]	[56]
	Real stimulation with no task	[20, 30]	
	Distracting tasks	[53]	
Post-processing	Removing artifactual data (and reinterpolation)	[35, 46, 58]	
	Artifact-subtraction	[20, 30]	[43, 58]
	PCA	[19]	[31]
	ICA	[23]	
	Source reconstruction	[53]	
	Fitting functions	[19]	
	Filtering	[29]	

2 The TMS-EEG Artifact(s)

2.1 Artifact Types

2.1.1 The Electromagnetic Artifact

TMS is based on the phenomenon of electromagnetic induction. The transient, time-varying magnetic field (with a pulse-duration of around 100 ms) creates an electric field in the brain and subpopulations of neuron discharge in response. Unfortunately, the same phenomenon also affects every electrical circuit in the coil vicinity. For example, high voltage (on the order of several volts, [15])

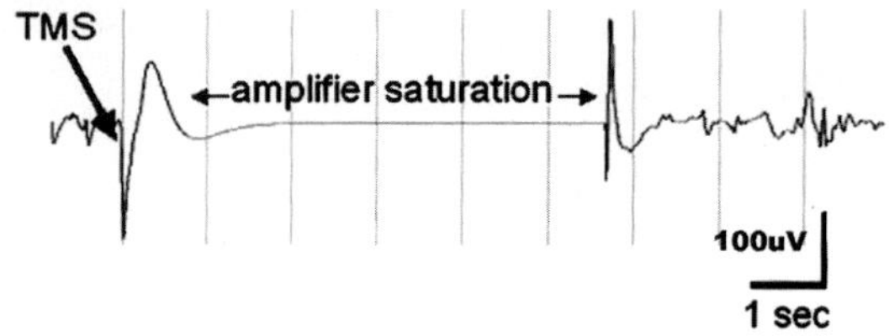

Fig. 1 Illustration of amplifier saturation after a single TMS pulse in a traditional EEG recording system. (Adapted with permission [16])

might be induced in the loops formed by disk- or ring-electrodes, the electrode leads, and amplifier circuits. This high voltage is a source of two problems: the first concerns safety, as traditional EEG electrodes overheat in the presence of strong induced (eddy) currents. The second concerns the acquisition of exploitable data. Traditional amplifiers, because of their design and wide-band characteristics, can respond to high-energy electric fields by saturating for many seconds or even minutes. This saturation conceals a large portion of the brain response to TMS ([16], see Fig. 1). Some amplifiers may even become damaged due to the TMS pulse [16].

In the last decade, new EEG hardware has been designed that allows for safe EEG recording in the magnetic field generated by TMS (or magnetic resonance imaging) without the risk of electrode overheating and/or amplifier saturation. However, even with these new systems, recordings will be contaminated by TMS-induced electric artifacts to different degrees, depending upon the setup (system, electrode preparation, etc.). See Fig. 2 illustrating the variability in electric artifact size and duration over studies.

Induced Voltage Peaks and Charges

The primary electromagnetic artifact generated by TMS consists of an immediate, brief, high-voltage peak. This artifact has a very high amplitude and can reach thousands of microvolts (see e.g., [17] and below). In addition, the electric field may induce charge-differences in the hardware circuitry, at the interface between the electrodes and the skin, and between skin layers (see [18] for the latter). Such electrical charges will gradually decay over time which may contribute to a longer lasting artifactual deflection that can follow the initial voltage peak (e.g., [19, 20]).

Recharge Artifact with Repetitive TMS

After the initial voltage peak coincident with the TMS discharge, a second artifactual peak might appear 5–70 ms after the TMS pulse when TMS pulses are applied in rapid succession during repetitive TMS (rTMS) [17]. This is created by the need for rapid recharge of the TMS device in rTMS protocols [17]. This artifact is of relatively low amplitude (on the order of 10 μV) and occurs consistently at a fixed latency after the TMS pulse, depending on TMS intensity (longer latency for higher intensities, Fig. 3).

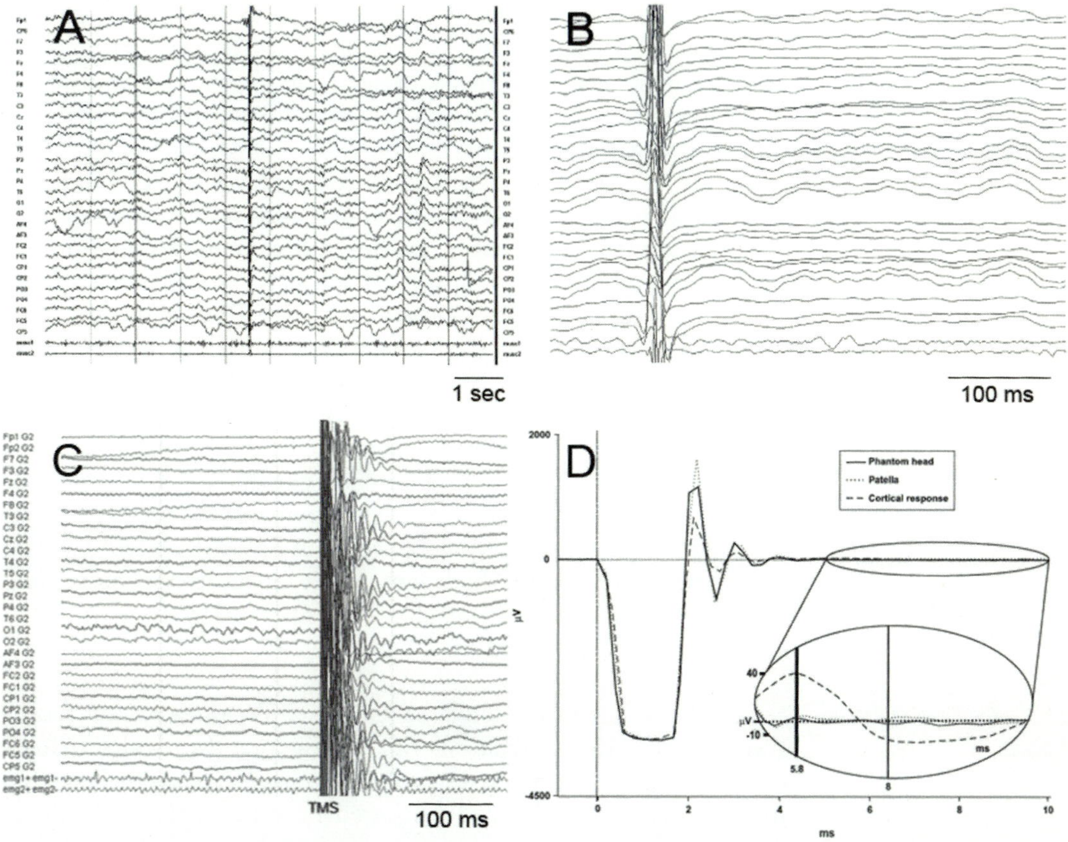

Fig. 2 Illustrations of immediate artifact recorded after TMS pulse. (**a**, **b**) Saturation of amplifiers designed to work in MR environment after stimulation of M1 at 120 % of MT. (Adapted with permission [46]). (**c**) Artifacts recorded with the same system. (Adapted with permission [29]). (**d**) Artifacts recorded with a DC amplifier allowing to adjust the amplifier sensitivity and operational range to TMS. Stimulation of the phantom head was performed at 50 % of MSO, stimulation of the patella at 40 % and stimulation of the cortex at 57 % (MT). (Adapted with permission [17])

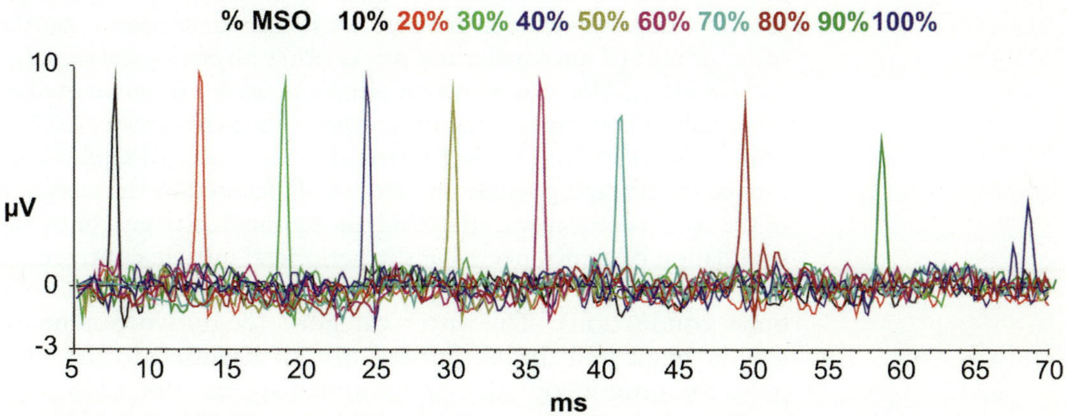

Fig. 3 Illustration of the amplitude and latency of the recharge artifact in function of stimulus intensity. (Adapted with permission [17])

2.1.2 The Mechanical Artifact Resulting from Electrode Motion

During EEG recordings, any movement of an electrode with respect to the skin surface mechanically disturbs the distribution of charge that is created at the skin-electrode interface. This results in a momentary change of the potential and is known as the motion artifact or movement artifact [21].

During TMS-EEG experiments, especially if the TMS coil is in contact with one or several electrodes, movement of those electrodes is almost unavoidable, causing a change in electric potential on the order of several millivolts. Even if the coil is not touching the electrodes, the electromagnetic effect of TMS might cause the movement of standard electrodes. According to Lorenz's law, "an induced current is always in such a direction as to oppose the motion or change causing it". Thus, the primary current in the coil, which creates the primary magnetic field responsible for brain stimulation, also creates a secondary current in the electrodes. This, in turn, creates a secondary magnetic field, opposed to the primary field. As a consequence, the coil and the electrodes exercise a repulsive force which can result in electrode movement. Additionally, electrode motion can result indirectly from TMS-induced muscle contraction.

2.1.3 The Physiological Artifacts

Scalp muscles may be activated by the TMS pulse directly or in response to the activation of nerves on the scalp. Muscle contractions cause strong artifacts (from hundreds to thousands of microvolts) in the electrodes located near the activated muscles. These artifacts are particularly prominent when stimulating frontal and temporal areas. Because muscle relaxation takes several tens of milliseconds [22], these artifacts often contaminate the brain signal of interest.

TMS might also trigger eye movements or blinks as a consequence of a startle caused by the loud click or discomfort associated with the pulse. Eye movements and blinks create a deflection of the potential mainly across channels of the forehead (ocular artifacts) depending upon the reference. This deflection is on the order of tens of microvolts and masks other physiological signals.

Finally, TMS can provoke sensory- and cognitive-evoked potentials. The former category consists of somatosensory-evoked potentials (SEPs) and auditory-evoked potentials (AEPs). SEPs are evoked by the taping sensation on the scalp (caused by the vibration of the coil in contact with the head during the discharge), or by the stimulation of cranial nerves or the peripheral muscle contraction. AEPs are elicited by the loud click (transmitted via both air and bone conduction). The latter category (cognitive potentials) may be generated through orientation of attention to the TMS pulse for a more cognitive or emotional appraisal of TMS (e.g., monitoring of outcome). Such cognitive artifacts would be characterized by responses to TMS that originate in non stimulated sites and are invariant to where over the scalp TMS is applied (see e.g., fronto-central theta-response in [23, 24], but see [25]). These artifacts can be mistaken for genuine TMS-induced activity as they reflect true brain signals and are, consequently, in the range of TMS-induced brain responses.

2.2 Artifact Size(s), Form(s), and Duration(s)

Electromagnetic artifacts induced by TMS are easily discernable in raw continuous EEG data (Fig. 2). Their peak voltages can be estimated using Faraday's law of electromagnetic induction.

These artifacts are generated in every loop of the EEG circuit where an electromotive force ε might be induced. According to Faraday's law, this "force" (measure in volts) is:

$$\varepsilon = \oint_c E \cdot dl = -\frac{\partial \phi_B}{\partial t}$$

where E is the electric field, c the perimeter of the loop through which the magnetic field enters, ϕ_B the magnetic flux and t the time. The magnetic flux ϕ_B through the loop of the surface S, created by the magnetic field B, can be calculated from the following equation:

$$\phi_B = \iint_S \vec{B} \cdot d\vec{S}$$

The tension (voltage) between two electrodes, separated by a distance d can be estimated as the product between the electric field E and the distance d. An estimation of the TMS-induced voltage U is thus:

$$U \propto \frac{\Delta B}{\Delta t} . \frac{S}{c} . d = \frac{\Delta B}{\Delta t} . \frac{R}{2} . d$$

where ΔB is the variation of the magnetic field (≈ 1 T), Δt the duration of the TMS pulse (≈ 100 μs), and d the distance between two electrodes (≈ 2 cm). With a radius of the loop R of 1 cm, the voltage U would be in the order of 1,000 mV. This is several orders of magnitude above physiological signals (such as spontaneous scalp EEG, task-evoked potentials, and TMS-evoked responses/potentials), which are generally inferior to 100 mV (see e.g., [26]).

Accordingly, most of the TMS-induced voltage is artifact and not brain signal. This can also be demonstrated by a simple experiment: recording EEG while TMS is applied with the coil touching the scalp (condition 1) or with a 1 cm thick sheet of plastic between the scalp and coil (condition 2). Discharging the coil over the hand motor cortex with TMS just above motor threshold evokes clear muscle twitches in condition 1 but rarely evokes any twitches in condition 2. This suggests more effective brain stimulation in the first as compared to the second case. Yet, the strong TMS-induced

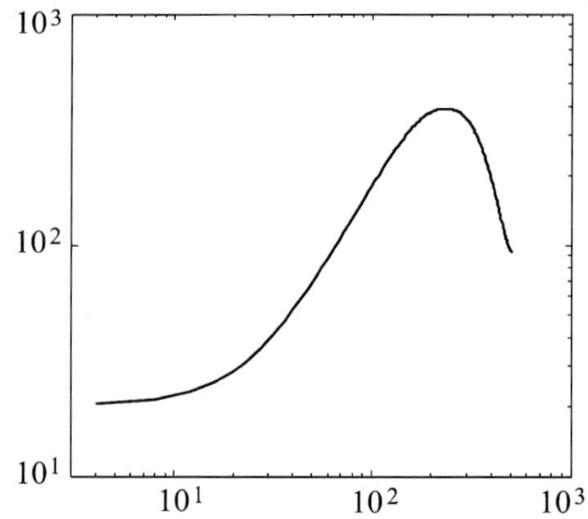

Fig. 4 Description of the artifact recorded on a manikin in the spectral domain. (Adapted with permission [29])

EEG-deflections do not show any clear difference between the two conditions (see [20] using a low slew-rate limiting amplifier, see also Sect. 3.1.2). As the measured voltages in the EEG are of similar amplitude in both cases, it has been concluded that many TMS-induced effects result from magnetically elicited contamination of EEG components (electrical contamination) rather than being of a physiological nature [20].

The electromagnetic portion of the artifact has also been evaluated through stimulation of a phantom head [17, 27–29], allowing characterization of the artifact in the absence of brain and muscular response (Fig. 4). This artifact depends on the TMS pulse form (e.g., monophasic or biphasic, see [17, 18]) but clearly differs from brain activity evoked by TMS. This discrepancy can be used to disentangle artifact from signal, e.g., when using independent component analysis (ICA) for post-processing if required (see Sect. 3.4.4).

The artifact size, form, and duration may vary across different EEG settings. For instance, the exponentially decreasing DC shift which often follows the immediate voltage peak (and which may reflect decay of TMS-induced charges, see Sect. 2.1.1.1) can span over several tens [19] to several hundreds of milliseconds ([20, 27]; Fig. 5), and will have to be eliminated by post-processing [30, 31]. With some TMS-EEG hardware, however, this slow decay is virtually absent and only the initial voltage peak is present, thereby shortening the TMS-EEG artifact considerably. The shortest artifact described thus far is of 5–8 ms duration, measured at a TMS intensity of about 60 % of maximum stimulator output [17, 24]. Therefore, depending upon the system employed, different levels of artifact control will be required.

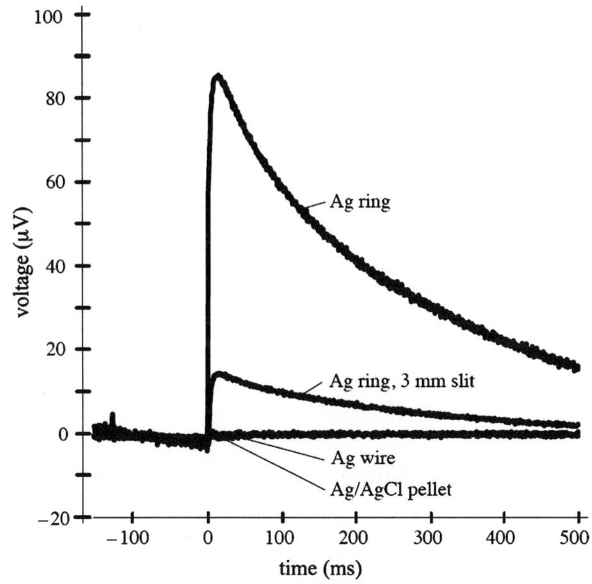

Fig. 5 Illustration of the slow decay of induced charges recorded with four types of electrode on a phantom. (Adapted, with permission [27])

It has to be kept in mind, however, that there is currently no system which allows a completely artifact-free recording. Plus, mechanical and physiological artifacts might be present. Carefully choosing the hardware and designing the experiment is crucial to optimally minimize the recorded artifacts.

3 Methodology: How to Minimize the TMS-EEG Artifact

Procedures for the reduction of electromagnetic artifacts aim to minimize artifact size and duration by adapting the electrical properties of the circuits and/or electrodes while keeping TMS intensity constant. This can be achieved through appropriate amplifier and electrode designs, wire arrangement, and skin preparation for optimizing electrode-skin impedance (see Sects. 3.1.1, 3.1.2, 3.2.1, and 3.2.2). These designs also reduce the risk of electrode heating and therefore meet safety requirements (Sect. 3.1.1). The recharge artifacts, on the other hand, may be eliminated by controlling the TMS machine (for delaying recharge artifacts see Sect. 3.2.3). Physiological artifacts can be reduced by experimental setup (sound masking, distraction task), or, if reduction is not possible, should at least be monitored (control conditions) (see Sects. 3.2.4–3.2.7). Finally, residual artifacts can be dealt with offline via post-processing (see Sect. 3.4).

3.1 Before the Experiment: Choosing Suitable Hardware

3.1.1 Electrodes

Note on safety: Electrodes may heat up after a few pulses, which is associated with a risk of skin burns. In standard disk electrodes, each pulse can increase temperature by 5 °C [21]. Thus, in a room at 20 °C, 5 pulses are sufficient to exceed the recommended safety limit of 41 °C for medical equipment (IEC-601, see [21]). Therefore, even in experiments where artifacts do not impede analysis (such as when the focus of the study is on EEG signals prior to TMS or signals several hundreds of milliseconds after TMS), one should be aware of the risk of burns and choose electrodes accordingly.

TMS-compatible electrodes should satisfy several criteria: they must (1) allow for a safe recording (2), limit TMS artifacts (both achieved through electrode design in regard to diameter, type, and material), and (3) allow for effective brain stimulation despite the offset of the coil from its ideal scalp position by the mounted electrodes (electrode-thickness). At the same time, the electrode should form a good electrical contact and impedance (resistance) should be kept as low as possible to allow for measuring of the electric potential on the skin.

It has been shown that both heating and force are proportional to the conductivity of the electrode material [22, 32]. The challenge is, therefore, to maintain high conductivity (for the accurate recording of scalp potentials) and simultaneously to prevent heating and force. To achieve this, one should ideally use small electrodes. This is because heat induction is proportional to the square of the electrode diameter [32]. Likewise, the electromagnetic force (responsible for electrode movement) is proportional to the thickness of the electrode and the cube of its diameter [22]. In addition, heating and force can be further reduced by the choice of special electrode designs and materials that minimize the current-loop area, the conductive mass of the electrodes, or both. This reduces TMS-induced eddy-currents in the electrodes, a source of heating and force. The current-loop area can be reduced by cutting a section out of silver-ring-electrodes [32]. A slit in an annulus-shaped electrode reduces both heating and the DC-offset by an order of magnitude (see e.g., [27] for a DC-offset reduction from 90 to 12 μV). The conductivity mass can be reduced by using conductive plastic pellet electrodes coated with a thin layer of silver epoxy to create an Ag/AgCl surface [33]. Pellet electrodes lead to a drop in the DC shift below 1 μV ([27]; see Fig. 5). TMS-compatible electrodes are commercially available.

3.1.2 Amplifiers

A TMS pulse might saturate one or several amplifiers in a standard EEG system. The amplifiers might then take hundreds of milliseconds to recover, preventing the recording of brain signals. Different amplifiers or preamplifiers have been designed to tackle this issue.

Sample-and-Hold Circuits

The sample-and-hold solution (developed by [27]) uses amplifier-gain control and sample-and-hold circuits to prevent the strong TMS artifact from being transmitted to the filter circuits. A sample-and-hold circuit consists of at least one switch and one capacitor. When the switch opens, the capacitor keeps tension at the level prior to opening. In the system developed by Virtanen et al. [27], gain control and sample-and-hold circuits protect the amplifier at different input stages. The gain of a first amplifier is reduced during the TMS pulse while a switch sets the entrance of a second amplifier simultaneously to zero (limiting voltage inputs). Further downstream, two sample-and-hold circuits are blocked shortly before the TMS pulse and are released shortly after the pulse in order to maintain voltage at a constant level during TMS. To test performance of this system, Virtanen et al. [27] have measured artifacts for a 2.5 ms-gating period from electrodes attached to a phantom (styrofoam dummy head covered by a leather sheet moistened with physiological saline solution). During the gating period, the voltage peak was smaller than ±1.7 μV in all channels. After gating, DC shifts were below ±1.3 μV and ±0.5 μV at 3 ms and 10 ms post-TMS respectively, with the exceptions of three channels directly under the coil. This system has been commercialized. Typically, gating periods start at 0.05–0.1 ms pre-pulse and end at 2–20 ms post-pulse [23, 34–39]. The shortest gating period ever tested was 100 μs by Ilmoniemi et al. [14]. The circuit may require additional time to recover [19].

A similar design has been implemented by Iramina and Maeno [40], but an attenuator was inserted between the first and the second amplifier and before the switch that sets the signal during the pulse to zero. The output of the second amplifier was filtered (high pass and anti-aliasing) before the sample-and-hold circuit. The amplifiers were turned off 10 ms before TMS and turned on 1 ms after. Artifact-free EEG signals could be measured 5 ms after the TMS pulse.

Low Slew-Rate Amplifiers

Another solution for preventing amplifier saturation is to use a simple amplifier/attenuator module that can be inserted between the recording electrodes and any conventional EEG recording device [16, 20, 30]. The module consists of a low slew-rate operational amplifier (0.07 V/μs) with a relatively low-gain bandwidth product (200 kHz). By adjusting the gain to 2,200 (downstream complimentary attenuation of 2,200 is incorporated in the demultiplexing unit), the high-frequency cutoff is equal to 90 Hz, preventing the circuit from responding to the very high slew-rate of the TMS pulse. The slew-rate limiting properties prevent amplifier blocking and significantly reduce the artifact, while most biological signals of interest (up to 90 Hz, i.e., the mid gamma range) are still captured. Note that although only frequencies up to 90 Hz are recorded, this contrasts with using low pass filters which modify the data and can induce artifacts themselves (i.e., filter artifacts) in the presence of the high-frequency, spike-shaped TMS artifact ([16]; the cutoff frequency was 70 Hz in [20, 30], see also [41]).

The advantage of this method is the simplification of the setup. Also, it can be used as an add-on with any EEG instrument [16]. Furthermore, this method allows for a continuous registration of the signal. Uncorrected artifact durations were initially reported to be in the order of 200–300 ms (reducible to 5–10 ms with post-processing [20], see also Sect. 3.4.2) and were later reduced to 30 ms at 100 % of stimulator output in the electrode directly under the coil [16]. In some electrodes located further than 5 cm from the coil, there was virtually no artifact [16].

Other Amplifiers

More recently, additional types of amplifier circuits have become commercially available which allow for continuous recording without the need for clamping the signal received by the amplifier (although this option remains available with many of these systems). In such systems, the artifact is entirely recorded. A high acquisition rate is recommended such that artifacts can be accurately modeled to allow for its elimination during post-processing (if required).

One option is to prevent amplifier saturation by using modern direct current (DC) amplifiers with wide dynamic ranges that do not contain an initial capacitor subject to saturation by the TMS pulse [42–45]. In such systems, the artifact lasts about 30 ms. Another system with wide dynamic range has been used by Levit-Binnun et al. [31]. The first 10 ms had to be removed from the data and artifacts were present up to 50 ms post-TMS, increasing the need to use off-line artifact removal procedure.

Another option is to use amplifiers and electrodes designed to work inside magnetic resonance (MR) scanners. These systems have been successfully used with TMS [17, 24, 29, 46, 47]. In a study by Fuggetta et al. [46], the MR-compatible amplifiers saturated at approximately 15 ms and recovered during the next 5 ms. The artifact thus remained large during the initial 20 ms, but data could be recorded from 30 ms post-TMS onwards (see Fig. 2). In addition, adjustment of the amplifier sensitivity and operational range may aid in minimizing the TMS artifact [17, 47]. Selecting high sensitivity (0.1 or 0.5 μV) over a large range (±3.277 mV or ±16.385 mV, respectively) is effective with regard to rapid signal recovery [17]. When stimulating a phantom (melon) or non-brain tissue (knee), Veniero et al. [17] showed that the signal returned to baseline at about 5–5.6 ms after the TMS pulse. With cortical stimulation, additional peaks occurred at 5.8 and 8 ms; which were interpreted to reflect the earliest recordable cortical responses following EEG contamination by the artifacts (see Fig. 2).

3.1.3 Synchronization Boxes

It may be advantageous to gate the signal for TMS discharge to the clock of the EEG data acquisition system, thereby allowing the recording of TMS artifacts without aliasing problems. Such a synchronization box, connected to the EEG unit, has only been used so far by Thut et al. [20, 30]. As a consequence, TMS pulses and induced artifacts were perfectly synchronized to the sampling rate of the EEG device. At low sampling rate (200 Hz) such as used by Thut et al. [20, 30], it is essential to avoid aliasing (variable artifact reconstruction due to under-sampling), especially if template artifacts need to be calculated for subtraction purposes (see Sect. 3.4.2). With higher sampling rates, this gating might not be necessary, but no research comparing artifact reduction with and without such synchronization has been conducted so far.

3.2 The Day of the Experiment: Experimental Setup and Subject Preparation

The choice of suitable hardware should be supported by careful experimental setup and participant preparation. This will help to further reduce or at least monitor the electric and the physiological artifacts (muscle activity, eye blinks/movements, somatosensory- and auditory-evoked potentials). It can also help to minimize several sources of artifact, such as electrode movement and TMS-induced electric charges.

3.2.1 Electrodes and Skin Preparation

In addition to having optimal properties to prevent overheating, electrodes should conserve good contact with the skin and impedance should be kept as low as possible (generally below 5 kΩ). Comparing recovery-times and amplitudes of TMS artifact under high impedance (21–25 kΩ) and low impedance conditions (<3 kΩ) (stimulation of knee), Veniero et al. [17] found differences of a factor of 2–3 between the two conditions (recovery-time at 15–20 ms versus 6 ms; amplitude-differences more than doubled in electrodes not directly stimulated).

To achieve low electrode impedance, the skin at the electrode-skin contact has to be cleaned using conventional skin preparation (scrubbing with alcohol and a wooden stick with a cotton tip before applying electrode paste; [21]). This will efficiently reduce the impedance in the outer epithelial layers of the skin. However, deeper layers of the skin epithelium will remain intact. Their capacitive properties will lead to a charging of the skin after the TMS pulse resulting in a slow after-discharge that will contribute to the long-lasting TMS artifact [18]. In order to reduce the skin capacitance and resistance, Julkunen et al. [18] proposed to short-circuit the epithelial layers via a mini-puncturing technique. In their study, the skin at electrode contact was punctured using holes within the electrodes (four per electrode). A custom-made mini-puncturing instrument was used. The optimal length of the tip (i.e., penetration depth) was not separately studied, but the authors recommended a length of about 0.5 mm. An adequate mini-puncture should not cause any bleeding. If applied correctly, mini-puncturing may even cause less discomfort and less skin damage than rubbing [18]. However, EEG caps should go through disinfection, and the mini-puncture tools should be made from disposable components [18].

In comparison to the traditional rubbing procedure, the mini-puncture preparation led to a similar spatial spreading of the artifact, but artifact amplitude and recovery-time were significantly reduced. Twenty-one ms post-TMS, the difference in the size of the artifact between the two preparation techniques was larger than twice the pre-stimulus standard deviation ($2 \times 1.35\ \mu V$). This difference increased with increasing TMS intensity [18].

The mini-puncture technique was tested with a biphasic stimulator. With a monophasic stimulator, the problem of accumulated charge is worse, so the recording should benefit even more from the mini-puncturing [18].

3.2.2 Wire and Coil placement

It is of primary importance to avoid any contact between the coil and the electrodes to minimize electrode movements [21]. In addition, whenever possible, reorientation of the coil (coil rotation) should be considered in order to keep peripheral muscle activation as low as possible and minimize muscle artifacts, especially when stimulating a frontal or temporal area where muscle artifacts can be considerable [22]. Once coil position and orientation have been determined, reorientation of the wires can help to further reduce artifacts, as discussed below.

The artifact can be reduced in duration if the electrode wires are kept loop free and arranged in a radial way from the center of the magnetic field [17]. When recordings are performed with high-density EEG, it is recommended that the loose part of the electrode wires are grouped together toward the amplifier and oriented away from the coil cable to avoid additional interference [17].

Sekiguchi et al. [48] have modeled and systematically studied the effects of electrode wire positioning on artifact size. It was argued that one can create a configuration (electrode wire perpendicular to the coil handle) in which the currents, running in opposite direction in the two rings of a figure-of-eight coil, should counteract each other. In a circuit model with one recording electrode and a reference electrode stimulated by a figure-of-eight coil, this supposition proved accurate. Over a human scalp, the real artifact was considerably reduced by arranging the wires perpendicular to the coil handle, although it could not be totally eliminated [48]. To optimize artifact reduction, Sekiguchi et al. [48] proposed a stepwise procedure of wire arrangement consisting of recording TMS-induced artifacts, then rearranging the wires for artifact-contaminated channels.

3.2.3 Controlling TMS Boosters

The TMS recharge artifact can be controlled at the TMS device. By controlling the power supply unit and using only one booster to recharge the TMS circuit (e.g., in Magstim Super Rapid), it is possible to delay recharge and associated artifacts to periods of less physiological interest [17]. The alternative of inserting a recharging delay circuit in the stimulation device could be used to the same effect [17]. If this is not possible, the artifact can be dealt with by post-processing (in particular Sect. 3.4.1), as it is of short duration and low amplitude.

3.2.4 Sound Masking

Several studies have pointed to the problem of contaminations by the auditory response to the coil-click (e.g., [26, 38, 46, 49–52]). In this regard, it is important to note that the coil-click can be conducted not only by air but also by bone (vibration of the temporal bone), bypassing the middle ear and reaching the cochlea directly. Nikouline et al. [49] recorded the auditory potentials created by the coil-click (100 dB at 0.5 cm from the coil) under three conditions. In the first, the center of the coil was held 2 cm above the scalp over the left motor cortex. In the second, the coil was held at the same position, but with a plastic piece rigidly connecting the head and the coil. In the last condition, normal stimulation was performed with the coil pressed against the head. In all three cases, the negative–positive (N1–P2) auditory complex was clearly identifiable in the EEG. The average N1–P2 peaks (maximal over central and parieto-temporal regions) were 9 mV, 14 mV, and 23 mV, respectively. To eliminate this artifact, the following procedures have been adopted.

Although simple hearing protection will help to reduce air conduction, it will not be sufficient for its elimination. More effective for masking the coil-generated click is to play a loud 90 dB white-noise through insert earphones. Participants have reported that this noise is generally sufficient to mask the auditory input and no AEP have been recorded when the coil is discharging 2 cm above the head under this condition [26, 46]. However, such sound masking may be insufficient to prevent the auditory artifact resulting from bone conductance.

Other studies limited the intensity of the masking noise by designing the noise so that the pressure level at each frequency matched the pressure levels of the time-varying frequencies from the TMS click. The volume was then adjusted until the subject did not perceive any TMS click. Bone conduction was attenuated by placing a thin layer of foam between the coil and the scalp [38, 50–52]. These procedures effectively abolished the auditory stimulation associated with TMS as demonstrated by the absence of any evoked response to sham TMS stimulation [50, 51].

3.2.5 Avoiding Orientation to TMS

In addition to auditory and tactile sensory brain responses, TMS may induce cognitive EEG components, e.g., such as those associated with attentional orientation towards TMS. In order to avoid such orientation effects, participants might be asked to participate in a distracting task. In a study by Huber et al. [53], subjects were engaged in a simple oddball task. Interspersed within a noise mask, tones were played at irregular interval and the participants had to respond as fast as possible with a mouse button click each time a tone was played. However, in many experimental designs, such distracting tasks might be a problem as reactivity and connectivity depend on current brain state and therefore also on mental activity [54].

3.2.6 Further Control Conditions

In some experimental designs (and when artifacts cannot be avoided), one might benefit from including control conditions to characterize some of the physiological artifacts: such as the residual AEP and tactile-evoked potential created by the vibration of the coil during coil discharge. Modeling these artifacts can help to tease apart which components of the TMS-induced potentials reflect a genuine TMS effect and which are unspecific, i.e., due to the associated sounds and coil taps. These artifacts can then be eliminated through subtraction. To model these artifacts, a sham condition [55] or real stimulation in the absence of the tested task [20, 30, 43] have been used, the latter additionally emulating the electrical artifacts that can then also be subtracted (see Sect. 3.4). Finally, assessing TMS effects as a function of changing coil orientation relative to the most effective stimulation position (perpendicular to the underlying gyral folding pattern [56]) might also help to differentiate which components are specific to the optimal orientation (and therefore reflect genuine TMS-evoked brain responses) and which are not [24, 42].

3.2.7 Familiarization with the TMS and EEG Setting

If the participants are TMS- and EEG-naïve, it will be advantageous to familiarize the participants with the particular setting by running training sessions, especially for TMS. It is important that the participant feel comfortable with the procedure, which will help to reduce eye movements, blinks, and muscle contraction. It will also likely reduce startle and orientation towards TMS.

3.2.8 EOG

It is recommended that eye movements and blinks be recorded to determine and discard contaminated trials offline. TMS artifacts render automatic artifact detection or ocular correction procedures difficult [28], but oculomotor artifacts can still be rejected by visual inspection.

3.3 During the Recordings

For some of the artifact reduction procedures (see post-processing below), it is important to obtain reproducible artifacts (besides reproducible signals). Reproducible artifacts are crucial for the calculation of a reliable artifact template that can then be subtracted from the data. Additionally, reproducible artifacts are easier to localize via topographical maps (see Sect. 3.4). To obtain reproducible artifacts, the positioning of the coil relative to EEG hardware should be kept as constant as possible. A movement as small as 5 mm might cause large changes in artifact morphology (in addition to large changes in brain response) ([57]—see below). Further, any coil displacement might increase the risk of displacing an electrode. Coil movements might be prevented by the use of a coil holder or by MRI-guided neuronavigation, both of which allow for precise and stable coil placement and orientation.

For the same reason, wire orientation, TMS intensity, EEG impedances, etc. should be kept as stable as possible. For longer experiments (>30 min), the impedances should be checked regularly to avoid size increase as the electrode paste dries [19].

3.4 After the Experiment: Post-Processing

Even when all the above steps are meticulously followed, TMS-EEG artifacts are unlikely to be eliminated completely. For instance, even with sample-and-hold circuits to fully prevent any brief peak-activity from entering into the amplifier during the hold-interval, there is likely some induced charge that does not decay completely before the end of the hold-period. This can result in residual (but significant) EEG contaminations [19], which requires post-processing. To remove residual artifacts, several off-line analysis approaches have been designed which, presumably, do not distort the physiological signal.

It has to be highlighted here that any removal of TMS-induced artifact comes at the risk of removing TMS-induced brain signals. We therefore recommend that post-processing for artifact removal only be used if unavoidable and that special care be taken if the corrected artifact stretches into the time-window of interest where EEG measurements are interpreted (often after 15 ms post-TMS).

3.4.1 Removing the Channels or Period Containing the TMS Artifacts

The first possible way to deal with the artifacts is to simply exclude the channels contaminated by systematic artifacts [35–37]. Unfortunately, the most contaminated channels are usually those closest to the stimulation site, which also contain the most information regarding brain reactivity to TMS. In addition, in the case of source reconstruction, such removals introduce bias in source localization [19]. Given that late signal recovery is not only channel-dependent but also varies across trials, rejecting the most affected trials in a channel-specific way might still conserve an acceptable number of trials per channel for subsequent analysis [58].

Another possibility is to simply ignore the initial tens of milliseconds containing the main part of the artifact and analyze later-evoked responses. Usually, TMS-EEG analysis starts 6–40 ms after the pulse, coincident with the onset of many signals of interest [17, 24, 39, 43, 44, 46, 47]. In order to also eliminate characteristic auditory and sensory responses, the analysis could focus on a very restricted period of time (e.g., from 20 to 50 ms after the pulse), if this is the period of interest [35].

After the removal of data from the artifact period, it is possible to reconstruct a continuous data set (however, without any meaningful information in the discarded section). For instance, Fuggetta et al. [43] cut out 40 ms segments (from −2 to 38 ms) and joined data points from before and after each removed segment. The random voltage step between the joined data points of each single trial was eliminated via trial-averaging. In a study on brain oscillations, Fuggetta et al. [55] concatenated artifact-free periods extracted from an rTMS session. This created edge artifacts in the EEG data at the junction between the segments. The frequency bands around the frequency of the edge and their harmonics did not reflect brain oscillations, but any intermediate frequency band was reliable [55]. When the first tens of milliseconds post-TMS are eliminated, another possibility is to set all values to zero (zeropadding, see, e.g., [39, 53, 59]) or to interpolate the missing data by third-order polynomial curves followed by appropriate filtering [58].

Thus, for many studies, simply discarding channels, trials, or periods of time containing the artifact might be sufficient, particularly if artifact durations are short (<10 ms). Information concerning the local early response to the TMS pulse might consequently be lost.

3.4.2 Creating TMS Template Artifacts and Subtraction

In order to ensure that a TMS-evoked component is not reflective of an artifact, it might be useful to compare EEG responses obtained from scalp electrodes with those recorded from a phantom head (e.g., [17, 24, 28]). Bender et al. [28] compared the N100, a negative component around 100 ms post-TMS (supposed to reflect inhibitory processes [60]) to the TMS-evoked potential that was produced in the same time-window when a glass dummy head was stimulated (this head was covered by a cloth soaked with water, so that impedances of about 5 kΩ were obtained). The short, sharp EEG artifact produced by the TMS pulse was transformed by the amplifier's hardware and filter characteristics to a slightly delayed high-frequency oscillation with decreasing amplitude. Even though this artifact was still present 100 ms after the pulse, the N100 recorded in children was significantly larger (-130.6 ± 71.9 μV) than the "N100" recorded in the dummy (4.8 ± 0.9 μV), supporting a neuronal origin of the N100.

While TMS-EEG on a phantom head elicits purely electrical artifacts, others have created a template of the artifact by recording EEG during TMS at rest [16, 20, 30, 58]. Obtained through multichannel recordings for each subject, this template can then be used to isolate the neuronal signal from artifact by subtraction (experimental condition minus rest). In these studies, participants were performing a task, and the purpose was to explore the changes to task-related activity as a consequence of TMS (not the activity evoked by the TMS pulse per se). By subtraction of the template artifacts, one will not only subtract artifacts, but also neuronal activity that has been induced by the TMS pulse in the brain tissue underlying the coil, isolating task-related potentials (TMS-modified or unaffected) from any TMS-locked event.

Thut et al. [20, 30] tested this procedure using a visual task, applying TMS over the occipital pole and recording visual-evoked potentials (VEPs) (using low slew-rate EEG amplifiers). The template artifact calculated in a TMS-only condition (rest) lasted up to 200–300 ms. Despite this considerable contamination, VEPs could be retrieved by subtracting the template artifact (TMS only) from the experimental condition (visual stimulation + TMS). The VEPs resulting from subtraction were virtually identical to control VEPs (recorded during a visual task without TMS) when TMS was applied at visual stimulus onset (see Fig. 6), but were modified by TMS (in P1- and N1-amplitude) when applied around 100 ms after visual stimulus onset (i.e., at the moment of most consistent visual suppression). Using a similar procedure, Reichenbach et al. [58] found a supra-additive interference between occipital TMS- and VEPs (P1 for 6 over 7 subjects and N1 for the last subject) dependent upon the TMS intensity. Subtraction methodology was also applied with TMS to other areas, e.g., posterior parietal cortex and vertex [43] and the cerebellum [20].

The advantage of subtracting TMS-artifact templates is that it deals with both electrical artifacts and auditory/somatosensory contaminations (although at the expense of eliminating TMS-induced activity that is of interest in many applications). Subtraction of template artifacts is therefore useful for studying single-pulse TMS interference with the generation of functional, task-related signals (e.g., event-related potentials) or with spontaneously generated (patho-) physiological activity (e.g., epileptic spikes). In exchange, it does not allow for the study of brain reactivity and connectivity because TMS-induced activity, which is confounded in the artifact, is also subtracted [20].

3.4.3 Source Analysis with PCA Around TMS Administration

This artifact correction option proposed by Litvak et al. [19] is based on a method for correcting eye-movement artifacts using scalp topographies and source analysis [61]. It is implemented in the BESA software and also available with the open-source MEEGTools toolbox (SPM8).

The main feature of this algorithm is that it models artifact topography and brain signals separately, which allows for disentanglement despite spatial and temporal correlation. To model the artifact, a principal component analysis (PCA) is performed during the time period in which the artifacts dominate (typically the first 15 ms after the pulse for the immediate TMS artifact). PCA is a linear transformation, from the original EEG data collected in channels to data expressed in an orthogonal coordinate system, in which the greatest variance is accounted for by the first component, the next largest variance by the second, etc. Thus, the first few components of the PCA should reflect the major part of the artifact.

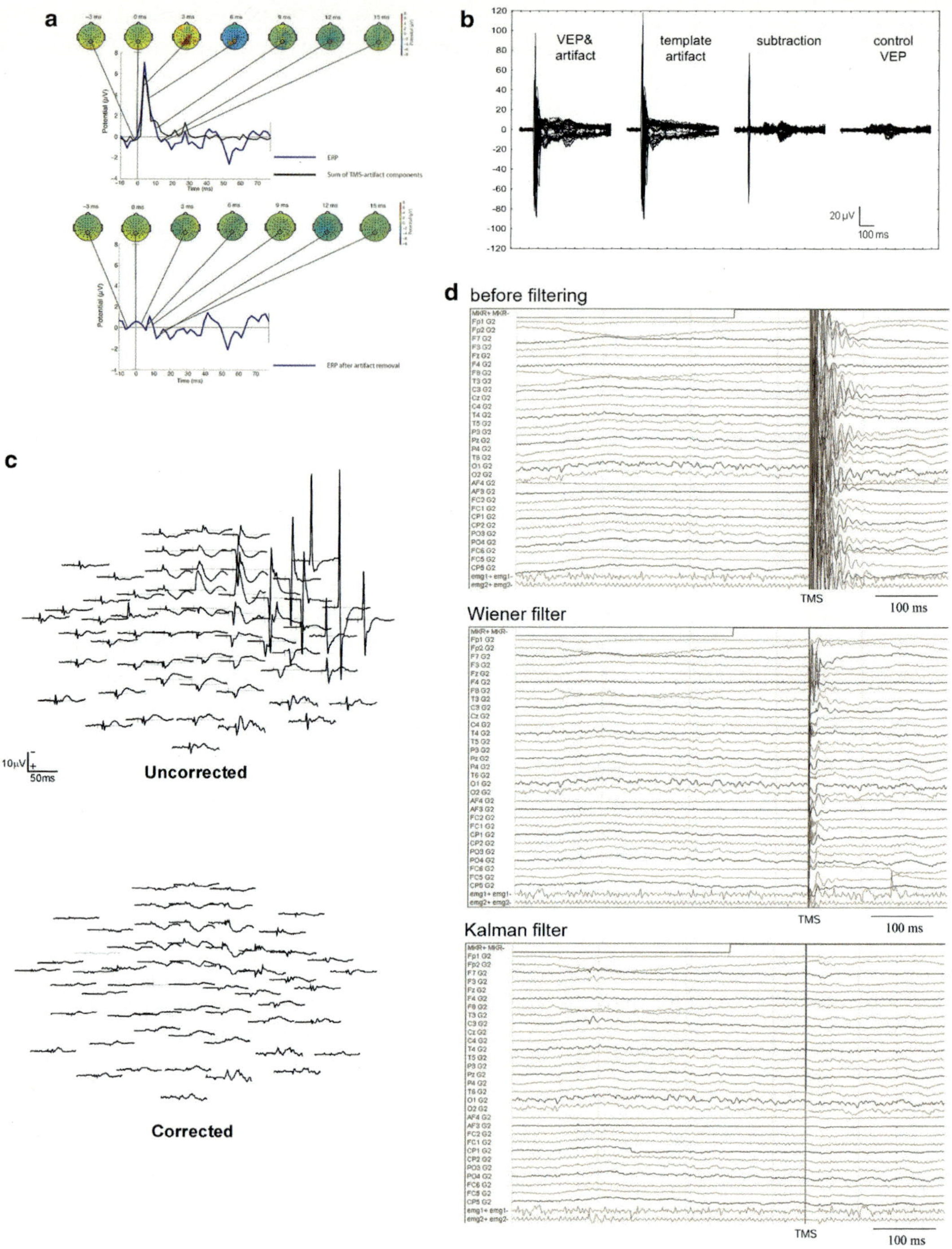

Fig. 6 Illustration of artifact removal efficiency. (**a**) ICA method. (Adapted with permission [23]). (**b**) Artifact-subtraction method. (Adapted with permission [20]). (**c**) Brain topography and PCA method. (Adapted with permission [19]). (**d**) Filtering methods. (Adapted with permission [29])

To model the brain signals, a surrogate model of the brain containing 15 sources is used, capable of reasonably approximating activity generated in the cerebral cortex [62]. A general model of brain activity (using the surrogate model) and artifacts (entering the first PCA component) is used to calculate, from the original EEG signals, the activity of both brain and artifact sources. Performing the inverse calculation—after removal of the artifact source—leads to cleaner EEG signals. Further PCA artifact components are then added one by one to the general model until no artifact can be seen in the EEG signals. Using the resulting artifact-corrected data set (single-subject data) and based on the grand-mean averaged data, a better source model accounting for brain response to TMS is then built (grand-mean model). The artifact correction procedure outlined above is then repeated on single-subject data, with the grand-mean based model replacing the surrogate model. For an illustration of the outcome of this algorithm see Fig. 6.

Litvak et al. [19] applied this artifact correction procedure to retrieve early TMS-evoked activity (10–60 ms latency) which was contaminated with residual artifacts despite the use of TMS-EEG dedicated hardware. Later parts of TMS-evoked responses were not affected. The results revealed multiple source activity already in this early phase of TMS-evoked responses.

Although the application of the method was first shown with EEG signals recorded with a dedicated sample-and-hold system [19], it is also applicable with standard EEG systems so long as saturation is prevented [31].

3.4.4 ICA Approach

ICA has been proposed for TMS-artifact removal by Hamidi et al. [23] (on data recorded through a sample-and-hold system). Decomposing EEG data by ICA involves a linear transformation from data collected in channels to data expressed in "virtual channels" (components). The independent components are chosen to produce the maximally independent signals (in the time domain). Because each EEG electrode records a linear combination of activity from distinct (physiological or artifactual) sources, the ICA procedure should ideally allow for the isolation of these sources of activity.

As with other procedures, the difficulty lies with distinguishing between artifactual ICA-components and physiological ICA-components (and to identify potential mixing). To this end, the artifact detection and removal procedure proposed by Hamidi et al. [23] relies on three main principles of artifact characteristics: first, the artifacts should fulfill the criteria of being spatially localized. In contrast to physiological EEG signals that will be detected by many electrodes on the scalp due to volume conductance (even from a very localized source), TMS-related artifacts originating from outside the skull are recorded mainly in the electrodes closest to the TMS coil. Second, TMS artifacts should fulfill the criteria of being temporally predictable, occurring systematically at pulse onset and lasting only for 10–15 ms after the pulse. Third, during repeated application of TMS at a fixed frequency (rTMS, as in [23]), the power spectra of artifact components should clearly be of a non-physiological nature. The artifact component has a strong peak at the stimulation frequency and several harmonics superimposed on a flat power spectrum. Any physiological signal, on the other hand, should cover a wider bandwidth, not contain strong harmonics, and maintain a 1/*f* pattern in the power spectrum.

For artifact-identification and reduction, Hamidi et al. [23] applied ICAs at two stages of pre-processing (first on continuous EEG data and then on epoch data). To make sure that only ICA-components reflecting artifacts are discarded, Hamidi et al. recommend to inspect the data for mixed components (showing elements of both physiological data and the above artifact criteria). In the case of mixed components, only those that explain relatively little variance in the data should be eliminated. As a final step, comparing the results with or without these ambiguous components might help to further validate the identification of artifact components [23]. Using this method, Hamidi et al. [23] studied the evolution of TMS-evoked activity over the course of a 10 Hz train of multiple successive pulses. The amplitude of evoked activity within each inter-pulse-interval(IPI) (100 ms) followed a quadratic relationship with pulse number, revealing complex dynamics of neuronal recruitment over an rTMS train.

It should also be noted that the ICA procedure outlined above partially relies on the TMS pulses being applied regularly and stimulation conditions remaining constant during the experiment.

3.4.5 Filtering Methods

It is not advisable to use conventional filters during recording or upon uncorrected (artifact-contaminated) signals as these might interact with the residual artifact and produce ringing artifacts (especially when low pass filters are applied) thereby increasing artifact duration (e.g., [63]). In addition, despite the fact that low frequencies usually reflect physiological signals and high frequencies usually reflect artifact, the large overlap between spectral components prevents any simple filter from removing the artifact without distorting the physiological signal [29]. To remedy this, specific off-line filter strategies have been tested.

Morbidi et al. [29] contrasted different filtering methods aimed at removing the electrical artifact from a recorded signal (data recorded with an MR-compatible EEG system). Of the two-filter strategies tested (Wiener filter versus Kalman filter), the Kalman filter approach (working properly on a non-stationary signal) yielded better results and is summarized below.

If one assumes that the electrophysiological signal is a stationary, zero-mean signal, then it is possible to obtain the average artifact by averaging the signal (electrophysiological plus artifact) over several trials. The average of the artifact can then be subtracted from individual trials in order to obtain the individual physiological signal. However, the artifactual signal is not necessarily perfectly reproducible and might contain a stochastic portion, thus preventing perfect results by simple subtraction. The Kalman filter does not rely on artifacts and signal being jointly stationary. Two dynamic models describing EEG and TMS signal generation are established and the Kalman filter is applied to their linear combination. The Kalman filter is a recursive (two-step) method. In the first step, the signal at time t can be estimated through the model from the estimate signal at the time t-1. The uncertainty relative to "real data" is estimated through the variance given to the model. In the second step, the estimation is corrected by averaging the previously estimated signal and measured signal, with weights dependent upon the uncertainty of the model.

For an example of Kalman filtering taken from Morbidi et al. [29], see Fig. 6.

3.4.6 Source Reconstruction to Identify Artifacts

In order to remove artifact from electrical contamination, Huber et al. [53] used a source reconstruction method. The origin of the average electrical activity of each peak of the TMS-evoked response was localized. A failure to locate a source within brain was considered as an evidence for artifactual contamination.

3.4.7 Modeling Artifacts by Exponential Functions

Artifacts can take the form of an initial, steep amplitude increase (offset) followed by a slow exponential decrease. This slow decay likely reflects the slow discharge of accumulated charges during the pulse and can be fitted with exponential functions [19] for removal from the data.

3.4.8 Final Automatic Rejection (After the Major Portion of the Artifact Has Been Deleted)

After the main electrical artifacts have been removed, traditional algorithms can be applied to remove residual TMS, muscle, or EOG artifacts. Two criteria are generally used: amplitude of the signal and slope (first derivative) which should both be kept below chosen thresholds [19, 55].

4 Benefits from Combining TMS and EEG

To illustrate the scope of this unique method combination, we outline some of its applications below.

4.1 Systems and Cognitive Neuroscience

In the fields of systems and cognitive neuroscience, the combination of TMS and EEG has opened new ways to study brain connectivity and function. Our overview of some key findings in these fields is organized according to the three main TMS-EEG approaches (as identified in [54]): (1) the "inductive" approach, during which TMS-induced brain activity is measured in the absence of any task (see also [64]) (2), the "interactive" approach, during which either TMS-induced modulation of task-related activity, or task-related modulation of TMS-evoked activity is assessed (see also [63]), and (3) the "rhythmic" approach, during which TMS-EEG interactions are studied within the perspective of brain rhythms (see also [65]).

4.1.1 The Inductive Approach

TMS can be used independently (without EEG) to probe the excitability of the primary motor cortex (M1) or the excitability of the visual cortex (V1/V2) via the measuring of TMS-induced motor-evoked potentials (MEPs) from target muscles or via the assessment of TMS-induced illusory visual percepts (phosphenes). Other, "silent" brain areas have also been stimulated without concurrent recordings of brain activity, but with other goals in mind, i.e., to explore the role of the stimulated area in a given task. Through the recording of TMS-evoked potentials (TEP) across the scalp, TMS-EEG can be used to quantify cortical excitability beyond motor and visual areas. By analyzing local brain responses, the excitability of stimulated areas can be probed. In addition, studying the subsequent spread of activity to anatomically connected areas provides information on brain connectivity.

Cortical Excitability

TEP-amplitude can be used to quantify local excitability. In doing so, different TEP peaks should be distinguished. Upon stimulation of the motor cortex, evoked activity lasts up to 200 ms post-TMS [42]. Successive positive and negative TMS-evoked deflections have been consistently identified over the scalp (e.g., [26, 37, 42, 64, 66]). These include peaks usually defined as N15/N18, P30, N40/N45, P55/P60, N100, P180/P190, and N280, which originate from different areas [26]. While some reflect the activation of the stimulated cortex (N18, N45, P55), others are related to a spreading of activity toward other regions (P30) or to a widespread brain response (N100). The N100-P190 complex was initially associated with the sound emitted by the discharging coil [49]. Recent studies using white-noise sound masking and manipulating coil orientation as controls, and comparing the N100-P190 with AEP- and SEP-topographies, have determined that part of this component is a genuine TMS effect [28, 42].

To characterize the excitability of the cortex in more detail, the shape of the input–output curve (i.e., stimulus intensity-response curve) has been assessed for different peaks of the EEG-response [36, 37, 51]. TEP-amplitudes increase with stimulus intensity, but this relationship depends on the peak examined and the area stimulated. TMS of M1 often evokes measurable brain activity at low stimulus intensities, probably significantly below 60 % motor threshold (MT) [37], whereas the prefrontal cortex shows less excitability [36]. This provides useful information for fine-tuning of TMS intensity during an experiment or for therapeutic treatment, e.g., for depression. Because the function of motor and prefrontal cortex can be altered by neurological disease, one should be aware that any relation between the excitability of the two areas in healthy subjects cannot be directly transferred to patients [36, 37].

In addition to defining local excitability via TEPs , the inductive approach can also substitute for MEPs during the study of short or long intracortical inhibition (SICI or LICI) and intracortical facilitation (ICF) via the paired-pulse paradigm [26, 44, 67]. Briefly, in the paired-pulse paradigm, a conditioning motor cortex TMS pulse can either promote or decrease the MEP evoked by a second test pulse, depending upon IPI [68–70]. MEPs typically decrease with IPIs of 3-ms or 100-ms (SICI and LICI respectively) and increase with an IPI of 20 ms (ICF). Integration of TMS and EEG allows one to explore ICF and ICI in neural substrates other than M1 [26, 44, 67]. Daskalakis et al. [67] showed, for instance, that both within the motor cortex and the prefrontal cortex, the LICI paradigm resulted in significant suppression of mean TMS-evoked brain activity. This suppression was correlated with MEP-suppression across subjects.

Neural Connectivity

Examining the time-course of TMS-evoked scalp potentials and their possible neural origin has also contributed to the study of neural connectivity. To date, neural connectivity has been probed using TMS over the cerebellum (e.g., [40]), the sensori motor and occipital cortices (e.g., [14]), and the motor and posterior parietal cortices (e.g., [19, 71]). Spreading of activation across areas ipsilateral to stimulation was recorded within 5–15 ms [14, 19] and across homologous regions in the opposite hemisphere within 20 ms [14, 71]. Both transcallosal and subcortical pathways might explain this contralateral hemispheric response.

Examining neuronal connectivity as a function of different states of consciousness, namely sleep and wakefulness, has revealed the dynamic nature of neural connectivity [38, 50]. During wakefulness, TMS evoked a low-amplitude wave with a complex, differentiated pattern including long-range spreading of cortical activity. Contrarily, during sleep, TMS produced either a local slow-wave (after stimulation of the sensori motor area) or a widespread slow-wave (after stimulation of the premotor area). These findings were interpreted to be reflective of a sleep-related loss of ability of the thalamocortical system to enter states that are both integrated and differentiated: the system either breaks down into independent modules or bursts into an undifferentiated response [38, 50]. Both features may be important to account for the fading of consciousness during the early phases of sleep (see the integrated information theory of consciousness in [72]).

Finally, analyzing the spread of activation can also reveal the existence of distinct "functional" connections. For instance, stimulation of the posterior parietal cortex has confirmed the existence of an "attentional" circuit (revealed by co-activation of prefrontal cortex) as well as a connection between dorsal and ventral streams (revealed by co-activation in temporal-occipital regions) [73].

In the absence of any tasks, the functional relevance of changes in excitability and connectivity remain hypothetical. Below, we describe what the TMS-EEG combination has achieved in the study of the "active" brain.

4.1.2 The Interactive Approach: Exploring Network Dynamics of the Active Brain

TMS-EEG can provide insight into how TMS acts on the brain to alter behavior. Specifically, it allows one to establish how modifications in brain reactivity and connectivity are reflected at the behavioral level. One approach consists of studying the modification of task-related brain activity (e.g., event-related potentials) by TMS, while another consists of studying the modification of TMS-evoked potentials as a function of task.

Modification of Task-Related Activity by TMS

In tasks requiring the allocation of visual attention, alterations of VEPs or of anticipatory visual activity over posterior sites (e.g., alpha activity) have been described after stimulation of remote areas, namely the right parietal cortex [43, 74], and the frontal eye field (FEF) [74, 75]. These alterations were absent after stimulation of control sites. The study by Fuggetta et al. [43] targeting parietal cortex showed impairment in task performance associated with a VEP-change, causally linking the parietal cortex with attentional control mechanisms. Likewise, the EP-findings of Taylor et al. [75] highlight the causal role of the FEF in the attention-mediated regulation of brain activity evoked over posterior sites. The findings of Capotosto et al. [74] revealed a causal role of FEF and of parietal cortex but in the control of ongoing posterior (alpha) activity (recorded during attention orienting in anticipation of a forthcoming visual event). All these studies therefore reveal cortico-cortical interaction in the regulation of input control.

Modification of TMS-Probed Excitability and Connectivity by Task

Because the amplitude and configuration of the TMS-evoked response depend on vigilance states [38, 50], TMS-EEG also offers a new opportunity to explore brain reactivity and effective connectivity in the active brain during different tasks or stages of task performance.

Task-dependent changes in brain reactivity were illustrated by Bender et al. [28], who showed that the amplitude of the N100 potentials evoked by motor cortex TMS is reduced during motor preparation when participants anticipate a forthcoming stimulus prompting a manual response. Because this involved pre-activation of the cortical structures necessary for a fast response, the attenuation of the TMS-evoked N100 indicates that this component may reflect an inhibitory response to the sudden, TMS-induced synchronization of neuronal discharge [28]. The link between inhibitory processes and the N100 was later confirmed by studies in which participants were asked either to resist or assist TMS-induced movements [76] or mechanically-induced movements [77]. In both instances, subjects were able to intentionally modulate their motor response to the external stimulus. EEG recordings revealed decreased N100 amplitude during preparation to assist as compared to resist, suggesting anticipatory reduction of intracortical inhibitory processes [76, 77].

Task-dependent change in effective connectivity was illustrated by Taylor et al. [78] who showed that TMS of the medial frontal cortex (close to the area usually described as the pre-supplementary motor area) impaired behavioral and EEG-measures of motor preparation, the latter reflected in changes of activity over M1 (remote effect). Notably, these effects were only observed when a conflict should be resolved, but TMS had no effect in the absence of conflict.

These studies underscore how different tasks can shape the balance between excitation and inhibition, or functional connectivity between anatomically connected areas.

4.1.3 The Rhythmic Approach

TMS-EEG combination has also been used to elucidate the role of brain oscillations in brain function. Here, again, TMS-EEG can be used in two ways: the first consists of studying the modification of ongoing brain oscillations by TMS, while the second consists of studying the modification of TMS outcome measures (excitability and connectivity) as a function of background oscillatory activity.

TMS Influences on Rhythmic Brain Activity

In addition to analyzing TMS-induced potentials in terms of components (or peaks), responses can be analyzed in terms of spectral composition. Single pulses over the motor cortex lead to synchronization in the α- and/or β-bands [26, 46] as well as enhanced coherence between motor and fronto-central areas, mainly in the α band [46]. It was suggested that either the pulse had activated “idling neurons” that, in turn, began to oscillate at α- and/or β-frequencies, or more likely, that the pulse led to a phase-reset of ongoing spontaneous activity in a population of neurons [26, 46]. Such phase-resetting of ongoing oscillations might be established in cortical networks (local pacemaker). However, the recruitment of a thalamic pacemaker through thalamocortical interactions is also conceivable [46, 79]. The phase-resetting hypothesis was supported by Van der Werf and Paus [59] who studied phase-locking of single-trial beta-responses over motor cortex. This phase-locking was highly significant, whereas the amplitude modulation was not. It is noteworthy that, while TMS is an external and artificial event, the oscillations induced by TMS are likely of physiological origin. Rosanova et al. [52] showed that when stimulated, each region tended to preserve its own natural frequency (e.g., α-oscillations over occipital cortex, β over parietal, and fast β/γ-oscillations over frontal areas). Hence, single-pulse TMS reveals the “natural rhythms” of different regions.

An interesting corollary of inducing brain oscillation by a single pulse is the prospect of entrainment when further TMS pulses are applied in phase-alignment with the induced oscillation. For instance, when stimulating an area oscillating spontaneously at α (i.e., an α-generator), one would expect to induce an α-oscillation by one TMS pulse [52] and to further enhance (entrain) such an oscillation with further TMS pulses applied at α-frequency in phase with the induced oscillation. The hypothesis that entrainment of physiological rhythms with frequency-tuned rhythmic TMS is feasible has been supported by TMS of an α-generator at α-frequency and concurrent EEG recordings [24]. Note that these entrainment effects have been measured during application of only a few pulses (5 pulse TMS-trains in [80]) and corroborate earlier studies inferring entrainment of brain oscillations by short rhythmic TMS-trains from mainly behavioral measures [81–84]. For instance, Klimesch et al. [81], who introduced the idea of entraining brain oscillations at the frequency of stimulation, showed that parietal TMS tuned to the individual α-frequency benefitted mental rotation while stimulation at other frequencies had no effect. This finding is in accord with EEG-findings that high posterior α-power predicts good performance in mental rotation tasks. More recently, frequency-specific behavioral effects of rhythmic TMS have been reported after parietal stimulation for performance in location-based and feature-based attention tasks as well as working memory tasks [82–84].

Other modifications of spectral content have been reported after prolonged rTMS (usually >5 min of stimulation). Various rTMS-parameters, such as duration of stimulation, frequency, and intensity, will influence the local oscillatory activity as well as synchronization between different areas [47, 55]. For instance, trains of 5 Hz rTMS that increase cortical excitability as reflected in enhanced MEP amplitude have been associated with a decrease of coupling between brain areas, possibly reflecting reduced efficacy of inhibitory cortico-cortical and/or cortico-subcortico-cortical projections [55]. On the contrary, 1 Hz rTMS promoting cortical inhibitory mechanisms as reflected in a decrease in MEPs amplitude translates into a local increase of power in the α-band [47]. But, see Veniero et al. [85] for changes in brain oscillations that are independent of rTMS protocol (e.g., 5 Hz vs. 1 Hz). As this chapter is focused on EEG signals recorded during TMS, any reader interested in EEG aftereffects of rTMS is invited to refer to the review by Thut and Pascual-Leone [86].

Rhythmic Background EEG Activity Influencing TMS Outcomes

When stimulating motor cortex with constant parameters, variability of TMS outcome (MEP amplitude) can be attributed to variability in the excitability of the spinal motoneuron pool recruited by the cortical efferent volley induced by TMS. In accordance with this view, MEP amplitude is only inconsistently or weakly correlated with the amplitude of TMS-evoked potentials, such as the N100 amplitude ([26], but see [28]), the peak-to-peak amplitude of the N15-P30 complex [87], or the integrated activity of the TMS-evoked response at electrodes over motor, pre motor, and sensori motor cortices [53]. In other words, variability in MEP amplitude is only weakly related to variability in cortical responses to TMS. From this, it stands to reason that an important source of MEP variability is subcortical. However, variability in TMS outcomes may also be related to spontaneous fluctuations in ongoing oscillatory activity.

Romei et al. [88, 89] showed that subsecond fluctuations in α-band power at occipital electrodes just prior to stimulation of occipital cortex were significantly linked to the presence or absence of an illusory visual percept (phosphene), both within and across individuals. These spontaneous fluctuations in brain excitability could be functionally relevant: the brain possibly maximizes the prospective gathering of visual information by periodically changing sensitivity for a given location [88]. Similarly, power in the alpha-, beta-, and/or gamma-bands just before stimulation of M1 has been shown to correlate with MEP-amplitude [90–93]. Finally, the phase of spontaneous oscillation also plays a role in shaping TMS-probed cortical excitability [93, 94].

4.2 Clinical Applications

A growing understanding of the normal brain constitutes a basis for the development of new clinical tools, both in diagnosis and treatment. For instance, the dependence of brain response on TMS intensity [37] or the triggering of natural oscillations in response to stimulation of different areas [52] may have diagnostic potential for several pathologies such as depression, epilepsy, or disorders of consciousness [21]. Similarly, TMS-EEG may provide a window into cortical connectivity, which could be useful for prognosis, for instance following brain injuries [64]. Regarding treatment, studies on sleep [38, 50, 53] have opened new perspectives for increasing sleep efficiency [72]. Similarly, the possibility to entrain brain rhythms [24, 80, 81] and concurrently modify cognitive performance [81–84] or promote synchronized oscillations in two areas by applying synchronous bifocal stimulation [95] might be further exploited for the treatment of various cognitive deficits. Below, we review some TMS-EEG studies on the understanding, diagnosis, and treatment of neurological conditions.

4.2.1 Pathophysiology

The TMS-EEG combination has been used to explore abnormalities in brain reactivity and connectivity in several conditions.

Alcohol intoxication modifies connectivity [35] as well as reactivity of the prefrontal cortex [96]. These changes may be linked to an interaction of alcohol with a large variety of inhibitory and excitatory mechanisms and may be associated with symptoms of alcohol intoxication such as mood lability and impairments in attention, memory, and judgment [96].

Abnormalities could also be reflected in the spectral content of brain response to TMS. In schizophrenia patients, for instance, TMS-evoked γ-activity and the spreading of γ-oscillations are impaired over motor, premotor, and prefrontal cortices [51, 97]. This might suggest that frontal thalamocortical circuits in these individuals are impaired in the capacity to effectively produce and synchronize γ-rhythms in response to external stimulation [51]. Alternatively, it is possible that it is modulation ($GABA_B$ mediated) rather than the generation of γ-oscillations that is dysfunctional in schizophrenia [97]. Modulation of γ-oscillations might be useful to shape the temporal profile of incoming information during different phases of cognitive tasks such as working memory [97]. Thus, selective deficits in the ability to modulate γ-oscillations may represent an important pathophysiological trait, which may be responsible for some aspects of deficits commonly found in schizophrenia [97].

4.2.2 Diagnosis

The TMS-EEG combination may also be useful for diagnostic purposes. Julkunen et al. [98] showed that cortical reactivity and connectivity were decreased in patients with Alzheimer's disease (AD). For patients with mild cognitive impairment (MCI), brain responses were frequently halfway between the values of controls and AD patients, consistent with the concept of MCI being a transitional state between normal aging and AD. Therefore, TMS-EEG might be sensitive to early functional changes in the diagnosis of MCI and AD.

Valentin et al. [99] showed that, unlike in healthy subjects, TMS induces late responses (0.1–1 s) in patients with focal epilepsy which either resemble epileptiform discharges or the onset of a new rhythm clearly different from the background EEG. The induction of these late responses could be used to identify the epileptogenic region and diagnose the existence of a focal epilepsy (except if originating in deep insular or mesiotemporal cortex). It has therefore been suggested that the TMS-EEG methodology might be included in the routine diagnosis of focal epilepsy, complementary to interictal scalp EEG [99].

4.2.3 Monitoring Treatment Outcome

The combination of TMS and EEG may also allow for the evaluation of the efficiency of antiepileptic rTMS treatment. Brodbeck et al. [100] used EEG during rTMS treatment in patients with focal epilepsy to measure changes in spike-patterns and/or rate. The examined protocol (6Hz-primed rTMS at 1 Hz) failed to show consistent results in more than one patient. Rotenberg et al. [101] suggest that the TMS-EEG combination might be used to fine-tune rTMS treatments in epilepsy, e.g., by adjusting the intensity to the TMS-EEG probed excitability of the stimulated area or by delivering TMS pulses at an appropriate timing relative to ongoing epileptiform activity. Concerning the latter "real-time" treatment approach, seizure identification triggering an anticonvulsive rTMS protocol have indeed proven to be realistic in rodents, which might pave the way for designing new tools for seizure detection triggering brain stimulation in humans [101].

Real-time detection of an EEG-signature for triggering TMS pulses was also tested with the aim to improve rTMS treatment for depression. Price et al. [102] designed an interactive rTMS method, whereby TMS pulses were applied in response to a selected pattern of background EEG analyzed in real-time. Patients treated with this interactive method showed a trend toward greater recovery than the group treated with the standard 10 Hz rTMS method.

Although these studies remain exploratory and the to-be-targeted EEG-patterns leading to the best outcome are largely unknown, such results encourage the study of EEG-guided TMS for clinical application.

5 Improving TMS Methodology

Finally, EEG measurements can provide the basis for more informed choices of TMS parameters, i.e., where, when, how much, and how long to stimulate. First, concerning the localization of stimulation, direct tracking of TMS outcome in brain space (e.g., via source estimates) could help to determine appropriate areas to stimulate when trying to disturb/enhance a brain area and/or task performance. Second, concerning the timing of stimulation, TMS could be triggered according to different time-points of a previously recorded, task-related brain response [30]. Third, EEG-guided TMS in real-time could base the triggering of stimulation upon a predefined pattern recorded in the electroencephalogram [103]. Fourth, the intensity of stimulation, traditionally based on motor threshold (MT) or phosphene threshold (PT), could be adjusted on the basis of TEP-amplitude in "silent" brain areas. Fifth, the potentiation of brain response during rTMS [47] might help to determine an appropriate duration of stimulation for reaching a clinical goal or avoiding critical stimulation safety levels. Interestingly, very low frequency rTMS (<0.2 Hz), for which no cumulative effects on MEP amplitude have been described, has shown a potentiation of the N100 amplitude, suggesting some plasticity may still occur at this low frequency [28]. Hence, the sensitivity of EEG could be used to track effects that otherwise might have been overlooked. This might be beneficial for improving TMS methodology in both neuroscientific and clinical research.

6 Concluding Remarks

The use of EEG and TMS raises numerous technical challenges, many of which can be resolved with careful consideration. When thoughtfully designed, performed, and analyzed, TMS-EEG experiments can make important contributions to the fields of cognitive and clinical neurosciences. Ultimately, TMS-EEG can contribute to questions regarding how TMS interacts with the brain to modify functions. TMS is a versatile neuroscientific tool and there is a virtually infinite space of TMS parameters that remain to be explored [104]. Combination of TMS with other neurophysiological techniques, e.g., EEG, fMRI, and PET, can provide a better idea of how TMS interacts with brain activity, and more informed choices as to effective TMS parameters. This may, in turn, aid in the development of better tools for diagnosis and rehabilitation. We believe that TMS-EEG will provide important new insights in the years to come.

References

1. Barker AT, Jalinous R, Freeston IL (1985) Non-invasive magnetic stimulation of human motor cortex. Lancet 1(8437):1106–1107
2. Walsh V, Cowey A (2000) Transcranial magnetic stimulation and cognitive neuroscience. Nat Rev Neurosci 1(1):73–79
3. Hummel FC, Cohen LG (2006) Non-invasive brain stimulation: a new strategy to improve neurorehabilitation after stroke? Lancet Neurol 5(8):708–712
4. Ridding MC, Rothwell JC (2007) Is there a future for therapeutic use of transcranial magnetic stimulation? Nat Rev Neurosci 8(7):559–567
5. Fregni F, Pascual-Leone A (2007) Technology insight: noninvasive brain stimulation in neurology-perspectives on the therapeutic potential of rTMS and tDCS. Nat Clin Pract Neurol 3(7):383–393
6. Miniussi C, Cappa SF, Cohen LG, Floel A, Fregni F, Nitsche MA et al (2008) Efficacy of repetitive transcranial magnetic stimulation/transcranial direct current stimulation in cognitive neurorehabilitation. Brain Stimul 1(4):326–336
7. Sack AT, Linden DE (2003) Combining transcranial magnetic stimulation and functional imaging in cognitive brain research: possibilities and limitations. Brain Res Brain Res Rev 43(1):41–56
8. Sack AT (2006) Transcranial magnetic stimulation, causal structure-function mapping and networks of functional relevance. Curr Opin Neurobiol 16(5):593–599
9. Bestmann S, Ruff CC, Blankenburg F, Weiskopf N, Driver J, Rothwell JC (2008) Mapping causal interregional influences with concurrent TMS-fMRI. Exp Brain Res 191(4):383–402
10. Driver J, Blankenburg F, Bestmann S, Vanduffel W, Ruff CC (2009) Concurrent brain-stimulation and neuroimaging for studies of cognition. Trends Cogn Sci 13(7):319–327
11. Ruff CC, Driver J, Bestmann S (2009) Combining TMS and fMRI: from 'virtual lesions' to functional-network accounts of cognition. Cortex 45(9):1043–1049
12. Silvanto J, Pascual-Leone A (2008) State-dependency of transcranial magnetic stimulation. Brain Topogr 21(1):1–10
13. Cracco RQ, Amassian VE, Maccabee PJ, Cracco JB (1989) Comparison of human transcallosal responses evoked by magnetic coil and electrical stimulation. Electroencephalogr Clin Neurophysiol 74(6):417–424
14. Ilmoniemi RJ, Virtanen J, Ruohonen J, Karhu J, Aronen HJ, Naatanen R et al (1997) Neuronal responses to magnetic stimulation reveal cortical reactivity and connectivity. Neuroreport 8(16):3537–3540
15. Siebner HR, Bergmann TO, Bestmann S, Massimini M, Johansen-Berg H, Mochizuki H

et al (2009) Consensus paper: combining transcranial stimulation with neuroimaging. Brain Stimul 2(2):58–80
16. Ives JR, Rotenberg A, Poma R, Thut G, Pascual-Leone A (2006) Electroencephalographic recording during transcranial magnetic stimulation in humans and animals. Clin Neurophysiol 117(8):1870–1875
17. Veniero D, Bortoletto M, Miniussi C (2009) TMS-EEG co-registration: on TMS-induced artifact. Clin Neurophysiol 120(7):1392–1399
18. Julkunen P, Paakkonen A, Hukkanen T, Kononen M, Tiihonen P, Vanhatalo S et al (2008) Efficient reduction of stimulus artefact in TMS-EEG by epithelial short-circuiting by mini-punctures. Clin Neurophysiol 119(2): 475–481
19. Litvak V, Komssi S, Scherg M, Hoechstetter K, Classen J, Zaaroor M et al (2007) Artifact correction and source analysis of early electroencephalographic responses evoked by transcranial magnetic stimulation over primary motor cortex. Neuroimage 37(1):56–70
20. Thut G, Ives JR, Kampmann F, Pastor MA, Pascual-Leone A (2005) A new device and protocol for combining TMS and online recordings of EEG and evoked potentials. J Neurosci Methods 141(2):207–217
21. Ilmoniemi RJ, Kicic D (2010) Methodology for combined TMS and EEG. Brain Topogr 22(4):233–248
22. Ilmoniemi RJ, Karhu J (2008) TMS and electroencephalography: methods and current advances. In: Wasserman EM, Epstein CM, Ziemann U, Walsh V, Paus T, Lisanby SH (eds) Oxford handbook of transcranial stimulation. Oxford University Press, Oxford, pp 593–608
23. Hamidi M, Slagter HA, Tononi G, Postle BR (2010) Brain responses evoked by high-frequency repetitive TMS: an ERP study. Brain Stimul 3(1):2–17
24. Thut G, Veniero D, Romei V, Miniussi C, Schyns P, Gross J (2011) Rhythmic TMS causes local entrainment of natural oscillatory signatures. Curr Biol 21(14):1176–1185
25. Garcia JO, Grossman ED, Srinivasan R (2011) Evoked potentials in large-scale cortical networks elicited by TMS of the visual cortex. J Neurophysiol 106(4):1734–1746
26. Paus T, Sipila PK, Strafella AP (2001) Synchronization of neuronal activity in the human primary motor cortex by transcranial magnetic stimulation: an EEG study. J Neurophysiol 86(4):1983–1990
27. Virtanen J, Ruohonen J, Naatanen R, Ilmoniemi RJ (1999) Instrumentation for the measurement of electric brain responses to transcranial magnetic stimulation. Med Biol Eng Comput 37(3):322–326
28. Bender S, Basseler K, Sebastian I, Resch F, Kammer T, Oelkers-Ax R et al (2005) Electroencephalographic response to transcranial magnetic stimulation in children: Evidence for giant inhibitory potentials. Ann Neurol 58(1):58–67
29. Morbidi F, Garulli A, Prattichizzo D, Rizzo C, Manganotti P, Rossi S (2007) Off-line removal of TMS-induced artifacts on human electroencephalography by Kalman filter. J Neurosci Methods 162(1–2):293–302
30. Thut G, Northoff G, Ives JR, Kamitani Y, Pfennig A, Kampmann F et al (2003) Effects of single-pulse transcranial magnetic stimulation (TMS) on functional brain activity: a combined event-related TMS and evoked potential study. Clin Neurophysiol 114(11):2071–2080
31. Levit-Binnun N, Litvak V, Pratt H, Moses E, Zaroor M, Peled A (2010) Differences in TMS-evoked responses between schizophrenia patients and healthy controls can be observed without a dedicated EEG system. Clin Neurophysiol 121(3):332–339
32. Roth BJ, Pascual-Leone A, Cohen LG, Hallett M (1992) The heating of metal electrodes during rapid-rate magnetic stimulation: a possible safety hazard. Electroencephalogr Clin Neurophysiol 85(2):116–123
33. Ives JR, Keenan JP, Schomer DL, Pascual-Leone A (1998) EEG recording during repetitive transcranial magnetic stimulation (rTMS). Neurology Suppl 4(50):A167
34. Tiitinen H, Virtanen J, Ilmoniemi RJ, Kamppuri J, Ollikainen M, Ruohonen J et al (1999) Separation of contamination caused by coil clicks from responses elicited by transcranial magnetic stimulation. Clin Neurophysiol 110(5):982–985
35. Kahkonen S, Kesaniemi M, Nikouline VV, Karhu J, Ollikainen M, Holi M et al (2001) Ethanol modulates cortical activity: direct evidence with combined TMS and EEG. Neuroimage 14(2):322–328
36. Kahkonen S, Komssi S, Wilenius J, Ilmoniemi RJ (2005) Prefrontal TMS produces smaller EEG responses than motor-cortex TMS: implications for rTMS treatment in depression. Psychopharmacology (Berl) 181(1):16–20
37. Komssi S, Kahkonen S, Ilmoniemi RJ (2004) The effect of stimulus intensity on brain responses evoked by transcranial magnetic stimulation. Hum Brain Mapp 21(3):154–164
38. Massimini M, Ferrarelli F, Esser SK, Riedner BA, Huber R, Murphy M et al (2007) Triggering sleep slow waves by transcranial

magnetic stimulation. Proc Natl Acad Sci U S A 104(20):8496–8501

39. Esser SK, Huber R, Massimini M, Peterson MJ, Ferrarelli F, Tononi G (2006) A direct demonstration of cortical LTP in humans: a combined TMS/EEG study. Brain Res Bull 69(1):86–94
40. Iramina K, Maeno T (2003) Measurement of evoked electroencephalography induced by transcranial magnetic stimulation. J Appl Phys 93(10):6718–6720
41. Epstein CM (1995) A simple artifact-rejection preamplifier for clinical neurophysiology. Am J EEG Technol 35:64–71
42. Bonato C, Miniussi C, Rossini PM (2006) Transcranial magnetic stimulation and cortical evoked potentials: a TMS/EEG coregistration study. Clin Neurophysiol 117(8): 1699–1707
43. Fuggetta G, Pavone EF, Walsh V, Kiss M, Eimer M (2006) Cortico-cortical interactions in spatial attention: a combined ERP/TMS study. J Neurophysiol 95(5):3277–3280
44. Fitzgerald PB, Maller JJ, Hoy K, Farzan F, Daskalakis ZJ (2009) GABA and cortical inhibition in motor and non-motor regions using combined TMS-EEG: a time analysis. Clin Neurophysiol 120(9):1706–1710
45. Taylor PC, Walsh V, Eimer M (2010) The neural signature of phosphene perception. Hum Brain Mapp 31(9):1408–1417
46. Fuggetta G, Fiaschi A, Manganotti P (2005) Modulation of cortical oscillatory activities induced by varying single-pulse transcranial magnetic stimulation intensity over the left primary motor area: a combined EEG and TMS study. Neuroimage 27(4):896–908
47. Brignani D, Manganotti P, Rossini PM, Miniussi C (2008) Modulation of cortical oscillatory activity during transcranial magnetic stimulation. Hum Brain Mapp 29(5): 603–612
48. Sekiguchi H, Takeuchi S, Kadota H, Kohno Y, Nakajima Y (2011) TMS-induced artifacts on EEG can be reduced by rearrangement of the electrode's lead wire before recording. Clin Neurophysiol 122(5):984–90
49. Nikouline V, Ruohonen J, Ilmoniemi RJ (1999) The role of the coil click in TMS assessed with simultaneous EEG. Clin Neurophysiol 110(8):1325–1328
50. Massimini M, Ferrarelli F, Huber R, Esser SK, Singh H, Tononi G (2005) Breakdown of cortical effective connectivity during sleep. Science 309(5744):2228–2232
51. Ferrarelli F, Massimini M, Peterson MJ, Riedner BA, Lazar M, Murphy MJ et al (2008) Reduced evoked gamma oscillations in the frontal cortex in schizophrenia patients: a TMS/EEG study. Am J Psychiatry 165(8): 996–1005
52. Rosanova M, Casali A, Bellina V, Resta F, Mariotti M, Massimini M (2009) Natural frequencies of human corticothalamic circuits. J Neurosci 29(24):7679–7685
53. Huber R, Maatta S, Esser SK, Sarasso S, Ferrarelli F, Watson A et al (2008) Measures of cortical plasticity after transcranial paired associative stimulation predict changes in electroencephalogram slow-wave activity during subsequent sleep. J Neurosci 28(31):7911–7918
54. Miniussi C, Thut G (2010) Combining TMS and EEG offers new prospects in cognitive neuroscience. Brain Topogr 22(4):249–256
55. Fuggetta G, Pavone EF, Fiaschi A, Manganotti P (2008) Acute modulation of cortical oscillatory activities during short trains of high-frequency repetitive transcranial magnetic stimulation of the human motor cortex: a combined EEG and TMS study. Hum Brain Mapp 29(1):1–13
56. Thielscher A, Opitz A, Windhoff M (2011) Impact of the gyral geometry on the electric field induced by transcranial magnetic stimulation. Neuroimage 54(1):234–243
57. Komssi S, Aronen HJ, Huttunen J, Kesaniemi M, Soinne L, Nikouline VV et al (2002) Ipsi- and contralateral EEG reactions to transcranial magnetic stimulation. Clin Neurophysiol 113(2):175–184
58. Reichenbach A, Whittingstall K, Thielscher A (2011) Effects of transcranial magnetic stimulation on visual evoked potentials in a visual suppression task. Neuroimage 54(2): 1375–1384
59. Van Der Werf YD, Paus T (2006) The neural response to transcranial magnetic stimulation of the human motor cortex. I. Intracortical and cortico-cortical contributions. Exp Brain Res 175(2):231–245
60. Nikulin VV, Kicic D, Kahkonen S, Ilmoniemi RJ (2003) Modulation of electroencephalographic responses to transcranial magnetic stimulation: evidence for changes in cortical excitability related to movement. Eur J Neurosci 18(5):1206–1212
61. Berg P, Scherg M (1994) A multiple source approach to the correction of eye artifacts. Electroencephalogr Clin Neurophysiol 90(3):229–241
62. Scherg M, Ille N, Bornfleth H, Berg P (2002) Advanced tools for digital EEG review: virtual source montages, whole-head mapping,

correlation, and phase analysis. J Clin Neurophysiol 19(2):91–112
63. Taylor PC, Walsh V, Eimer M (2008) Combining TMS and EEG to study cognitive function and cortico-cortico interactions. Behav Brain Res 191(2):141–147
64. Komssi S, Kahkonen S (2006) The novelty value of the combined use of electroencephalography and transcranial magnetic stimulation for neuroscience research. Brain Res Rev 52(1):183–192
65. Thut G, Miniussi C (2009) New insights into rhythmic brain activity from TMS-EEG studies. Trends Cogn Sci 13(4):182–189
66. Fitzgerald PB (2010) TMS-EEG: a technique that has come of age? Clin Neurophysiol 121(3):265–267
67. Daskalakis ZJ, Farzan F, Barr MS, Maller JJ, Chen R, Fitzgerald PB (2008) Long-interval cortical inhibition from the dorsolateral prefrontal cortex: a TMS-EEG study. Neuropsychopharmacology 33(12): 2860–2869
68. Claus D, Weis M, Jahnke U, Plewe A, Brunholzl C (1992) Corticospinal conduction studied with magnetic double stimulation in the intact human. J Neurol Sci 111(2):180–188
69. Kujirai T, Caramia MD, Rothwell JC, Day BL, Thompson PD, Ferbert A et al (1993) Corticocortical inhibition in human motor cortex. J Physiol 471:501–519
70. Ziemann U (1999) Intracortical inhibition and facilitation in the conventional paired TMS paradigm. Electroencephalogr Clin Neurophysiol Suppl 51:127–136
71. Iwahashi M, Arimatsu T, Ueno S, Iramina K (2008) Differences in evoked EEG by transcranial magnetic stimulation at various stimulus points on the head. Conf Proc IEEE Eng Med Biol Soc 2008:2570–2573
72. Massimini M, Tononi G, Huber R (2009) Slow waves, synaptic plasticity and information processing: insights from transcranial magnetic stimulation and high-density EEG experiments. Eur J Neurosci 29(9):1761–1770
73. Zanon M, Busan P, Monti F, Pizzolato G, Battaglini PP (2010) Cortical connections between dorsal and ventral visual streams in humans: evidence by TMS/EEG co-registration. Brain Topogr 22(4):307–317
74. Capotosto P, Babiloni C, Romani GL, Corbetta M (2009) Frontoparietal cortex controls spatial attention through modulation of anticipatory alpha rhythms. J Neurosci 29(18):5863–5872
75. Taylor PC, Nobre AC, Rushworth MF (2007) FEF TMS affects visual cortical activity. Cereb Cortex 17(2):391–399
76. Bonnard M, Spieser L, Meziane HB, de Graaf JB, Pailhous J (2009) Prior intention can locally tune inhibitory processes in the primary motor cortex: direct evidence from combined TMS-EEG. Eur J Neurosci 30(5): 913–923
77. Spieser L, Meziane HB, Bonnard M (2010) Cortical mechanisms underlying stretch reflex adaptation to intention: a combined EEG-TMS study. Neuroimage 52(1):316–325
78. Taylor PC, Nobre AC, Rushworth MF (2007) Subsecond changes in top down control exerted by human medial frontal cortex during conflict and action selection: a combined transcranial magnetic stimulation electroencephalography study. J Neurosci 27(42): 11343–11353
79. Van Der Werf YD, Sadikot AF, Strafella AP, Paus T (2006) The neural response to transcranial magnetic stimulation of the human motor cortex. II. Thalamocortical contributions. Exp Brain Res 175(2):246–255
80. Thut G, Schyns PG, Gross J (2011) Entrainment of perceptually relevant brain oscillations by non-invasive rhythmic stimulation of the human brain. Front Psychol 2:170
81. Klimesch W, Sauseng P, Gerloff C (2003) Enhancing cognitive performance with repetitive transcranial magnetic stimulation at human individual alpha frequency. Eur J Neurosci 17(5):1129–1133
82. Romei V, Gross J, Thut G (2010) On the role of prestimulus alpha rhythms over occipito-parietal areas in visual input regulation: correlation or causation? J Neurosci 30(25):8692–8697
83. Romei V, Driver J, Schyns PG, Thut G (2011) Rhythmic TMS over parietal cortex links distinct brain frequencies to global versus local visual processing. Curr Biol 21(4): 334–337
84. Sauseng P, Klimesch W, Heise KF, Gruber WR, Holz E, Karim AA et al (2009) Brain oscillatory substrates of visual short-term memory capacity. Curr Biol 19(21): 1846–1852
85. Veniero D, Brignani D, Thut G, Miniussi C (2011) Alpha-generation as basic response-signature to transcranial magnetic stimulation (TMS) targeting the human resting motor cortex: a TMS/EEG co-registration study. Psychophysiology 48(10):1381–1389
86. Thut G, Pascual-Leone A (2010) A review of combined TMS-EEG studies to characterize lasting effects of repetitive TMS and assess their usefulness in cognitive and clinical neuroscience. Brain Topogr 22(4):219–232
87. Maki H, Ilmoniemi RJ (2010) The relationship between peripheral and early cortical

activation induced by transcranial magnetic stimulation. Neurosci Lett 478(1):24–28

88. Romei V, Brodbeck V, Michel C, Amedi A, Pascual-Leone A, Thut G (2008) Spontaneous fluctuations in posterior alpha-band EEG activity reflect variability in excitability of human visual areas. Cereb Cortex 18(9):2010–2018
89. Romei V, Rihs T, Brodbeck V, Thut G (2008) Resting electroencephalogram alpha-power over posterior sites indexes baseline visual cortex excitability. Neuroreport 19(2):203–208
90. Zarkowski P, Shin CJ, Dang T, Russo J, Avery D (2006) EEG and the variance of motor evoked potential amplitude. Clin EEG Neurosci 37(3):247–251
91. Lepage JF, Saint-Amour D, Theoret H (2008) EEG and neuronavigated single-pulse TMS in the study of the observation/execution matching system: are both techniques measuring the same process? J Neurosci Methods 175(1):17–24
92. Sauseng P, Klimesch W, Gerloff C, Hummel FC (2009) Spontaneous locally restricted EEG alpha activity determines cortical excitability in the motor cortex. Neuropsychologia 47(1):284–288
93. Maki H, Ilmoniemi RJ (2010) EEG oscillations and magnetically evoked motor potentials reflect motor system excitability in overlapping neuronal populations. Clin Neurophysiol 121(4):492–501
94. Dugué L, Marque P, Vanrullen R (2011) The phase of ongoing oscillations mediates the causal relation between brain excitation and visual perception. J Neurosci 31(33): 11889–11893
95. Plewnia C, Rilk AJ, Soekadar SR, Arfeller C, Huber HS, Sauseng P et al (2008) Enhancement of long-range EEG coherence by synchronous bifocal transcranial magnetic stimulation. Eur J Neurosci 27(6):1577–1583
96. Kahkonen S, Wilenius J, Nikulin VV, Ollikainen M, Ilmoniemi RJ (2003) Alcohol reduces prefrontal cortical excitability in humans: a combined TMS and EEG study. Neuropsychopharmacology 28(4):747–754
97. Farzan F, Barr MS, Levinson AJ, Chen R, Wong W, Fitzgerald PB et al (2010) Evidence for gamma inhibition deficits in the dorsolateral prefrontal cortex of patients with schizophrenia. Brain 133(Pt 5):1505–1514
98. Julkunen P, Jauhiainen AM, Westeren-Punnonen S, Pirinen E, Soininen H, Kononen M et al (2008) Navigated TMS combined with EEG in mild cognitive impairment and Alzheimer's disease: a pilot study. J Neurosci Methods 172(2):270–276
99. Valentin A, Arunachalam R, Mesquita-Rodrigues A, Garcia Seoane JJ, Richardson MP, Mills KR et al (2008) Late EEG responses triggered by transcranial magnetic stimulation (TMS) in the evaluation of focal epilepsy. Epilepsia 49(3):470–480
100. Brodbeck V, Thut G, Spinelli L, Romei V, Tyrand R, Michel CM et al (2010) Effects of repetitive transcranial magnetic stimulation on spike pattern and topography in patients with focal epilepsy. Brain Topogr 22(4):267–280
101. Rotenberg A (2010) Prospects for clinical applications of transcranial magnetic stimulation and real-time EEG in epilepsy. Brain Topogr 22(4):257–266
102. Price GW, Lee JW, Garvey CA, Gibson N (2010) The use of background EEG activity to determine stimulus timing as a means of improving rTMS efficacy in the treatment of depression: a controlled comparison with standard techniques. Brain Stimul 3(3): 140–152
103. Price GW (2004) EEG-dependent ERP recording: using TMS to increase the incidence of a selected pre-stimulus pattern. Brain Res Brain Res Protoc 12(3):144–151
104. Reithler J, Peters JC, Sack AT (2011) Multimodal transcranial magnetic stimulation: using concurrent neuroimaging to reveal the neural network dynamics of noninvasive brain stimulation. Prog Neurobiol 94(2):149–165

Part V

Clinical Considerations

Chapter 12

Transcranial Magnetic Stimulation (TMS) Clinical Applications: Therapeutics

Jared Cooney Horvath, Umer Najib, and Daniel Press

Abstract

The number and type of neuropsychiatric conditions being treated by transcranial magnetic stimulation (TMS) is ever increasing. In addition, the parameters utilized to treat these conditions are ever evolving. In this chapter, we explore the evidence concerning the efficacy of TMS to treat a number of psychiatric and physiological conditions: including depression, obsessive compulsive disorder, schizophrenia, post traumatic stress disorder, chronic pain, tinnitus, stroke, and Parkinson's disease. In addition, we outline the most commonly utilized protocols and parameters to treat each of the above conditions.

Key words Transcranial magnetic stimulation (TMS), Depression, Obsessive compulsive disorder, Schizophrenia, Post traumatic stress disorder, Chronic pain, Tinnitus, Stroke, Parkinson's disease

1 Introduction

According to the Institute of Medicine's (IOM) 2001 *Healthcare Quality Initiative* report, three of the six dimensions of "perfect care" are safety, effectiveness, and efficiency [1]. Extrapolated to reflect the field of neuropsychiatric therapeutics, a *perfect care* paragon would contain mild or no health-related side-effects, high levels of symptom alleviation, and specific action with little or no dispersive effect. A cursory examination of the modern state of pharmaceutical therapies makes it clear these goals are going largely unmet: although efficacy rates are often high for many drugs, effects are usually diffuse across the entire CNS and the practice of creating pill-regimens to combat medication side-effects is ever expanding.

Using the IOM goals as a guideline, focal noninvasive brain stimulation represents a potential step in a promising direction. More specifically, repetitive transcranial magnetic stimulation (rTMS) provides clinicians with a therapeutic tool that allows for effective temporal and spatial specificity whilst avoiding many systemic side-effects. Unfortunately, despite these assets, the matter of therapeutic efficacy still remains nebulous and largely unknown.

Alexander Rotenberg et al. (eds.), *Transcranial Magnetic Stimulation*, Neuromethods, vol. 89,
DOI 10.1007/978-1-4939-0879-0_12, © Springer Science+Business Media New York 2014

In this chapter, we will explore the evidence concerning rTMS efficacy across a number of psychiatric and physiological conditions. As you read, it is important to remember two points: first—the use of rTMS for therapeutic purposes (with the exclusion of a single protocol) is off-label. As such, risks and ethical questions must be addressed accordingly. Second—proof of principle does not automatically translate to proof of clinical efficacy. Research showing a treatment *may* work is not the same as research showing a treatment *does* work. Continued exploration is encouraged and practitioners are advised to keep abreast of current research as the therapeutic landscape continues to grow and shift.

The parameters outlined for each illness below are in NO WAY meant to represent accepted or even efficacious treatment. We are including them only to help readers decipher what previous researchers have attempted so as to place results and/or typical outcomes of discussed protocols in a more comprehensive (and comprehendible) framework.

2 Three Therapeutic Paradigms

Just as clinicians may prescribe several different classes of drugs to treat the same condition across varied individuals (depending upon specific patient profiles), so too can rTMS be applied in diverse fashions to treat similar symptoms. Within the realm of therapeutics, rTMS treatment protocols generally fall under three distinct categories:

Direct Targeting: This paradigm targets the specific area of dysfunction and attempts to modulate cortical excitability (increase or decrease) to normalize abnormal levels of neural activity.

Distance Effect: This paradigm targets a neural area functionally linked to the area of dysfunction and attempts to normalize abnormal activity levels via paradoxical facilitation or inhibition.

Distributed Modulation: This paradigm attempts to induce distributed modulation of brain activity via the network-specific release of neurotransmitters which serve to normalize abnormal activity levels.

Which protocol a clinician determines appropriate depends not only upon the results of efficacy studies, but also upon individual patient reactions, tolerances, and comfort levels. Therefore, it is important practitioners not only stay abreast of emerging evidence but also develop effective screening and assessment measures to aid in the determination of proper therapeutic protocols.

3 Psychiatric Applications

3.1 Depression

As rTMS treatment for depression will be explored in great depth elsewhere in this volume, we will keep our examination rather cursory and focused on therapeutic trajectory and the current state of affairs.

Although the complete psychopathology of clinical depression is not yet fully understood, a commonality amongst many sufferers is prefrontal cortex dysfunction. Although mildly debated, evidence from several imaging and behavioral studies suggests hypoactivity of the left dorsolateral prefrontal cortex (LDLPFC) is linked to symptom severity in both primary and secondary depression ([2, 3]—for debate, see: [4]). Accordingly, the goal of most rTMS depression protocols is to modulate cortical excitability across specific neural networks to normalize any abnormal activity levels.

Most research conducted looking at the effects of TMS on depression fall under the category of Direct Targeting (recently: [5, 6]). Typical parameters for this application are:

Frequency	10–20 Hz (Excitatory)
Location	LDLPFC
Intensity	Suprathreshold
Duration	Varied

Several researchers have explored the outcomes of Distance Effect modulation (recently: [7, 8]). Typical parameters for this application are:

Frequency	1 Hz (Inhibitory)
Location	RDLPFC
Intensity	Suprathreshold
Duration	Varied

Although the distributed modulation paradigm has been explored quite vigorously in animal models, only two research groups have looked at the possibility of utilizing it in humans for the treatment of depression, with mixed-results [9, 10]. The parameters utilized were:

Frequency	10 Hz (excitatory)
Location	LDLPFC
Intensity	At threshold
Duration	750–3,500 pulses
Affected NT	Dopamine

There have been several open and sham-controlled randomized studies looking at the effects of rTMS on symptoms of depression. Arguably the two most influential to date have been O'Reardon and colleague's [11] industry sponsored study ($N=301$) and George and colleague's [12] NIH sponsored study ($N=109$). Both studies demonstrated that daily prefrontal rTMS produced a significant and clinically meaningful antidepressant effect greater than sham, and the former study led to the FDA-approved TMS depression protocol (FDA approval K061053).

A review of several meta-analyses supports the theory that rTMS can alleviate many symptoms of medication-resistant depression (Table 1). The most recent analysis [13] concluded that low frequency stimulation over RDLPFC (distance effect) showed a trend toward better response than high frequency stimulation over LDLPFC, although full statistical significance was not achieved. In addition, Slotema and colleagues determined the highest effect size was obtained from studies utilizing rTMS as a monotherapy (as opposed to an add-on to psychopharmaceutical treatment).

Table 1
A review of select meta-analyses exploring the efficacy of rTMS for the treatment of depression

Meta-analysis	*N* trials	Mean effect size	Conclusion
[39]	5	NNT=2–3*	rTMS demonstrates beneficial effects, but the extent and duration have yet to be defined
[40]	12	0.81*	Active stimulation shows a significant difference compared to sham. However, the clinical significance is modest
[41]	16	0.67*	High frequency stimulation shows a significant difference compared to sham
[42]	13	SMD=−0.35*	High frequency stimulation shows a significant difference compared to sham following 2 weeks of treatment. However, effect is nonsignificant at 2-week follow-up
[43]	6	WMD=−1.1	High frequency rTMS shows no difference from sham stimulation. However, greater power is needed for conclusive results
[44]	33	0.71*	Active stimulation shows a significant difference compared to sham, though variability is high
[45]	30	0.39*	High frequency stimulation shows a significant difference compared to sham
[13]	34	0.55*	Active stimulation shows a significant difference compared to sham. Low frequency rTMS trends toward better response compared to high frequency

WMD weighted mean difference, *SMD* standardized mean difference, *NNT* number needed to treat
* Asterisk denotes a significant difference between stimulation and control conditions

These data suggest that rTMS is, indeed, an efficacious tool in the treatment of medication-resistant depression. Although further clinical data is needed, the prognosis seems favorable.

3.2 Obsessive Compulsive Disorder

Obsessive compulsive disorder (OCD) is marked by intrusive and recurrent thoughts that lead to repetitive behaviors that are ego dystonic and cause significant distress to the sufferer and those related [14]. Current models (based on animal studies, imaging studies, and associative symptomology) emphasize the importance of abnormal cortico-striatal-thalamo-cortical (CSTC) circuitry in OCD pathophysiology (for review: [15, 16]). Due to this extended functional network dynamic theory, all studies looking at the effect of rTMS on OCD fall under the Distance Effect paradigm. More specifically, the orbitofrontal cortex (which is accessible to the TMS pulse and has been singled-out in imaging studies and stereotactic lesioning procedures as a region important to OCD dynamics) is often targeted with the hopes of modulating the entire dysfunctional CSTC network.

To date, studies exploring the therapeutic value of rTMS in OCD populations have utilized widely varying parameters and have reported widely varying results (Table 2). To compound, replication attempts of any single protocol have yet to enter the literature. As such, any conclusions we attempt to draw would be cursory and largely unsubstantiated. Perhaps the only thing that is safe to say is that more research in this area is needed to address the contradictory claims thus far made and determine whether OCD is an illness amenable to the tool of TMS.

3.3 Schizophrenia

Schizophrenia is a debilitating illness that's treatment has proven rather resistant to treatment (at least holistically). The symptomology of schizophrenia is typically divided into two categories: positive symptoms (auditory hallucinations, delusions, etc.) and negative symptoms (blunted affect, anhedonia, etc. [17]).

Typical pathophysiological models make reference to patient cortical excitability and inhibition abnormalities. TMS motor-evoked potential (MEP) and pre-pulse inhibition studies reveal an inability for sufferers of positive symptoms to suppress responses to inputs, a state known as hyperexcitability [18–20]. Intuitively, sufferers of negative symptoms often show reduced activation in frontal and deep brain regions, a state known as hypoactivation [21, 22]. As rTMS has been linked to LTP- and LTD-like cortical modulation, many researchers have explored its therapeutic potential across several schizophrenic symptoms.

Several studies have been conducted recently looking at the effects of rTMS on auditory verbal hallucinations (AVH). As AVH is a positive symptom and is, therefore, linked to cortical

Table 2
A review of select studies exploring the efficacy of rTMS for the treatment of obsessive compulsive disorder (parameters and results)

Study	Stimulation details	Conclusion	N
[46]	Frequency: 20 Hz (excitatory) Location: RLPFC/LLPFC Intensity: Subthreshold Duration: 800/1 session	Significant decrease in compulsive urges for 8 h after RLPFC stimulation (compared to LLPFC and sham)	12 (crossover)
[47]	Frequency: 1 Hz (inhibitory) Location: RDLPFC Intensity: Suprathreshold Duration: 1,200/18 sessions	No significant effect (compared to sham)	10 Active 8 Sham
[48]	Frequency: 10 Hz (excitatory) Location: LDLPFC/RDLPFC Intensity: Suprathreshold Duration: 1,500/10 sessions	Significant improvement in obsessions and compulsions from both conditions (open trial)	6 LDLFPFC 6 RDLPFC
[49]	Frequency: 1 Hz (inhibitory) Location: LDLPFC Intensity: Suprathreshold Duration: 1,800/10 sessions	No significant effect (compared to sham)	33
[50]	Frequency: 1 Hz (inhibitory) Location: SMA Intensity: At threshold Duration: 1,200/10 sessions	Significant improvement in overall symptomology (open study)	10
[51]	Frequency: 10 Hz (excitatory) Location: LDLPFC Intensity: Suprathreshold Duration: 1,500/10 sessions	No significant effect (compared to sham)	10 Active 8 Sham
[52]	Frequency: 1 Hz (inhibitory) Location: RDLPFC—SMA Intensity: Suprathreshold—at threshold Duration: 1,200–1,200/10 sessions	No significant effect (compared to sham)	10 Active 10 Sham
[53]	Frequency: 10 Hz (excitatory) Location: RDLPFC Intensity: Suprathreshold Duration: 800/10 sessions	No significant effect (compared to sham)	21 Active 21 Sham
[54]	Frequency: 1 Hz (inhibitory) Location: SMA (bilateral) Intensity: At threshold Duration: 1,200/4–8 weeks	Nonsignificant trend towards better response in active stimulation (as compared to sham)	9 Active 9 Sham
[55]	Frequency: 10 Hz (excitatory) Location: rDLPFC Intensity: Suprathreshold Duration: 2,000/6 weeks	No significant effect (compared to sham)	15 Active 15 Sham

hyperexcitability, the majority of protocols have utilized low frequency suppressive TMS over the left temporoparietal cortex (Direct Targeting). Typical parameters are:

Frequency	1 Hz (Inhibitory)
Location	LTPC
Intensity	Subthreshold
Duration	Varied

A review of several meta-analyses suggests rTMS has a moderate effect upon the severity of AVH (Table 3). Although more evidence is needed, the lack of efficacious treatment options for schizophrenic patients suffering from AVH makes rTMS a viable consideration.

Although apparently effective in alleviating AVH symptoms, low frequency stimulation over LTPC does not appear to have any statistical affect on other positive symptoms associated with schizophrenia (Table 4). Although Freitas and colleagues were able to find a significant effect in their meta-analysis for AVH, when all non-sham studies were eliminated from analysis, all therapeutic effects of TMS on other positive symptoms disappeared. This alludes to the possibility that placebo played a role in positive symptom alleviation and, as such, inhibitory stimulation of temporoparietal cortex does not seem to be an effective treatment for symptoms beyond AVH. Additional research exploring the stimulation of different cortical areas using different stimulatory paradigms is recommended (perhaps distance effect or distributed modulation).

In addition to positive symptoms, there have been a number of studies exploring the effects of rTMS on negative symptomology.

Table 3
A review of select meta-analyses exploring the efficacy of suppressive rTMS over the left temporoparietal cortex for the treatment of auditory verbal hallucinations in schizophrenia

Meta-analysis	*N* trials	Mean effect size	Conclusion
[56]	10	0.76*	rTMS appears to be an efficacious treatment for auditory hallucinations
[57]	10	0.51*	rTMS may prove useful as a complementary treatment to schizophrenia and AVH
[23]	9	1.04*	Larger control trials are warranted
[13]	7	0.54*	TMS shows promise and might prove an efficacious clinical tool

* Asterisk denotes a significant difference between stimulation and control conditions

Table 4
A review of select meta-analyses exploring the efficacy of suppressive rTMS over the left temporoparietal cortex for the treatment of positive symptoms associated with schizophrenia

Meta-analysis	*N* trials	Mean effect size	Conclusion
[56]	10	0.21	rTMS does not appear to be an efficacious treatment for positive symptoms (beyond auditory hallucinations)
[23]	12	Controlled and uncontrolled = 0.50* Controlled only = 0.17	More controlled studies need to be undertaken to determine true effect

* Asterisk denotes a significant difference between stimulation and control conditions

Table 5
A review of select meta-analyses exploring the efficacy of excitatory rTMS over the left dorsolateral prefrontal cortex for the treatment of the negative symptoms of schizophrenia

Meta-analysis	*N* trials	Mean effect size	Conclusion
[23]	8	Controlled and Uncontrolled = 0.49* Controlled Only = 0.21	More controlled studies need to be undertaken to determine true effect
[13]	7	0.39	It is uncertain if TMS would prove efficacious in the treatment of negative symptoms
[58]	9	0.43*	TMS could prove efficacious in the treatment of negative symptoms

* Asterisk denotes a significant difference between stimulation and control conditions

As negative symptoms have been linked to hypoactivation in frontal regions, many researchers have explored the effects of excitatory stimulation to DLPFC (Direct Effect). Typical parameters are:

Frequency	10 Hz (Excitatory)
Location	LDLPFC
Intensity	Suprathreshold
Duration	Varied

Meta-analysis data from this type of protocol is mixed (Table 5). Although statistically significant effects were found in two meta-analyses, the consideration of placebo effects eliminated the significant results of one [23]. As such, it is too early to make any informed determination on the effects of excitatory rTMS to frontal regions on the negative symptomology of schizophrenia. More research is needed to parse out true effect from placebo.

3.4 Post Traumatic Stress Disorder

As classified by the Diagnostic and Statistical Manual of Mental Disorders, Fourth Edition (DSM-IV), post traumatic stress disorder (PTSD) is an anxiety disorder characterized by the development of intrusive thoughts, hyperarousal, flashback, nightmares, sleep disturbances, emotional numbing, withdrawal, and other clinical symptoms after exposure to a frightening ordeal involving physical harm or the threat of physical harm. This disorder, which affects an estimated 7.7 million American adults (NIH), may cause considerable interference in sufferers' occupational, social, and interpersonal functioning.

Neural models emphasize the role of the limbic and paralimbic dysfunction in this disorder. Although an exact pathogenesis is not yet known (and probably varies across patients), recent imaging studies have implicated faulty communication between the amygdale (threat assessment), the hippocampus (contextual information encoding), and the anterior cingulated gyrus (fear extinction) in the formation of PTSD symptomology (for review: [24]).

As with OCD, because rTMS can only reach superficial neural layers and cannot selectively stimulate deep brain structures, all PTSD studies have attempted to modulate network dynamic via a Distance Effect paradigm. A review of the, admittedly, sparse evidence suggests high frequency rTMS over the RDLPFC may generate a potential therapeutic benefit for several symptoms of PTSD (Table 6). Further research in this domain is needed and arguably warranted.

Table 6
A review of select studies exploring the efficacy of rTMS for the treatment of post traumatic stress disorder (parameters and results)

Study	Stimulation details	Conclusion	*N*
[59]	Frequency: 0.3 Hz (inhibitory) Location: Motor cortex (bilateral) Intensity: 100 % Machine capacity Duration: 15 per side/30 total	Transient symptom improvement (open trial)	10
[60]	Frequency: 1 Hz (inhibitory) Location: RDLPFC Intensity: Subthreshold Duration: 1,200/17 or 30 sessions	Significant decrease of PTSD symptom severity from both conditions (open trial—only one patient per condition)	2
[61]	Frequency: 1 or 5 Hz (In/Ex) Location: LDLPFC Intensity: Subthreshold Duration: 600/10 sessions	Comorbid depression symptoms improved, PTSD symptomology improved only minimally (open trial)	12
[62]	Frequency: 1 or 10 Hz (In/Ex) Location: RDLPFC Intensity: subthreshold Duration: 100 or 400/10 sessions	Significant decrease of PTSD symptom severity from high frequency condition, but not from low (as compared to sham)	8 Low frequency 10 High frequency 6 Sham

(continued)

Table 6 (continued)

Study	Stimulation details	Conclusion	N
[63]	Frequency: 1 Hz (inhibitory) Location: RDLPFC Intensity: At threshold Duration: 1,800/30 sessions	No significant effects (as compared to sham)	9 (crossover)
[64]	Frequency: 20 Hz (excitatory) Location: LDLPFC/RDLPFC Intensity: Subthreshold Duration: 1,600/10 sessions	Significant decrease of PTSD symptom severity from both active conditions (as compared to sham). RDLPFC generated larger effects than LDLPFC	10 LDLPFC 10 RDLPFC 10 Sham

4 Physiological Applications

4.1 Chronic Pain

Recent imaging studies suggest a diverse network of neural regions is involved in chronic pain—including nuclei in the thalamus, somatosensory cortex, and limbic system [25, 26]. As imaging has revealed extensive projections between the motor cortex and specific thalamic nuclei [27, 28], many TMS studies have targeted M1 to explore the mechanisms of pain and therapeutic effects of rTMS (distance effect/distributed modulation). Several PET studies have revealed this to be a solid decision as motor cortex stimulation appears to increase regional cerebral flow in several neural structures related not only to the sensation of pain but also the emotional valence of pain (such as the insula, cingulate gyrus, and areas of the upper brain stem—[29–31]).

Although several early studies explored inhibitory rTMS protocols, many recent studies have switched to excitatory protocols in an attempt to mimic efficacious implanted motor cortex stimulation devices (which typically range from 20 to 55 Hz). Results thus far are encouraging, however questions of duration and sustainability still linger (Table 7). It appears single rTMS sessions can result in analgesic effects, but what is the feasibility of this type of therapy for chronic, long-term relief? This issue is paramount when considering the future of TMS therapy with this disorder.

4.2 Parkinson's Disease: Motor Function

Motor symptoms of Parkinson's disease (PD) (tremors, rigidity, postural instability, bradykinesia, etc.) are believed to arise due to insufficient levels of dopamine within the midbrain [32]. Although both subthalamic nucleus (STN) and globus pallidus internus (GPI) deep brain stimulation (DBS) have proven effective in the alleviation of many of these symptoms (for review: [33, 34]), the high cost and invasive nature of this treatment has led many

Table 7
A review of select studies exploring the efficacy of rTMS for the treatment of chronic pain (parameters and results)

Study	Stimulation details	Conclusion	*N*
[65]	Frequency: 0.5 or 10 Hz (In/Ex) Location: M1 Intensity: Subthreshold Duration: 600 or 1,000/1 session	Significant decrease in pain after 10 Hz but not after 0.5 Hz (as compared to sham)	18
[66]	Frequency: 10 Hz (excitatory) Location: M1 Intensity: Subthreshold Duration: 1,000/1 session	Significant decrease in pain (as compared to sham)	14
[67]	Frequency: 20 Hz (excitatory) Location: M1 Intensity: Subthreshold Duration: 800/1 session	No significant effects (as compared to sham)	12
[68]	Frequency: 10 Hz (excitatory) Location: M1 Intensity: Suprathreshold Duration: 120/1 session	Significant decrease in pain (as compared to sham)	10
[69]	Frequency: 10 Hz (excitatory) Location: M1 Intensity: Subthreshold Duration: 1,000/1 session	Significant decrease in chronic pain (ascompared to sham) as influenced by pain origin and pain site	60
[70]	Frequency: 20 Hz (excitatory) Location: M1 Intensity: Subthreshold Duration: 2,000/5 sessions	Significant decrease in pain (as compared to sham)	48
[71]	Frequency: 1 or 20 Hz (In/Ex) Location: M1 Intensity: Subthreshold Duration: 1,600/1 session	Significant decrease in pain after 20 Hz but not after 1 Hz (as compared to sham)	14
[72]	Frequency: 1 or 5 Hz (In/Ex) Location: M1 Intensity: Subthreshold Duration: 500/5 sessions	Trending, but no significant decrease in pain for either condition (at 0.05)	27
[73]	Frequency: 10 Hz (Ex) Location: M1 Intensity: Subthreshold Duration: 2,000/3 real, 3 sham	TMS was associated with significant improvements in average daily pain in 3 of 4 participants	4
[74]	Frequency: 10 Hz (Ex) Location: M1 Intensity: Subthreshold Duration: 1,500/14 sessions over 25 weeks	Active TMS significantly reduced pain intensity (compared to sham) for a period of 25 weeks	16 Active 14 Sham
[75]	Frequency: 10 Hz (Ex) with iTBS prime Location: M1 Intensity: Subthreshold Duration: 1,000/daily sessions	Analgesic effects of 10 Hz stimulation is enhanced with iTBS priming	14

researchers to search for stimulatory alternatives. Due to the peripheral nature of TMS, research utilizing this technique must rely upon either distance effects or distributed modulation to obtain results.

Although, to date, the majority of studies looking at the effects of TMS on PD motor symptoms have targeted the M1 region, the variation amongst stimulation intensities and durations has made it incredibly difficult to generate any generalizations regarding efficacy (Table 8). Fregni and colleagues [35] undertook a meta-analysis which revealed a significant (albeit modest—0.62) therapeutic effect of TMS over M1. However, the authors themselves cautioned that the heterogeneous data was difficult to synthesize and questions still abound. Perhaps the only solid conclusion researchers have been able to reach is that stimulation over the

Table 8
A review of select studies exploring the efficacy of rTMS for the treatment of motor function in Parkinson's disease (parameters and results)

Study	Stimulation details	Conclusion	*N*
[76]	Frequency: 5 Hz (excitatory) Location: M1 Intensity: Subthreshold Duration: 2,250/5 sessions	Significant decrease in global UPDRS score (as compared to sham). Large effects in contralateral limb	10
[77]	Frequency: 0.2 Hz (inhibitory) Location: M1 Intensity: Suprathreshold Duration: 100/1 session	No significant effects (as compared to sham)	85
[78]	Frequency: 5 Hz (excitatory) Location: M1 Intensity: Suprathreshold Duration: 2,000/10 sessions	Significant decrease in global UPDRS score (as compared to sham) Large effects in contralateral limb	36
[79]	Frequency: 0.5 or 10 Hz (In/Ex) Location: M1 Intensity: Subthreshold Duration: 600 or 2,000/1 session	Both groups showed significant improvements in motor function (as compared to sham)	12
[80]	Frequency: 15 Hz (excitatory) Location: LDLPFC Intensity: Suprathreshold Duration: 200/10 sessions	Trend towards stability in treatment group and increased UPDRS scores in sham group, but not significant (at 0.05)	42
[81]	Frequency: 15 Hz (excitatory) Location: LDLPFC Intensity: Suprathreshold Duration: 200/10 sessions	No significant effects (as compared to fluoxetine medication treatment)	25

(continued)

Table 8 (continued)

Study	Stimulation details	Conclusion	N
[82]	Frequency: 25 Hz (excitatory) Location: Bilateral DLPFC and bilateral M1 Intensity: At threshold Duration: 300 (per site)/8 sessions	Motor function times gradually decreased (as compared to sham)	18
[83]	Frequency: 10 Hz (excitatory) Location: DLPFC Intensity: Subthreshold Duration: 450/10 sessions	No significant effects on UPDRS global scores (as compared to sham)	13
[84]	Frequency: 1 Hz (inhibitory) Location: Vertex Intensity: Subthreshold Duration: 100/10 sessions	Vertex stimulation produced no therapeutic value and should be abandoned (recommendation of Arias et al.)	9 Active 9 Sham
[85]	Frequency: 1 Hz (inhibitory) Location: M1 Intensity: Subthreshold Duration: 1,800/4 sessions	No significant differences between active and sham stimulation were reported	10
[86]	Frequency: 1 Hz (inhibitory) Location: Cerebellum Intensity: Subthreshold Duration: 600/1 session	rTMS improved upper limb gross motor skills but reduced fine motor skills	20
[87]	Frequency: iTBS (excitatory) Location: M1 and DLPFC Intensity: Subthreshold Duration: 600/8 sessions	Although mood improved, iTBS generated no significant effects on motor performance	13 Active 13 Sham
[88]	Frequency: 25 Hz (excitatory) Location: bilateral M1 Intensity: Subthreshold Duration: 1,000/15 sessions	rTMS improved bradykinesia as measured via simple and complex taping tasks	10

DLPFC appears to have no effect on PD motor symptoms, regardless of intensity. It might be advisable to focus future research on replication rather than additional exploration.

4.3 Stroke Rehabilitation: Motor Dysfunction

Kolominsky-Rabas and colleagues [36] state that over 60 % of stroke survivors suffer from some form of pronounced and persistent deficit. The majority of these deficits are motor in nature. Despite decades of research, scientists and clinicians have not yet devised a consistent method whereby full motor recovery after stroke can be achieved.

With regards to TMS, arguably the most explored therapeutic approach involves increasing excitability within the ipsilesional motor cortex. The direct targeting influence of high frequency

Table 9
A review of select studies exploring the efficacy of ipsilateral rTMS for the treatment of motor function rehabilitation following stroke (parameters and results)

Study	Stimulation details	Conclusion	*N*
[89]	Frequency: 3 Hz (excitatory) Location: Ipsilesional M1 Intensity: At- or suprathreshold Duration: 300/10 sessions	Significant increase in physical therapy augmented motor function (as compared to sham)	52
[90]	Frequency: 10 Hz (excitatory) Location: Ipsilesional M1 Intensity: Subthreshold Duration: 160/2 sessions	Significant increase in MEP amplitude and motor function (as compared to sham)	15
[91]	Frequency: 3 Hz (excitatory) Location: Ipsilesional M1 Intensity: Suprathreshold Duration: 900/5 sessions	Significant increase in motor function (as compared to sham) Less pronounced than 1 Hz contralateral	24
[92]	Frequency: 5 Hz (excitatory) Location: Ipsilesional M1 Intensity: Subthreshold Duration: 750/10 sessions	Significant increase in motor function (as compared to sham)	40
[93]	Frequency: iTBS (excitatory) Location: Ipsilesional M1 Intensity: Subthreshold Duration: 600 followed by 10 days of rehab	There were no differences between the active and sham group	10 Active 10 Sham

stimulation (3–10 Hz) over the affected M1 has proven effective not only in increasing MEP size within the paretic hand, but also increasing motor function when used in tandem with a movement-based physical therapy (Table 9). Although questions still remain regarding the sustainability of these effects, initial data certainly looks promising.

A secondary paradigm employed by researchers examining motor recovery after stroke involves decreasing the excitability of the contralesional motor cortex. Due to interhemispheric inhibition, downregulation of one M1 region has been shown to generate a paradoxical excitation of the opposing M1 region in healthy subjects [37]. Exploiting these distance effects, many studies have utilized low frequency (1 Hz) stimulation over the healthy motor cortex. Again, this paradigm has proven effective in increasing motor function and control in the paretic hand (Table 10). These findings not only allude to the potential benefit of TMS as a therapeutic tool in the recovery of motor function after stroke, but also to the power of paradoxical facilitation in the human cortex. This ability to utilize distance effects to effect change may prove beneficial in conditions where direct targeting has been made difficult or impossible due to occlusion.

Table 10
A review of select studies exploring the efficacy of contralateral rTMS for the treatment of motor function rehabilitation following stroke (parameters and results)

Study	Stimulation details	Conclusion	*N*
[94]	Frequency: 1 Hz (inhibitory) Location: Contralesional M1 Intensity: Subthreshold Duration: 1,200/1 session	Transient improvement in dexterity but not grip strength (as compared to sham)	12 crossover)
[95]	Frequency: 1 Hz (inhibitory) Location: Contralesional M1 Intensity: At threshold Duration: 600/1 session	Significant improvement in motor function (as compared to sham)	15 (crossover)
[96]	Frequency: 1 Hz (inhibitory) Location: Contralesional M1 Intensity: subthreshold Duration: 100/10 sessions	Decreased spasticity, improvement in motor function from subjects with intact motor pathways only	64 (>5 years post-stroke)
[97]	Frequency: 1 Hz (inhibitory) Location: Contralesional M1 Intensity: Subthreshold Duration: 1,500/1 session	Significant improvement in motor learning and function (as compared to sham)	20
[91]	Frequency: 1 Hz (inhibitory) Location: Contralesional M1 Intensity: At threshold Duration: 900/5 sessions	Significant improvement in motor function (as compared to sham) More pronounced than 3 Hz ipsilateral	24
[92]	Frequency: 1 Hz (inhibitory) Location: Contralesional M1 Intensity: Suprathreshold Duration: 150/10 sessions	Significant increase in motor function (as compared to sham)	40
[98]	Frequency: 6 Hz (excitatory) prime with 1 Hz (Inhibitory) treatment Location: Contralesional M1 Intensity: Subthreshold Duration: 1,600 total/22 sessions	Positive results reported for upper limb hemiparesis following 15 days of stimulation and occupational therapy	11
[93]	Frequency: cTBS (inhibitory) Location: Contralesional M1 Intensity: Subthreshold Duration: 600 followed by 10 days of rehabilitation	There were no differences between the active and sham group	10 Active 10 Sham
[99]	Frequency: 1 Hz (inhibitory) Location: Contralesional M1 Intensity: Subthreshold Duration: 1,200/22 sessions	Positive results reported for upper limb hemiparesis following 15 days of stimulation and occupational therapy	204

4.4 Tinnitus

Tinnitus, the subjective perception of simple acoustical features in the absence of a physical source, has long been assumed to occur due to specific hearing damage which results in hyperexcitability of the auditory cortex. This hyperexcitability is reflected in increased spontaneous neuronal firing and synchronized firing across all levels of the auditory system [38]. The ability of TMS to inhibit cortical excitability has led many researchers to explore the direct therapeutic potential of low frequency stimulation over the temporal cortex.

Despite having the appearance of an ideal tool for the alleviation of the underlying cause of tinnitus, TMS data collected to date has been ambivalent (Table 11). A number of studies have reported significant results from low frequency stimulation, but the size, duration, and location of the benefit fluctuates between studies. In addition, several studies have attempted to generate distance effects via high frequency stimulation over the contralateral auditory cortex. Again, results from these studies vary widely and do not point to any clear benefit.

Table 11
A review of select studies exploring the efficacy of both low and high frequency rTMS for the treatment tinnitus (parameters and results)

Study	Stimulation details	Conclusion	N
Low frequency			
[100]	Frequency: 1 Hz (inhibitory) Location: Varied Intensity: Suprathreshold Duration: 2,000/5 sessions	Significant reduction of tinnitus (as compared to sham)	12
[101]	Frequency: 1 Hz (inhibitory) Location: Varied Intensity: Suprathreshold Duration: 1,200/1 session	8 patients reported significant reduction of tinnitus complaints	13
[102]	Frequency: 1 Hz (inhibitory) Location: LTC Intensity: Suprathreshold Duration: 2,000/10 sessions	Significant reduction of tinnitus complaints	28
[103]	Frequency: 1 Hz (inhibitory) Location: Varied Intensity: Suprathreshold Duration: 300, 900, 1,800/1 session	Only moderate reduction of tinnitus	9
[104]	Frequency: 1 Hz (inhibitory) Location: LTC Intensity: Suprathreshold Duration: 2,000/10 sessions	Significant reduction in tinnitus complaints from patients with recent-onset symptoms and normal hearing	45

(continued)

Table 11 (continued)

Study	Stimulation details	Conclusion	*N*
[105]	Frequency: 1 Hz (inhibitory) Location: LTC Intensity: Suprathreshold Duration: 2,000/10 sessions	Significant reduction in left-sided and bilateral tinnitus, but no significant effect for right sided tinnitus	194
[106]	Frequency: 1 Hz (inhibitory) Location: Varied Intensity: At Threshold Duration: 2,000/10 sessions	Significant reduction of tinnitus on contralateral side (as opposed to ipsilateral side)	31
[107]	Frequency: 1 Hz (inhibitory) Location: Varied Intensity: Suprathreshold Duration: 2,000/10 sessions	Significant reduction of tinnitus severity from both active and sham stimulation—longer duration from active group	52
[108]	Frequency: 1 Hz (inhibitory) Location: LTC Intensity: Suprathreshold Duration: 1,020/5 sessions	Significant reduction of tinnitus (as compared to sham)	20
High frequency			
[109]	Frequency: 10 Hz (excitatory) Location: Varied Intensity: Suprathreshold Duration: 450/10 sessions	Stimulation over LTC showed significant reduction of tinnitus – stimulation elsewhere had no effect	14
[110]	Frequency: 1–20 Hz (In/Ex) Location: Auditory cortex Intensity: Subthreshold Duration: 200/1 session	Short duration suppression across all conditions. Tinnitus duration negatively correlated with optimal TMS frequency	114
[111]	Frequency: 10 Hz (excitatory) Location: LTC Intensity: Subthreshold Duration: 30/1 session	Significant reduction of tinnitus (as compared to sham)	7
[101]	Frequency: 10 Hz (excitatory) Location: Varied Intensity: Suprathreshold Duration: 30/1 session	No significant effects	13
[112]	Frequency: 25 Hz (excitatory) Location: Varied Intensity: Subthreshold Duration: 2,000/10 sessions	Significant reduction of tinnitus on contralateral side (as opposed to ipsilateral side)	31

5 Conclusion

It is clear that, despite growing popularity and several instances of successfully derived evidence, the future of TMS as a clinical tool is rather ambivalent. Looking at the above tables, one almost feels an understandable fever-pitch excitement over the promises and application of this novel tool. However, after attempting to alleviate a plethora of disorders and ailments, it is clear TMS is not the panacea some may have hoped it would be. As any new device suffers its growing pains, it seems as though TMS as a therapeutic device may finally be in a mature position to narrow parametric experimentation to those few disorders amenable to stimulatory therapy (namely: depression, motor disorders, and rehabilitation). As we push forward into the next phase of TMS application, clinicians and researchers alike would do well to reflect upon possible therapeutic mechanisms of action (direct, distance, or distributed modulation) and develop appropriate therapies according to the best potential paradigmatic mode.

References

1. Institute of Medicine (2001) Crossing the quality chasm: a new health system for the 21st century. Retrieved from: http://www.iom.edu/Global/News%20Announcements/Crossing-the-Quality-Chasm-The-IOM-Health-Care-Quality-Initiative.aspx
2. Phillips ML, Drevets WC, Rauch SL, Lane R (2003) Neurobiology of emotion perception II: Implications for major psychiatric disorders. Biol Psychiatry 54:515–528
3. Grimm S, Beck J, Schuepbach D, Hell D, Boesiger P, Bermpohl F, Niehaus L, Boeker H, Northoff G (2008) Imbalance between left and right dorsolateral prefrontal cortex in major depression is linked to negative emotional judgment: an fMRI study in severe major depressive disorder. Biol Psychiatry 63(4):369–376
4. Fitzgerald PB, Oxley TJ, Laird AR, Kulkarni J, Egan GF, Daskalakis ZJ (2006) An analysis of functional neuroimaging studies of dorsolateral prefrontal cortical activity in depression. Psychiatry Res 148(1):33–45
5. Charnsil C, Suttajit S, Boonyanaruthee V, Leelarphat S (2012) An open-label study of adjunctive repetitive transcranial magnetic stimulation (rTMS) for partial remission in major depressive disorder. Int J Psychiatry Clin Pract 16(2):98–102
6. Galletly C, Gill S, Clarke P, Burton C, Fitzgerald PB (2012) A randomized trial comparing repetitive transcranial magnetic stimulation given 3 days/week and 5 days/week for the treatment of major depression: is efficacy related to the duration of treatment or the number of treatments? Psychol Med 42(5):981–988
7. Fitzgerald PB, Hoy KE, Herring SE, McQueen S, Peachey AV, Segrave RA, Maller J, Hall P, Daskalakis ZJ (2012) A double blind randomized trial of unilateral left and bilateral prefrontal cortex transcranial magnetic stimulation in treatment resistant major depression. J Affect Disord 139(2):193–198
8. Eche J, Mondino M, Haesebaert F, Saoud M, Poulet E, Brunelin J (2012) Low- vs high-frequency repetitive transcranial magnetic stimulation as an add-on treatment for refractory depression. Front Psychiatry 3:13
9. Kuroda Y, Motohashi N, Ito H, Ito S, Takano A, Takahashi H, Nishikawa T, Suhara T (2010) Chronic repetitive transcranial magnetic stimulation failed to change dopamine synthesis rate: preliminary L-[β-11C]DOPA positron emission tomography study in patients with depression. Psychiatry Clin Neurosci 64(6):659–662
10. Pogarell O, Koch W, Pöpperl G, Tatsch K, Jakob F, Zwanzger P, Mulert C, Rupprecht R, Möller HJ, Hegerl U, Padberg F (2006) Striatal dopamine release after prefrontal repetitive transcranial magnetic stimulation in

major depression: preliminary results of a dynamic [123I] IBZM SPECT study. J Psychiatr Res 40(4):307–314

11. O'Reardon JP, Solvason HB, Janicak PG, Sampson S, Isenberg KE, Nahas Z, McDonald WM, Avery D, Fitzgerald PB, Loo C, Demitrack MA, George MS, Sackeim HA (2007) Efficacy and safety of transcranial magnetic stimulation in the acute treatment of major depression: a multisite randomized controlled trial. Biol Psychiatry 62(11):1208–1216
12. George MS, Lisanby SH, Avery D, McDonald WM, Durkalski V, Pavlicova M, Anderson B, Nahas Z, Bulow P, Zarkowski P, Holtzheimer PE, Schwartz T, Sackeim HA (2010) Daily left prefrontal transcranial magnetic stimulation therapy for major depressive disorder: a sham-controlled randomized trial. Arch Gen Psychiatry 67(5):507–516
13. Slotema CW, Blom JD, Hoek HW, Sommer IE (2010) Should we expand the toolbox of psychiatric treatment methods to include repetitive transcranial magnetic stimulation (rTMS)? A meta-analysis of the efficacy of rTMS in psychiatric disorders. J Clin Psychiatry 71(7):873–884
14. Boileau B (2011) A review of obsessive-compulsive disorder in children and adolescents. Dialogues Clin Neurosci 13(4):401–411
15. Saxena S, Rauch SL (2000) Functional neuroimaging and the neuroanatomy of obsessive–compulsive disorder. Psychiatr Clin North Am 23:563–586
16. Ting JT, Feng G (2011) Neurobiology of obsessive-compulsive disorder: insights into neural circuitry dysfunction through mouse genetics. Curr Opin Neurobiol 21(6):842–848
17. Sims A (2002) Symptoms in the Mind: An Introduction to Descriptive Psychopathology. W. B. Saunders, Philadelphia
18. Puri BK, Davey NJ, Ellaway PH, Lewis SW (1996) An investigation of motor function in schizophrenia using transcranial magnetic stimulation of the motor cortex. Br J Psychiatry 169(6):690–695
19. Fitzgerald PB, Brown TL, Daskalakis ZJ, Kulkarni J (2002) A transcranial magnetic stimulation study of inhibitory deficits in the motor cortex in patients with schizophrenia. Psychiatry Res 114(1):11–22
20. Daskalakis ZJ, Christensen BK, Chen R, Fitzgerald PB, Zipursky RB, Kapur S (2002) Evidence for impaired cortical inhibition in schizophrenia using transcranial magnetic stimulation. Arch Gen Psychiatry 59(4): 347–354
21. Reid PD, Daniels B, Rybak M, Turnier-Shea Y, Pridmore S (2002) Cortical excitability of psychiatric disorders: reduced post-exercise facilitation in depression compared to schizophrenia and controls. Aust N Z J Psychiatry 36(5):669–673
22. Wobrock T, Kadovic D, Falkai P (2007) Cortical excitability in schizophrenia: studies using transcranial magnetic stimulation. Nervenarzt 78(7):753–763
23. Freitas C, Fregni F, Pascual-Leone A (2009) Meta-analysis of the effects of repetitive transcranial magnetic stimulation (rTMS) on negative and positive symptoms in schizophrenia. Schizophr Res 108(1–3):11–24
24. Rauch SL, Shin LM, Phelps EA (2006) Neurocircuitry models of posttraumatic stress disorder and extinction: human neuroimaging research—past, present, and future. Biol Psychiatry 60(4):376–382
25. Zippo AG, Storchi RM, Valente M, Caramenti GC, Biella GE (2011) Neural substrates of chronic pain in the thalamocortical circuit. Available from Nature Precedings http://hdl.handle.net/10101/npre.65481
26. May A (2007) Neuroimaging: visualising the brain in pain. Neurol Sci 28(Suppl 2): S101–S107
27. Ahn S, Lee SK (2011) Diffusion tensor imaging: exploring the motor networks and clinical applications. Korean J Radiol 12(6): 651–661
28. Guye M, Parker GJ, Symms M, Boulby P, Wheeler-Kingshott CA, Salek-Haddadi A, Barker GJ, Duncan JS (2003) Combined functional MRI and tractography to demonstrate the connectivity of the human primary motor cortex in vivo. Neuroimage 19(4):1349–1360
29. Laird AR, Robbins JM, Li K, Price LR, Cykowski MD, Narayana S, Laird RW, Franklin C, Fox PT (2008) Modeling motor connectivity using TMS/PET and structural equation modeling. Neuroimage 2:424–436
30. Hayashi T, Ohnishi T, Okabe S, Teramoto N, Nonaka Y, Watabe H, Imabayashi E, Ohta Y, Jino H, Ejima N, Sawada T, Iida H, Matsuda H, Ugawa Y (2004) Long-term effect of motor cortical repetitive transcranial magnetic stimulation [correction]. Ann Neurol 56(1):77–85, Erratum in: Ann Neurol, 56(2):311
31. Chouinard PA, Van Der Werf YD, Leonard G, Paus T (2003) Modulating neural networks with transcranial magnetic stimulation applied over the dorsal premotor and primary motor cortices. J Neurophysiol 90(2):1071–1083
32. Jankovic J (2008) Parkinson's disease: clinical features and diagnosis. J Neurol Neurosurg Psychiatry 79:368–376

33. Lozano AM, Snyder BJ (2008) Deep brain stimulation for parkinsonian gait disorders. J Neurol 255(Suppl 4):30–31
34. Andrade P, Carrillo-Ruiz JD, Jiménez F (2009) A systematic review of the efficacy of globus pallidus stimulation in the treatment of Parkinson's disease. J Clin Neurosci 16(7):877–881
35. Fregni F, Simon DK, Wu A, Pascual-Leone A (2005) Non-invasive brain stimulation for Parkinson's disease: a systematic review and meta-analysis of the literature. J Neurol Neurosurg Psychiatry 76(12):1614–1623
36. Kolominsky-Rbas PL, Weber M, Gefeller O, Neundoerfer B, Heuschmann PU (2001) Epidemiology of ischemic stroke subtypes according to TOAST criteria: incidence, recurrence, and long-term survival in ischemic stroke subtypes: a population-based study. Stroke 32(12):2735–2740
37. Théoret H, Kobayashi M, Valero-Cabré A, Pascual-Leone A (2003) Exploring paradoxical functional facilitation with TMS. Suppl Clin Neurophysiol 56:211–219
38. Kaltenbach JA (2009) Insights on the origins of tinnitus: An overview of recent research. Hearing J 62(2):26–29
39. McNamara B, Ray JL, Arthurs OJ, Boniface S (2001) Transcranial magnetic stimulation for depression and other psychiatric disorders. Psychol Med 31(7):1141–1146
40. Holtzheimer PE 3rd, Russo J, Avery DH (2003) A meta-analysis of repetitive transcranial magnetic stimulation in the treatment of depression. Psychopharmacol Bull 35(4):149–169, Erratum in: Psychopharmacol Bull, 37(2), 5
41. Burt T, Lisanby SH, Sackeim HA (2002) Neuropsychiatric applications of transcranial magnetic stimulation: a meta analysis. Int J Neuropsychopharmacol 5(1):73–103
42. Martin JL, Barbanoj MJ, Schlaepfer TE, Thompson E, Pérez V, Kulisevsky J (2003) Repetitive transcranial magnetic stimulation for the treatment of depression. Systematic review and meta-analysis. Br J Psychiatry 182:480–491
43. Couturier JL (2005) Efficacy of rapid-rate repetitive transcranial magnetic stimulation in the treatment of depression: a systematic review and meta-analysis. J Psychiatry Neurosci 30(2):83–90
44. Herrmann LL, Ebmeier KP (2006) Factors modifying the efficacy of transcranial magnetic stimulation in the treatment of depression: a review. J Clin Psychiatry 67(12):1870–1876
45. Schutter DJ (2009) Antidepressant efficacy of high-frequency transcranial magnetic stimulation over the left dorsolateral prefrontal cortex in double-blind sham-controlled designs: a meta-analysis. Psychol Med 39(1):65–75
46. Greenberg BD, George MS, Martin JD, Benjamin J, Schlaepfer TE, Altemus M, Wassermann EM, Post RM, Murphy DL (1997) Effect of prefrontal repetitive transcranial magnetic stimulation in obsessive-compulsive disorder: a preliminary study. Am J Psychiatry 154(6):867–869
47. Alonso P, Pujol J, Cardoner N, Benlloch L, Deus J, Menchón JM, Capdevila A, Vallejo J (2001) Right prefrontal repetitive transcranial magnetic stimulation in obsessive-compulsive disorder: a double-blind, placebo-controlled study. Am J Psychiatry 158(7):1143–1145
48. Sachdev PS, McBride R, Loo CK, Mitchell PB, Malhi GS, Croker VM (2001) Right versus left prefrontal transcranial magnetic stimulation for obsessive-compulsive disorder: a preliminary investigation. J Clin Psychiatry 62(12):981–984
49. Prasko J, Pasková B, Záleský R, Novák T, Kopecek M, Bares M, Horácek J (2006) The effect of repetitive transcranial magnetic stimulation (rTMS) on symptoms in obsessive compulsive disorder. A randomized, double blind, sham controlled study. Neuro Endocrinol Lett 27(3):327–332
50. Mantovani A, Lisanby SH, Pieraccini F, Ulivelli M, Castrogiovanni P, Rossi S (2006) Repetitive transcranial magnetic stimulation (rTMS) in the treatment of obsessive-compulsive disorder (OCD) and Tourette's syndrome (TS). Int J Neuropsychopharmacol 9(1):95–100
51. Sachdev PS, Loo CK, Mitchell PB, McFarquhar TF, Malhi GS (2007) Repetitive transcranial magnetic stimulation for the treatment of obsessive compulsive disorder: a double-blind controlled investigation. Psychol Med 37(11):1645–1649
52. Kang JI, Kim CH, Namkoong K, Lee CI, Kim SJ (2009) A randomized controlled study of sequentially applied repetitive transcranial magnetic stimulation in obsessive-compulsive disorder. J Clin Psychiatry 70(12):1645–1651
53. Sarkhel S, Sinha VK, Praharaj SK (2010) Adjunctive high-frequency right prefrontal repetitive transcranial magnetic stimulation (rTMS) was not effective in obsessive-compulsive disorder but improved secondary depression. J Anxiety Disord 24(5):535–539
54. Mantovani A, Simpson HB, Fallon BA, Rossi S, Lisanby SH (2010) Randomized sham-controlled trial of repetitive transcranial magnetic stimulation in treatment-resistant obsessive-compulsive disorder. Int J Neuropsychopharmacol 13(2):217–227

55. Mansur CG, Myczkowki ML, de Barros CS, Sartorelli Mdo C, Bellini BB, Dias AM, Bernik MA, Marcolin MA (2011) Placebo effect after prefrontal magnetic stimulation in the treatment of resistant obsessive-compulsive disorder: a randomized controlled trial. Int J Neuropsychopharmacol 14(10):1389–1397
56. Aleman A, Sommer IE, Kahn RS (2007) Efficacy of slow repetitive transcranial magnetic stimulation in the treatment of resistant auditory hallucinations in schizophrenia: a meta-analysis. J Clin Psychiatry 68(3): 416–421
57. Tranulis C, Sepehry AA, Galinowski A, Stip E (2008) Should we treat auditory hallucinations with repetitive transcranial magnetic stimulation? A metaanalysis. Can J Psychiatry 53(9):577–586
58. Dlabac-de Lange JJ, Knegtering R, Aleman A (2010) Repetitive transcranial magnetic stimulation for negative symptoms of schizophrenia: review and meta-analysis. J Clin Psychiatry 71(4):411–418
59. Grisaru N, Amir M, Cohen H, Kaplan Z (1998) Effect of transcranial magnetic stimulation in posttraumatic stress disorder: a preliminary study. Biol Psychiatry 44(1):52–55
60. McCann UD, Kimbrell TA, Morgan CM, Anderson T, Geraci M, Benson BE, Wassermann EM, Willis MW, Post RM (1998) Repetitive transcranial magnetic stimulation for posttraumatic stress disorder. Arch Gen Psychiatry 55(3):276–279
61. Rosenberg PB, Mehndiratta RB, Mehndiratta YP, Wamer A, Rosse RB, Balish M (2002) Repetitive transcranial magnetic stimulation treatment of comorbid posttraumatic stress disorder and major depression. J Neuropsychiatry Clin Neurosci 14(3): 270–276
62. Cohen H, Kaplan Z, Kotler M, Kouperman I, Moisa R, Grisaru N (2004) Repetitive transcranial magnetic stimulation of the right dorsolateral prefrontal cortex in posttraumatic stress disorder: a double-blind, placebo-controlled study. Am J Psychiatry 161(3): 515–524
63. Osuch EA, Benson BE, Luckenbaugh DA, Geraci M, Post RM, McCann U (2009) Repetitive TMS combined with exposure therapy for PTSD: a preliminary study. J Anxiety Disord 23(1):54–59
64. Boggio PS, Rocha M, Oliveira MO, Fecteau S, Cohen RB, Campanhã C, Ferreira-Santos E, Meleiro A, Corchs F, Zaghi S, Pascual-Leone A, Fregni F (2010) Noninvasive brain stimulation with high-frequency and low-intensity repetitive transcranial magnetic stimulation treatment for posttraumatic stress disorder. J Clin Psychiatry 71(8):992–999
65. Lefaucheur JP, Drouot X, Nguyen JP (2001) Interventional neurophysiology for pain control: duration of pain relief following repetitive transcranial magnetic stimulation of the motor cortex. Neurophysiol Clin 31(4): 247–252
66. Lefaucheur JP, Drouot X, Keravel Y, Nguyen JP (2001) Pain relief induced by repetitive transcranial magnetic stimulation of precentral cortex. Neuroreport 12(13):2963–2965
67. Rollnik JD, Wüstefeld S, Däuper J, Karst M, Fink M, Kossev A, Dengler R (2002) Repetitive transcranial magnetic stimulation for the treatment of chronic pain—a pilot study. Eur Neurol 48(1):6–10
68. Pleger B, Janssen F, Schwenkreis P, Völker B, Maier C, Tegenthoff M (2004) Repetitive transcranial magnetic stimulation of the motor cortex attenuates pain perception in complex regional pain syndrome type I. Neurosci Lett 356(2):87–90
69. Lefaucheur JP, Drouot X, Ménard-Lefaucheur I, Nguyen JP (2004) Neuropathic pain controlled for more than a year by monthly sessions of repetitive transcranial magnetic stimulation of the motor cortex. Neurophysiol Clin 34(2):91–95
70. Khedr EM, Kotb H, Kamel NF, Ahmed MA, Sadek R, Rothwell JC (2005) Longlasting antalgic effects of daily sessions of repetitive transcranial magnetic stimulation in central and peripheral neuropathic pain. J Neurol Neurosurg Psychiatry 76(6):833–838
71. André-Obadia N, Peyron R, Mertens P, Mauguière F, Laurent B, Garcia-Larrea L (2006) Transcranial magnetic stimulation for pain control. Double-blind study of different frequencies against placebo, and correlation with motor cortex stimulation efficacy. Clin Neurophysiol 117(7):1536–1544
72. Irlbacher K, Kuhnert J, Röricht S, Meyer BU, Brandt SA (2006). Central and peripheral deafferent pain: therapy with repetitive transcranial magnetic stimulation. Nervenarzt 77(10), 1196, 1198–203.
73. Borckardt JJ, Smith AR, Reeves ST, Madan A, Shelley N, Branham R, Nahas Z, George MS (2009) A pilot study investigating the effects of fast left prefrontal rTMS on chronic neuropathic pain. Pain Med 10(5):840–849
74. Mhalla A, Baudic S, Ciampi de Andrade D, Gautron M, Perrot S, Teixeira MJ, Attal N, Bouhassira D (2011) Long-term maintenance of the analgesic effects of transcranial magnetic stimulation in fibromyalgia. Pain 152(7): 1478–1485

75. Lefaucheur JP, Ayache SS, Sorel M, Farhat WH, Zouari HG, Ciampi de Andrade D, Ahdab R, Ménard-Lefaucheur I, Brugières P, Goujon C (2012) Analgesic effects of repetitive transcranial magnetic stimulation of the motor cortex in neuropathic pain: Influence of theta burst stimulation priming. Eur J Pain 16(10):1403–1413
76. Siebner HR, Rossmeier C, Mentschel C, Peinemann A, Conrad B (2000) Short-term motor improvement after sub-threshold 5-Hz repetitive transcranial magnetic stimulation of the primary motor hand area in Parkinson's disease. J Neurol Sci 178(2):91–94
77. Okabe S, Ugawa Y, Kanazawa I, Effectiveness of rTMS on Parkinson's Disease Study Group (2003) 0.2-Hz repetitive transcranial magnetic stimulation has no add-on effects as compared to a realistic sham stimulation in Parkinson's disease. Mov Disord 18(4):382–388
78. Khedr EM, Farweez HM, Islam H (2003) Therapeutic effect of repetitive transcranial magnetic stimulation on motor function in Parkinson's disease patients. Eur J Neurol 10(5):567–572
79. Lefaucheur JP, Drouot X, Von Raison F, Ménard-Lefaucheur I, Cesaro P, Nguyen JP (2004) Improvement of motor performance and modulation of cortical excitability by repetitive transcranial magnetic stimulation of the motor cortex in Parkinson's disease. Clin Neurophysiol 115(11):2530–2541
80. Fregni F, Santos CM, Myczkowski ML, Rigolino R, Gallucci-Neto J, Barbosa ER, Valente KD, Pascual-Leone A, Marcolin MA (2004) Repetitive transcranial magnetic stimulation is as effective as fluoxetine in the treatment of depression in patients with Parkinson's disease. J Neurol Neurosurg Psychiatry 75(8):1171–1174
81. Boggio PS, Fregni F, Bermpohl F, Mansur CG, Rosa M, Rumi DO, Barbosa ER, Odebrecht Rosa M, Pascual-Leone A, Rigonatti SP, Marcolin MA, Araujo Silva MT (2005) Effect of repetitive TMS and fluoxetine on cognitive function in patients with Parkinson's disease and concurrent depression. Mov Disord 20(9):1178–1184
82. Lomarev MP, Kanchana S, Bara-Jimenez W, Iyer M, Wassermann EM, Hallett M (2006) Placebo-controlled study of rTMS for the treatment of Parkinson's disease. Mov Disord 21(3):325–331
83. del Olmo MF, Bello O, Cudeiro J (2007) Transcranial magnetic stimulation over dorsolateral prefrontal cortex in Parkinson's disease. Clin Neurophysiol 118(1):131–139
84. Arias P, Vivas J, Grieve KL, Cudeiro J (2010) Controlled trial on the effect of 10 days low-frequency repetitive transcranial magnetic stimulation (rTMS) on motor signs in Parkinson's disease. Mov Disord 25(12): 1830–1838
85. Filipović SR, Rothwell JC, Bhatia K (2010) Low-frequency repetitive transcranial magnetic stimulation and off-phase motor symptoms in Parkinson's disease. J Neurol Sci 291(1–2):1–4
86. Minks E, Mareček R, Pavlík T, Ovesná P, Bareš M (2011) Is the cerebellum a potential target for stimulation in Parkinson's disease? Results of 1-Hz rTMS on upper limb motor tasks. Cerebellum 10(4):804–811
87. Benninger DH, Berman BD, Houdayer E, Pal N, Luckenbaugh DA, Schneider L, Miranda S, Hallett M (2011) Intermittent theta-burst transcranial magnetic stimulation for treatment of Parkinson disease. Neurology 76(7):601–609
88. González-García N, Armony JL, Soto J, Trejo D, Alegría MA, Drucker-Colín R (2011) Effects of rTMS on Parkinson's disease: a longitudinal fMRI study. J Neurol 258(7):1268–1280
89. Khedr EM, Ahmed MA, Fathy N, Rothwell JC (2005) Therapeutic trial of repetitive transcranial magnetic stimulation after acute ischemic stroke. Neurology 65(3):466–468
90. Kim YH, You SH, Ko MH, Park JW, Lee KH, Jang SH, Yoo WK, Hallett M (2006) Repetitive transcranial magnetic stimulation-induced corticomotor excitability and associated motor skill acquisition in chronic stroke. Stroke 37(6):1471–1476
91. Khedr EM, Abdel-Fadeil MR, Farghali A, Qaid M (2009) Role of 1 and 3 Hz repetitive transcranial magnetic stimulation on motor function recovery after acute ischaemic stroke. Eur J Neurol 16(12):1323–1330
92. Emara TH, Moustafa RR, Elnahas NM, Elganzoury AM, Abdo TA, Mohamed SA, Eletribi MA (2010) Repetitive transcranial magnetic stimulation at 1Hz and 5Hz produces sustained improvement in motor function and disability after ischaemic stroke. Eur J Neurol 17(9):1203–1209
93. Talelli P, Wallace A, Dileone M, Hoad D, Cheeran B, Oliver R, Vandenbos M, Hammerbeck U, Barratt K, Gillini C, Musumeci G, Boudrias MH, Cloud GC, Ball J, Marsden JF, Ward NS, Di Lazzaro V, Greenwood RG, Rothwell JC (2012) Theta burst stimulation in the rehabilitation of the upper limb: A semirandomized, placebo-controlled trial in chronic stroke patients. Neurorehabil Neural Repair 26(8):976–987

94. Liepert J, Zittel S, Weiller C (2007) Improvement of dexterity by single session low-frequency repetitive transcranial magnetic stimulation over the contralesional motor cortex in acute stroke: a double-blind placebo-controlled crossover trial. Restor Neurol Neurosci 25(5–6):461–465
95. Nowak DA, Grefkes C, Dafotakis M, Eickhoff S, Küst J, Karbe H, Fink GR (2008) Effects of low-frequency repetitive transcranial magnetic stimulation of the contralesional primary motor cortex on movement kinematics and neural activity in subcortical stroke. Arch Neurol 65(6):741–747
96. Málly J, Dinya E (2007) Recovery of motor disability and spasticity in post-stroke after repetitive transcranial magnetic stimulation (rTMS). Brain Res Bull 76(4):388–395
97. Takeuchi N, Tada T, Toshima M, Chuma T, Matsuo Y, Ikoma K (2008) Inhibition of the unaffected motor cortex by 1 Hz repetitive transcranial magnetic stimulation enhances motor performance and training effect of the paretic hand in patients with chronic stroke. J Rehabil Med 40(4):298–303
98. Kakuda W, Abo M, Kobayashi K, Momosaki R, Yokoi A, Fukuda A, Umemori T (2011) Application of combined 6-Hz primed low-frequency rTMS and intensive occupational therapy for upper limb hemiparesis after stroke. NeuroRehabilitation 29(4):365–371
99. Kakuda W, Abo M, Shimizu M, Sasanuma J, Okamoto T, Yokoi A, Taguchi K, Mitani S, Harashima H, Urushidani N, Urashima M (2012) A multi-center study on low-frequency rTMS combined with intensive occupational therapy for upper limb hemiparesis in post-stroke patients. J Neuroeng Rehabil 9(1):4
100. Kleinjung T, Steffens T, Langguth B, Eichhammer P, Marienhagen J, Hajak G, Strutz J (2006) Treatment of chronic tinnitus with neuronavigated repetitive Transcranial Magnetic Stimulation (rTMS). HNO 54(6): 439–444
101. Londero A, Lefaucheur JP, Malinvaud D, Brugieres P, Peignard P, Nguyen JP, Avan P, Bonfils P (2006) Magnetic stimulation of the auditory cortex for disabling tinnitus: preliminary results. Presse Med 35(2 Pt 1):200–206
102. Langguth B, Zowe M, Landgrebe M, Sand P, Kleinjung T, Binder H, Hajak G, Eichhammer P (2006) Transcranial magnetic stimulation for the treatment of tinnitus: a new coil positioning method and first results. Brain Topogr 18(4):241–247
103. Plewnia C, Reimold M, Najib A, Reischl G, Plontke SK, Gerloff C (2007) Moderate therapeutic efficacy of positron emission tomography-navigated repetitive transcranial magnetic stimulation for chronic tinnitus: a randomised, controlled pilot study. J Neurol Neurosurg Psychiatry 78(2):152–156
104. Kleinjung T, Steffens T, Sand P, Murthum T, Hajak G, Strutz J, Langguth B, Eichhammer P (2007) Which tinnitus patients benefit from transcranial magnetic stimulation? Otolaryngol Head Neck Surg 137(4):589–595
105. Frank G, Kleinjung T, Landgrebe M, Vielsmeier V, Steffenhagen C, Burger J, Frank E, Vollberg G, Hajak G, Langguth B (2010) Left temporal low-frequency rTMS for the treatment of tinnitus: clinical predictors of treatment outcome—a retrospective study. Eur J Neurol 17(7):951–956
106. Khedr EM, Etraby AE, Hemeda M, Nasef AM, Razek AA (2010) Long-term effect of repetitive transcranial magnetic stimulation on motor function recovery after acute ischemic stroke. Acta Neurol Scand 121(1):30–37
107. Anders M, Dvorakova J, Rathova L, Havrankova P, Pelcova P, Vaneckova M, Jech R, Holcat M, Seidl Z, Raboch J (2010) Efficacy of repetitive transcranial magnetic stimulation for the treatment of refractory chronic tinnitus: a randomized, placebo controlled study. Neuro Endocrinol Lett 31(2):238–249
108. Marcondes RA, Sanchez TG, Kii MA, Ono CR, Buchpiguel CA, Langguth B, Marcolin MA (2010) Repetitive transcranial magnetic stimulation improve tinnitus in normal hearing patients: a double-blind controlled, clinical and neuroimaging outcome study. Eur J Neurol 17(1):38–44
109. Plewnia C, Bartels M, Gerloff C (2003) Transient suppression of tinnitus by transcranial magnetic stimulation. Ann Neurol 53(2):263–266
110. De Ridder D, Verstraeten E, Van der Kelen K, De Mulder G, Sunaert S, Verlooy J, Van de Heyning P, Moller A (2005) Transcranial magnetic stimulation for tinnitus: influence of tinnitus duration on stimulation parameter choice and maximal tinnitus suppression. Otol Neurotol 26(4):616–619
111. Fregni F, Marcondes R, Boggio PS, Marcolin MA, Rigonatti SP, Sanchez TG, Nitsche MA, Pascual-Leone A (2006) Transient tinnitus suppression induced by repetitive transcranial magnetic stimulation and transcranial direct current stimulation. Eur J Neurol 13(9):996–1001
112. Khedr EM, Abo-Elfetoh N, Rothwell JC, El-Atar A, Sayed E, Khalifa H (2010) Contralateral versus ipsilateral rTMS of temporoparietal cortex for the treatment of chronic unilateral tinnitus: comparative study. Eur J Neurol 17(7):976–983

Chapter 13

Transcranial Magnetic Stimulation (TMS) Clinical Applications: Diagnostics

Josep Valls-Sole

Abstract

Since its introduction in 1985, transcranial magnetic stimulation (TMS) has become a powerful tool for research as a noninvasive and painless technique to effectively stimulate the human cortex. Over the years, TMS has proven useful not only for the assessment of motor cortex physiology but also for the diagnosis and prognosis of many neurological disorders involving the corticospinal tract. In this chapter, we explore the theoretical and practical considerations of several key diagnostic protocols, including those used to asses motor cortical and corticospinal layout, excitability, and functional integrity. In addition, we illustrate how each of these protocols can be utilized during the diagnostic process of a number of neurological diseases, including stroke, paralysis, myelopathy, spinal cord injury, multiple sclerosis, motoneuron disease, and varied movement disorders.

Key words Transcranial magnetic stimulation (TMS), Excitability, Integrity, Diagnosis, Clinical assessment

1 General Remarks

Since its introduction in 1985, transcranial magnetic stimulation (TMS) has become a powerful tool for research as a noninvasive and painless technique to effectively stimulate the human cortex [1–4]. Over the years, TMS has proven useful not only for the assessment of motor cortex physiology but also for the diagnosis and prognosis of many neurological disorders involving the corticospinal tract. These tests of clinical applicability, however, are relatively few when one considers the large amount of research being conducted using this technique. Indeed, only a handful of motor tract stimulation research paradigms are in use for diagnostic purposes. One reason for this discrepancy between research and clinical applications is the lack of sufficient standardization to allow for replication of results across a large number of centers. In addition, there are many technical factors hindering the growth of TMS in the clinical realm, such as variable effects due to coil position,

Alexander Rotenberg et al. (eds.), *Transcranial Magnetic Stimulation*, Neuromethods, vol. 89,
DOI 10.1007/978-1-4939-0879-0_13, © Springer Science+Business Media New York 2014

waveform type, and motor threshold variability. Due to these factors, the results of many TMS experiments are somewhat laboratory-dependent and, although the researchers have no doubt been careful enough to check whether results obtained in their patients were different from those obtained in a comparable group of healthy subjects using the same procedure, many procedures have not proliferated beyond the laboratories in which they were first developed.

Many points favor the use of TMS in clinical practice, at least when considering the functional assessment of conduction time in the corticospinal tract: it is a painless technique, it is noninvasive, and there is sufficient physiological knowledge about its effects on the motor cortex for an adequate interpretation of the results generated by a standardized exam. Stimulation of cortical tissue by TMS is produced without the passage of electrical current. Therefore, activation of cortical neurons occurs with no concomitant electrical stimulation of the skin or subcutaneous tissue. This makes the procedure well tolerated by most patients. Also, the magnetic field induced by TMS gradually attenuates with increasing distance from the coil—therefore, short latency muscle responses (motor evoked potentials, or MEPs) are generated via cortical rather than subcortical stimulation of the motor pathway, even at maximal TMS intensity.

Although the MEPs are generated via the excitation of cortical neurons, they are mediated by activation of spinal (alpha) motoneurons and motor axons. Therefore, the latency and size of the MEP modulates according to the excitability level of alpha motoneurons and the state of nerve conduction at root and peripheral nerves. For effects of longer latency than the MEP, it is necessary to consider the sound produced by the passage of the current through the TMS coil as well as the excitation of subcutaneous nerve terminals. These collateral effects may induce effects generated in trigeminal muscle and cutaneous afferents as well as in many scalp, facial, neck, and trunk muscle afferents. These added inputs can cause reflex effects at the spinal or supraspinal level which may induce muscle responses that can combine and modify the late effects generated by TMS.

2 Methodological Remarks

The TMS practitioner must know the basics of technique being attempted, the characteristics of the device being utilized, and the subtle tricks required for a skillful exam in clinical practice. As with many other neurophysiological procedures, the examiner must be aware of the objectives when applying TMS in a particular patient. These will necessarily shift and change depending upon each patient's unique clinical problems. Additionally, the examiner must know whether a patient is consuming any medication that could be a contraindication for receiving TMS or otherwise affect the results of the exam. A basic neurological clinical examination can help in recognizing clinical signs that can be helpful for interpretation of results. As such, the examiner should preferably be a certified physician with both technical and clinical knowledge of neurological disorders. It is helpful, also, if an assistant aids in the exam, the setup, the placement of electrodes, the positioning of the coil, and the monitoring of the patient. This will help ensure the practitioner does not overlook or omit any technical or procedural requirements.

2.1 Stimulation

The TMS apparatus is a high-voltage, high-current capacitor that is discharged in a fraction of a second [5] through an electrically isolated coil. The transient passage of the electrical current through the coil produces a strong time-varying perpendicular magnetic field for a very short duration (<1 ms). This, in turn, induces a secondary current of opposite direction in conductive tissue within the magnetic field (eddy current). In humans, the electric field generated in the brain is modified by regional differences in tissue conductivity. Because of this, excitable tissue in the cortex will be more likely activated in areas where fibers run perpendicular to the magnetic field and, where there are shifts in axonal alignment. At a subcortical level, if the coil is applied over the posterior fossa, the stimulus will preferentially activate the cranial nerves at their exit sites in the skull while, if the coil is applied over the spine, it will preferentially activate the spinal nerves at their entrance to the bone foramina [6].

There are various forms of coils. The most commonly used in clinical practice are the round coil (a circle of 9 cm diameter that is active at the edges), the figure-of-eight coil (two small, conjoined coils that produce a strong magnetic field at the central junction), and the double-cone coil (two circular coils joined at a 90° angle which produce the strongest magnetic field in the virtual point of convergence). The round coil is the most commonly used in the conventional clinical examination. It is important to note which face of the coil is in contact with the scalp as the current flows in only one direction only (usually counterclockwise when the "A" side is facing upwards—the manufacturer will provide the user with this information). Since the motor cortex is preferentially depolarized by currents flowing in the posterior to anterior direction and the current generated in the brain tissue goes in the opposite direction with respect to that in the coil, most coils will preferentially activate the left motor cortex when positioned over the vertex with side A facing up. The orientation of the coil handle has no relevance when TMS is applied with a round coil but it may be very relevant when using a "figure-of-eight" coil [7].

The electrical current inducing the magnetic field can either be monophasic or biphasic (although additional shapes have been explored for experimental purposes). A monophasic pulses consist of a strong initial current flow in one direction (peaking at approximately 50 μs after pulse onset) followed by a low amplitude slow return to baseline. This limits the efficacy of the initial current flow, producing a stimulation effect that is physiologically comparable to a real monophasic current pulse. Because of this, the stimulation effect of a monophasic pulse is directionally specific and induces a tissue current that is opposite to the initial rising phase of the monophasic current. If monophasic stimulation currents are used, the current in the coil will flow mainly in the direction marked by the handle. Since the current induced in the brain tissue is opposite to that in the coil and the most effective stimulation of M1 will be obtained with posterior to anterior directed currents, it turns out that the most effective position will be with the handle perpendicular to the central sulcus in an anterior position.

A biphasic pulse resembles a cosine waveform. Because of this, the direction-specific stimulation effect is less pronounced, but still present [8, 9]. During a biphasic pulse, the current direction reverses two times: the initial current rise is followed by a current reversal below zero and a subsequent rise back to zero. In contrast to the monophasic current waveform, the cosine current waveform induces physiologically significant tissue current flow in two directions, according to the two phases of the pulse. However, the second (reversal) phase has the largest amplitude and the longest duration and, accordingly, it is the one that induces the strongest tissue current. Since tissue current goes in opposite direction to the current in the coil and the second phase in the biphasic pulse is reversed with respect to the phase of monophasic stimulators, it results that, with the same position of the coil, the neural elements preferentially activated by a biphasic pulse are different from those activated by a monophasic pulse.

The maximum intensity of a TMS-generated magnetic field is on the order of 1.5–2.0 T. This translates into an induced current in the neural structures, which intensity is difficult to determine because of the inhomogeneity of the brain tissue. The effective intensity of TMS is usually measured in percentage of the maximum stimulator intensity (100 %). It depends upon many variables, including equipment parameters, subject's age, and target muscle. In spite of this, the minimum intensity at which the stimulus induces a muscle response is one of the most relevant variables in clinical application of TMS (see below). Finally, distance between the stimulating coil and the cortical target area is a critical variable when determining the intensity required for effective cortical stimulation. In order to minimize the effects of varying inter- and intra-individual distances, the coil should be placed in direct contact with the skin, with the horizontal plane parallel to the patient's head.

2.2 Recording

The descending corticospinal volleys generated by TMS induce a post-synaptic excitatory potential in the alpha motoneuronal membrane. This will transform into motoneuronal firing only if depolarization threshold is reached. Therefore, the local excitability level of the spinal motoneurons will, together with the strength of the descending input the efficacy of transsynaptic cortico-motoneuronal excitation, determine whether an MEP is induced in the target muscles. The MEP is usually recorded from a muscle after cleansing the skin to reduce the impedance. Surface electrodes are then placed in a belly-tendon arrangement, with the active electrode as near as possible to the muscle motor point. The ground electrode should be secured in a place where electronic artifacts are reduced. It might be important to avoid bending of the cables, which could induce more artifacts in the presence of the magnetic field. Despite the care taken in the positioning of electrodes, the MEP may pick up activity of neighboring muscles activated by the same stimulus. Indeed, the MEP is a mix of several action potentials generated in various muscles activated synchronously [10]. The output from cortical stimulation may also be based upon movements rather than muscles, with little specificity of responses.

Each laboratory should establish a standardized setup for routine examination and produce its own normative values in accordance with the technique being used. However, the recorded MEP should always be clearly visible on the monitor of the recording device. The optimum screen settings depend on many factors such as the stimulus intensity, the muscle being studied, and the voluntary pre-activation. However, the amplification factor and temporal resolution of the screen should be chosen so that the induced MEP is optimally displayed. Ideally, the recorded trace to TMS should start some time before the stimulus to assess the off-line level of relaxation or pre-activation in the target muscle. The time recorded after TMS depends on whether the examiner wants to measure just the MEP or longer latency effects (for instance, the silent period induced during a tonic contraction).

Whatever the stimulus type, intensity, coil, or exam purpose, it is a good idea to record MEPs from homologous muscles bilaterally (e.g., from both hands) simultaneously. With a round coil, one can evaluate the small side-to-side latency difference in the MEPs that is usually seen due to the preferential stimulation of one hemisphere with each single pulse of TMS. It is, of course, necessary to use the two sides of the coil for preferential stimulation of either the right or the left hemispheres, and record a number of MEPs in homonymous target muscles of each side for a proper characterization of amplitude and latency of the MEP.

3 TMS Tests of Clinical Applicability

3.1 Motor Threshold

Resting motor threshold (RMT) is defined as the minimum stimulus intensity at which TMS evokes a measurable MEP in the target muscle at rest. It is usually measured as the lowest stimulus intensity required to induce MEPs of a predetermined amplitude 50 % of the time (the MEP should be clearly distinguishable from background noise) [11]. The stimulus intensity used in clinical practice should be expressed as percentage of motor threshold and it is recommended to use a predetermined percentage of an individual's RMT when measuring MEP latency and amplitude. As stated above, the MEP is the product of activation not only of the motor cortex but also of the alpha motoneuron. Therefore, the excitability measured through evaluation of MEP amplitude is a function not only of the motor cortex but also of the excitability of the entire corticospinal tract.

Even though RMT is a basic and necessary parameter for the evaluation of MEP data, its mean value is unfortunately rather variable between individuals and even within individuals [12–14]. Normal values for RMT may range from 30 to 100 % of the stimulator intensity—therefore, intra-individual differences in RMT are usually not relevant for the assessment of dysfunction. Apart from genetic or other as yet unknown factors, distance of the magnetic stimulating coil from the motor cortex and geometrical folding of the motor cortex may influence RMT [15]. Differences in RMT among muscles are likely related to the number and density of cortico-cortical axons and corticospinal neurons innervating that muscle. This is highest for hand muscles. Accordingly, RMT is lowest for these muscles and higher for axial, proximal arm, and lower limb muscles [16]. The motor threshold measured while sustaining a tonic muscle contraction, known as active motor threshold (AMT), is less variable among individuals than RMT, probably because the increase in motoneuronal excitability induced by voluntary contraction overrides other concomitant changes.

The RMT can be considered a surrogate of motor tract excitability. However, it is a relatively unstable measure, influenced by mood, attention, fatigue, medication, hyperventilation, and other conditions [17–19]. Therefore, it is risky to assume that RMT will not change in a given individual during follow-up examinations. In spite of this, even if it does not differ between patients and healthy subjects, RMT may be a reliable measure of the effects of interventions such as medication [20] or repetitive TMS [21] and reflect changes in the excitability of the corticospinal tract due to certain neurological disorders. RMT is often significantly increased in cerebral stroke or in advanced cases of lateral sclerosis [22], and reduced in untreated epilepsy [23, 24].

3.2 Other Measures of Motor Cortex and Corticospinal Tract Excitability

Apart from the fact that its variability in healthy subjects may prevent finding abnormal results in groups of patients, assessing motor tract excitability through the RMT is a limited technique as it allows only for assessment of the highest excitable neurons and does not allow for assessment of inhibitory circuits. There are two techniques, however, which can be used for such purposes: The short-interval intracortical inhibition (SICI) technique allows experimenters to evaluate excitability of inhibitory circuits and the triple stimulation technique (TST) allows experimenters to more precisely assess excitability in a large population of motoneurons.

3.2.1 SICI

SICI is examined by applying two subsequent stimuli through the same coil [25]. The first stimulus is of subthreshold intensity for the targeted muscle at rest. Its intensity is determined as a percentage of AMT in such a way that it is practically impossible to, when delivered at rest, cause any descending activity through the corticospinal tract. The second stimulus is of suprathreshold intensity, enough to induce an MEP of rather stable amplitude. When the two stimuli are applied in quick succession (with an interval of between 1 and 4 ms), the amplitude of the MEP elicited by the second stimulus is dramatically reduced (an effect known also as intracortical inhibition).

Studies of the descending corticospinal volleys [26, 27] have demonstrated that the number of descending inputs after the second stimulus is reduced when it is preceded by the low intensity stimulus. Therefore, changes in the amount of SICI should reflect changes in corticomotor excitability. It has been shown that SICI is reduced in healthy subjects preparing to move [28–30], during a sustained movement [31] or even by just imagining movement preparation [32]. SICI has been found to be abnormal in many patients with neurological diseases, such as Parkinson's disease, dystonia, and similar disorders of movement [33–35].

3.2.2 MEP Amplitude. Recruitment Curves

The amplitude of the MEP to single pulse TMS is not usually employed as a measure of functional relevance because of its large variability and dependence on many technical factors. Theoretically, the amplitude of the MEP reveals the fraction of motor axons activated by TMS expressed as a percentage of innervation received at the target muscle by dividing the MEP amplitude by the compound muscle action potential (CMAP). This is known as the amplitude ratio. Accordingly, the MEP amplitude might be considered a rough measure of motor cortex excitability. However, MEP amplitude depends also on the amount of desynchronization of the efferent volley [36]. A rather objective way to quantify the amplitude of the MEP is to measure how it changes with progressive increases in stimulus intensity or the overall degree of muscle contraction [37]. Input–output or recruitment curves measure how cortical excitability changes according to demands [37–39].

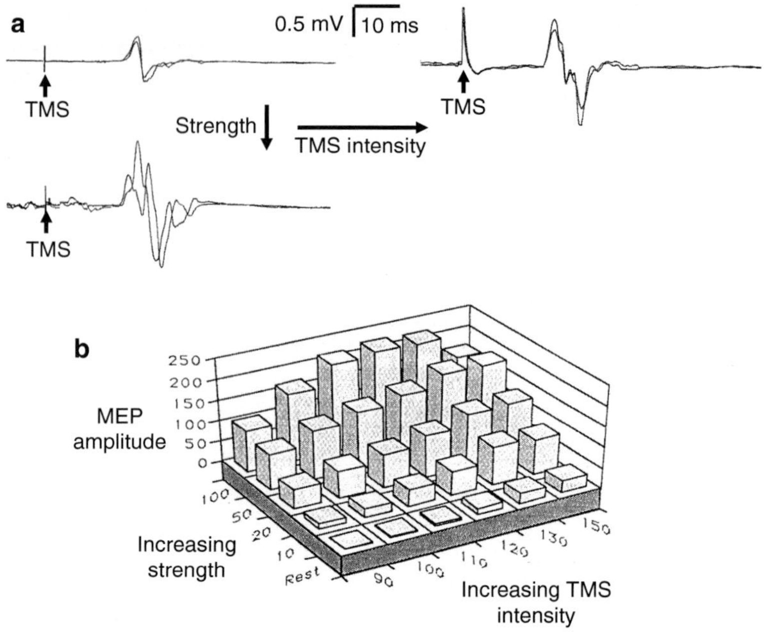

Fig. 1 (**a**) Examples from a healthy subject of MEP changes in latency and amplitude by increasing TMS intensity (compare *right* with *left side* recordings in the *upper part* of the figure) or strength of muscle contraction (compare *upper* and *lower* recordings on the *left side* of the figure). Two traces are superimposed in each recording to show consistency. (**b**) A graphic representation of the increase in MEP amplitude (*z* axis) when increasing TMS intensity (*x* axis) or strength of muscle contraction (*y* axis)

Usually, stimulus intensity is progressively increased in percentage steps out of RMT, while muscle activity is increased in steps based on percentages of the maximum force (Fig. 1).

3.2.3 Triple Stimulation Technique

The principle of the TST is the collision between ascending and descending volleys in the upper limb [40, 41]. For thenar muscle recordings, the TST requires one to apply a suprathreshold TMS pulse, followed by a supramaximal electrical stimulus at the wrist, followed by a supramaximal electrical stimulus at the Erb's point. The three stimuli must be timed appropriately. The one in the wrist should follow TMS by 10–12 ms in such a way that the antidromic volley generated at the wrist is able to collide with all the descending inputs generated in the motor axons by TMS. The electrical stimulus in the Erb's point should be applied 8–10 ms after the wrist stimulus for the orthodromic volley to collide with the remaining antidromic inputs (if there is any left) in the median nerve.

In healthy subjects, such combination of collisions should give rise to a CMAP recorded in the thenar muscles after Erb's point stimulation that is supramaximal—similar in shape and amplitude to a control action potential elicited by just the Erb's point stimulation alone. A reduction in the output from the motor cortex after TMS will cause an incomplete collision with the antidromic volley generated by the wrist stimulus. As a consequence, the remaining antidromic inputs will collide with the efferent volley generated after Erb's point stimulation and diminish the amplitude of the action potential recorded in the thenar muscle. As during conventional MEP recordings, patients should, if possible, perform a slight tonic contraction of the target muscle to facilitate TMS-induced corticospinal excitation. A pre-innervation of the target muscle should also be carried out during the peripheral control TST to ensure comparability. The TST is a stimulation technique that effectively "resynchronizes" corticomotor excitation at the level of the peripheral motor axon.

In patients with peripheral neuropathy, desynchronization along the peripheral motor axon may be substantial and, in turn, reduce the MEP amplitude due to phase cancellation. In addition, musculo-neural reexcitation may occur in the field of intramuscular nerve endings and induce antidromically ascending nerve action potentials that interfere with the collisions of the TST. Therefore, it is advisable to use a control measurement of TST, which is done simply by replacing of the first TMS pulse with a proximal peripheral nerve stimulus at the Erb's point. In this condition, since TST-control and TST-test are subjected to the same degree to peripheral desynchronization and musculo-neural reexcitation, the ratio between the CMAP induced by the TST-test and the one induced by the TST-control would cancel any influence from peripheral nerve lesions and leave the results influenced only by corticospinal function. TST, therefore, reflects the number of corticospinal neurons brought to discharge by TMS, expressed as a percentage of total number of neurons innervating the target muscle. In healthy persons, the ratio is close to 100 %. If the ratio is substantially under 100 %, it is recommended to gradually increase the TMS intensity to see whether an increase in MEP/CMAP ratio can be induced at higher intensities. Like the conventional MEP measurement, the TST cannot differentiate between a structural lesion of the axons or a functional conduction block with the axons still intact (neuropraxia). The corticospinal excitability measured by TST has been of clinical interest thus far in various disorders involving the motor tract, such as amyotrophic lateral sclerosis (ALS) [42], spinocerebellar ataxia type 6 [43], multiple system atrophy [44], and multiple sclerosis [45].

3.3 Central Motor Conduction Time

After deciding which coil to use, this exam begins by locating the motor "hot spot"—the site at which TMS elicits the largest MEP of shortest latency with lowest stimulus intensity. This search should begin with stimulus intensities above mean threshold for general population (e.g., 60–70 % of maximal stimulator output) and, when the largest response has been obtained, stimulus intensity should be diminished to find RMT (as outlined above).

The exam proceeds with stimulation at the hot spot using a predetermined intensity (usually, 110–120 % above RMT). Slight changes in coil orientation and stimulus intensity may be needed until the examiner feels confident with the results obtained.

To measure central motor conduction time (CMCT), it is necessary to subtract the peripheral conduction time from the latency of TMS-induced MEPs. Two methods are currently used in clinical routine to estimate conduction time in the peripheral motor nerve: (1) Magnetic stimuli can be applied at spinal level (spinal magnetic stimulation) to activate the nerve roots. This will give rise to a CMAP in the target muscle. This response latency is the peripheral conduction time. (2) The motor nerve can be stimulated electrically to induce an F wave. Peripheral conduction time is then measured according to the formula:

$$\text{PCT}[\text{ms}] = 0.5 \times (\text{F} + \text{M} - 1\ \text{ms}).$$

where F corresponds to the shortest F-wave latency and M equals the distal motor latency. One millisecond is subtracted to account for a central delay at the cell body of the motoneuron.

The two methods have different physiological mechanisms that the examiner must understand before deciding which method is more suitable for a given patient. Magnetic stimulation activates the spinal nerve while passing through the bone foramina, where the electrical resistance is lower in comparison to other points [6, 46]. This results in the conduction time along the very proximal section within the spinal canal being assigned to the CMCT. The resulting error in CMCT estimation depends on the length of the nerve section in the spinal canal and can range between 0.5 and 1.4 ms for the cervical spine and 3.0–4.1 ms for the lumbar spine. The stimulus intensity for spinal cord stimulation should be slightly suprathreshold for motoneuron activation. If higher intensities of stimulation are used, the point of nerve stimulation may move more distally, inducing an additional and uncontrolled error in the estimation of peripheral motor conduction time [47]. The F-wave latency can be rather variable. It is recommended to use the wave with the shortest latency among 10–20 consecutively collected F waves. Since the F wave is generated at the motoneuronal level, this method gives a better approximation of the CMCT than magnetic stimulation as it includes the intraspinal segment of the nerve roots in the peripheral conduction time. As a consequence, the CMCT will be shorter if calculated with the F-wave method than with the spinal stimulation method. However, the F-wave method can only be used in distal muscles and may give inconsistent results if a conduction block exists in the peripheral nerve [48]. The best recommendation is to use one method consistently and include this information on any report as the method utilized has very relevant practical implications (e.g., in the assessment of lesions affecting the nerve roots, as in Guillain-Barré syndrome, chronic inflammatory demyelinating polyneuropathy or polyradiculitis).

There is no universal agreement on which is the preferred state of the muscle for clinical studies; but authors tend to use a slight tonic contraction to increase the efficacy of TMS. In any case, any clinical report must state whether the MEPs were recorded with the muscle at rest or during contraction, as the facilitatory effect of muscle contraction on the MEPs to cortical stimulation does not occur in the F wave nor in spinal magnetic stimulation. Even if central conduction time is measured at rest, it is worth measuring a few MEPs during a weak voluntary contraction to test for the expected latency shortening and amplitude increase. However, if a patient cannot perform a contraction due to hemiparesis, the examiner should keep in mind that requesting voluntary contraction will facilitate the healthy side only, which will produce an inhibitory effect to the contralateral homologous muscles [49].

The CMCT, as is the case with many other electrophysiological measurements, can be judged abnormal if longer than 2.5 or 3 SD above the mean of the normative data acquired in a control group of healthy subjects matched for age, sex, and height, examined using the same technique and by the same examiner/s. It is crucial to build a set of control data that will match the specific population being explored. CMCT is probably the TMS-derived parameter most commonly used in clinical practice. It can evaluate the functional integrity of the corticospinal tract in conditions such as hemiplegia due to a stroke, multiple sclerosis, cervical spondylotic myelopathy, ALS, and other diseases potentially involving the motor pathway. Lengthening of CMCT is, in most instances, due to slowness of conduction in the corticospinal tract—as in demyelinating disorders [50]. It is worth noting, though, that CMCT can be lengthened due to the presence of axonal destruction, degeneration of fast-conducting cortico-motoneuronal fibers, or impaired temporal and spatial summation of excitatory postsynaptic potentials from weakened cortico-motoneuronal input at spinal motoneurons. Assessment of CMCT, along with nerve conduction studies, is useful for ensuring that the motor pathway is intact in patients with psychogenic disorders presenting with motor deficit [51].

3.4 Motor Mapping

Penfield and Boldrey [52] described for the first time in humans that direct low intensity stimulation of selective areas of the motor cortex generated movements in specific parts of the body. This was replicated first with electrical cortical stimulation [53] and, then, with magnetic cortical stimulation [54]. Point by point, cortical stimulation through the scalp allowed researchers to create maps of MEP representations across the motor cortex. One must keep in mind, however, that the MEP is a crude representation of cortical activation. As stated above, cortical stimulation generates movement patterns rather than muscle responses and, therefore, maps of muscles are only an approximation of the cortical representation. A resolution of 1 cm steps in all directions around Cz is the most commonly used method for conventional mapping studies. The amplitude of the MEP recorded to a given stimulation point can be represented in a two dimensional grid for quantitative purposes. More technical and practical details can be found in Thickbroom et al. [55] and Wolf et al. [10].

Cortical maps typically provide three pieces of quantitative data: the optimal position at which to obtain the largest MEP response (the so-called hot spot), the center of gravity (defined as the area in the grid with the largest concentration of spots from which TMS induces large amplitude MEPs), and the area of the scalp from which MEPs of more than a predefined threshold amplitude can be obtained (area of muscle representation). While the optimal position, or hot spot, presumably corresponds to the location of the most excitable population of neurons that project to a target muscle, the area of cortical muscle representation depends on the intensity of the stimulus and the area of the cortex that is considered [56]. Therefore, rather than using absolute values, as reference, it is wise to do comparative studies using the same parameters in control and test conditions.

With a few exceptions, mapping studies have shown that the area of muscle representation is more vulnerable than the hot spot to medical or experimental interventions [57]. In most instances, cortical maps are evaluated at rest. However, Darling et al. [58] showed that the variability of MEPs to TMS was lower if the patient maintained a low level tonic contraction (5–10 % of maximum). This observation suggests that comparisons of physiological changes during recovery in people with neuropathology may be influenced by the resting state of muscle contraction. The clinical interest in studying cortical maps relates to patients with mirror movements [59, 60], cortical dysgenesis [61], degenerative diseases [62], spinal cord injury [63], or amputees [64]. Relatively recently, technical advances have provided the possibility to detect the projection of the magnetic field into the brain tissue using sensor-based electromagnetic navigation systems (nTMS). This provide with a more accurate delineation of cortical maps for easier access to tumor resection near eloquent areas of the cortex [65, 66]. Picht et al. [67] found good correlation for the assessment of the hot spot between nTMS and direct peroperatory cortical stimulation.

3.5 Silent Period

Voluntary muscle contraction leads to an increase of MEP size and a shortening of MEP latency due to enhanced excitability at the corticospinal tract. Additionally, the EMG activity related to the voluntary contraction is suppressed during a period following the MEP. This is the so-called silent period [68–72]. It has a rather long duration in comparison to the silent period induced by peripheral

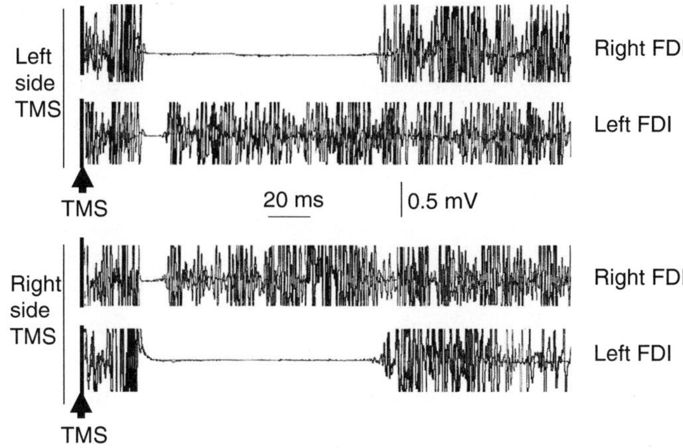

Fig. 2 Silent periods elicited in simultaneously contracting first dorsal interossei (FDI) of both sides by unilateral (focal) TMS. In the *upper two graphs*, TMS is applied over the motor cortex of the left hemisphere. See the induction of a long duration contralateral silent period in the right FDI and of a short ipsilateral silent period in the left FDI. In the *lower two graphs*, TMS is applied over the motor cortex of the right hemisphere. See the contralateral silent period in the left FDI and the ipsilateral silent period in the right FDI. Two traces are superimposed in each case for consistency

nerve electrical stimulation in the same muscle and Fuhr et al. [68] demonstrated that, while the period originates from post-activation depression of spinal cord alpha motoneurons, it sustains due to activation of inhibitory neurons at the brain. It is possible to induce a silent period in hand muscles at rather low stimulus intensities independently from the presence of an MEP [70, 73]. However, to avoid variability in the duration of the silent period, which is the variable most useful in the clinical context, it is recommended to use a higher intensity than that used to measure the MEP latency (150 % if possible). The TMS-induced silent period duration in hand muscles is about 200 ms, calculated from the delivery of the stimulus to the resumption of the EMG activity (Fig. 2). This seems not to be related to the size of the MEP as identically sized MEPs evoked by cervical cord stimulation [72] or by brainstem stimulation [74] are accompanied by silent periods of different duration than those induced by TMS.

The silent period is an inhibitory phenomena attributed to cortical functions (another is the SICI, referred to above). Werhahn et al. [75] found that the cortical silent period is likely mediated by GABA-B receptors, while the GABA-A receptors mediate SICI. As such, the two forms of cortical inhibition may be used in different conditions or diseases to test the function of GABA receptors. Testing the silent period might be of clinical interest in patients with Parkinson's disease (typically shortened when compared to control subjects) [76]. Interestingly, the silent period returns to values closer to "normality" with dopaminergic drugs [77, 78]. Group comparison of cortical silent period duration has a similar drawback as RMT clinical assessment: namely, it has a large variability in healthy persons. However, comparison between the hemispheres may be of value because of little side-to-side differences in normal subjects.

As shown in Fig. 2, focal unilateral cortical TMS induces a silent period not only in the contralateral hand muscles but also in the ipsilateral muscles [79, 80]. This is often attributed to transcallosal conduction as it is often absent or delayed in patients with congenital, acquired, or surgical lesions of the corpus callosum. This could represent a simple test for callosal function [80]. However, deep brain stimulation of the subthalamic nucleus in patients with Parkinson's disease has revealed that activation of the corticospinal tract at a point caudal to the corpus callosum can also generate a well-defined ipsilateral silent period.

3.6 Modulation of Segmental Reflexes

MEPs are generated by excitatory postsynaptic potentials in alpha motoneurons triggered by inputs from the descending corticospinal tract. However, the corticospinal tract is likely not the only descending tract influencing spinal motoneurons after TMS. Descending tracts likely only provide for a small portion of direct excitatory inputs to spinal motoneurons. The reticulospinal tract and other descending tracts from the brainstem could become activated after TMS by cortico-subcortical connections and, in this way, contribute to the arrival of descending inputs at the motoneuron. In most alpha motoneurons, however, there is a denser network of connections (both, excitatory and inhibitory) from propriospinal interneurons than from cortico-motoneuronal neurons. Therefore, there is often spinal modulation of inputs to the alpha motoneuron. Currently, there is little evidence for the effects of TMS on propriospinal interneurons. Taylor and coworkers [81] demonstrated an effect of TMS on the ipsilateral stretch reflex, and Kato et al. [82] showed that TMS produced subliminal inhibition in the circuit of the Ia inhibitory interneuron in forearm muscles. Rothwell et al. [83] showed that muscle spindle afferents changed their frequency of firing after TMS, indicating that the descending volley has reached fusimotor neurons. Indirect evidence of TMS effects at the presynaptic alpha motoneuron level was reported by Valls-Sole et al. [84] who showed that vibration-induced inhibition of the H reflex was transiently decreased by TMS at an intensity subthreshold that needed to evoke a soleus MEP. This is suggestive of effects of the TMS volley on presynaptic inhibitory neurons or circuits.

After inducing excitatory postsynaptic potentials in the alpha motoneuron via direct (cortico-motoneuronal) or indirect (after activation of propriospinal interneurons) inputs, TMS will cause an enhancement of excitability in the motoneurons for the duration of the excitatory postsynaptic potential. If this rise in excitability is not sufficient to cause them to fire, the alpha motoneurons will be primed. If additional excitatory inputs reach the motoneuron during such priming, responses will be larger than expected for the intensity of the input. This effect can be easily studied using the H reflex or the F wave. A very early study of the effects of TMS on segmental excitability was carried out by Cowan et al. [85] using cortical electrical stimulation. In this chapter, the authors examined only short intervals between cortical and segmental stimulation and reported a general facilitatory effect in various muscles, except in the soleus where they reported inhibition. Later studies using TMS and longer conditioning periods revealed that there is in fact facilitation of the soleus H reflex by TMS beginning at about 5–10 ms and lasting for about 20 ms followed by a second phase of facilitation beginning at about 60–70 ms and lasting up to 100–110 ms [86, 87]. TMS modulation of the soleus H reflex may vary while standing and during the different phases of walking [88]. Interestingly, a study of the effects of suprathreshold TMS on the H reflex recorded in the abductor pollicis brevis muscle showed a reduction of the segmental reflex response at about 30 ms [89]. In that study, the authors reported similar effects for the F wave recorded in the same muscle using TMS of higher intensity (200 % of RMT).

Modulation of spinal reflexes of leg muscles may have some clinical interest for the physiological analysis of gait disturbances, although this presents some technical difficulties [88]. When studied at rest, different patterns of abnormalities have been reported in different disorders (Fig. 3). In patients with Parkinson's disease, Valls-Sole and Valldeoriola [87] reported a significant decrease of the second facilitatory phase, whereas a different type of abnormality was reported in patients with idiopathic spastic paraparesis by Serranova et al. [90]. Wolfe et al. [91] found that modulation of the H reflex by TMS was preserved in some patients with spinal cord injury in whom no MEPs could be elicited, suggesting that the technique could be used to improve detection of preserved conduction in the descending motor tracts. Benito Penalva et al. [92] reported a correlation between the amplitude of the first facilitatory phase and the scores of clinical scales of improvement on gait during the first 3 months after the lesion.

3.7 Integration of Afferent Inputs

The excitability of the motor cortex and that of the whole motor pathway, including the alpha motoneurons, is modified by inputs from the sensory system. The profile and size of that modulation, which is measured as the change in amplitude or area of the MEP, depends upon the type of sensory afferents activated. The effects of trains of electrical stimuli were first shown by Mariorenzi et al., [93] in hand muscles and later by Kasai et al. [94] in foot muscles.

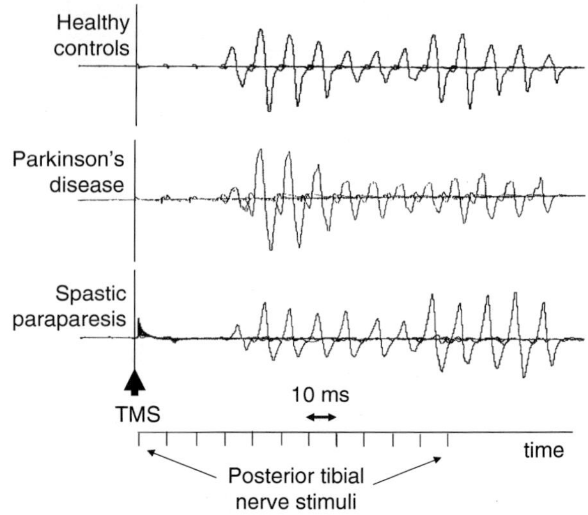

Fig. 3 Effects of TMS on the soleus H reflex in a healthy subject, a patient with Parkinson's disease and a patient with spastic paraparesis. TMS of subthreshold intensity for eliciting a soleus MEP is applied at the vertical line (*arrow*). A posterior tibial nerve electrical stimulus of an intensity inducing a H reflex of an amplitude of about 1 mV in the soleus muscle is applied at every 10 ms step after time 0. Traces obtained up to the interval of 110 ms are superimposed to better demonstrate the systematic changes in amplitude of the H reflex in each case

Further studies used vibration [95], high intensity electrical stimuli [96], or laser inputs [97]. An indirect way of providing evidence for the effects of sensory inputs was devised by inducing deafferentation through ischemic block [57, 98]. This contributed to further understanding the mechanisms of plasticity after amputation or spinal cord injury [99–101].

Progress in the elucidation of the effects of sensory inputs on MEP modulation was made by Stefan et al. [102], who reported that pairing peripheral nerve stimulus with TMS at an appropriate interval generated a consistent increase in MEP amplitude lasting longer than the period of stimulation. The mechanisms involved in this long lasting facilitation of the MEP after paired associative stimulation relate to synaptic long-term potentiation at the cortical level [103]. Modulation of the MEP by sensory afferents has been found to be abnormally reduced in diseases with defective sensorimotor integrations, such as dystonia [104, 105]. Interestingly, paired associative stimulation has been found to induce a larger amplitude increase of the MEP in patients with Parkinson's disease who are off therapy, a condition that was normalized after levodopa intake [106]. The effects of paired associative stimulation have been reported to differentiate between patients with organic dystonia and those with psychogenic dystonia [107].

4 Diseases in Which TMS Can Be of Help in Clinical Practice

4.1 Stroke

Vascular lesions involving the motor cortex or, more commonly, subcortical regions occupied by the motor tract cause a conduction block and, possibly, ischemic damage in motor axons of the corticospinal tract. The logical correlate of this condition is the absence of TMS-induced MEPs in the muscles receiving innervation from motoneurons supplied by the damaged axons. The preservation of MEPs in paralyzed muscles suggests a relatively good prognosis for functional recovery [108–111]. If the MEP is elicited during the acute phase, its latency is usually delayed. Latency shortening over repeated testing correlates positively with muscle strength and hand function test scores in patients recovering function [112, 113].

Unilateral lack of function in the corticospinal tract after stroke is likely to induce excitability changes in the contralateral motor cortex and, possibly, in other structures of the motor pathway. Interhemispheric asymmetry in regard to excitability of the motor cortex may imply that the unaffected motor cortex exerts an unwanted enhanced inhibition over the affected hand and, reciprocally, the affected hemisphere exerts reduced inhibitory action over the unaffected hand. One way to enhance motor function in the paretic hand after stroke may be through the down-regulation of activity in the intact motor cortex [114]. Ipsilateral MEPs to focal unilateral TMS are found with a higher prevalence in the non-affected side post-stroke more so than in healthy subjects, a manifestation of enhanced excitability [115]. No significant contribution was found of such hyperexcitability of the hemisphere contralateral that affected during the recovery of hand function in stroke patients [116]. However, the unaffected hemisphere may participate in the recovery of motor function in axial muscles, such as is the case in lingual paresis [117].

Training programs might enhance plastic change in the affected hemisphere. This was observed by Liepert et al. [118] who described an increase in the cortical representation of the thenar muscles after training. This increase correlated with improved dexterity. Interestingly, although the total area of the motor map increased, the center of gravity and the hot spot remained the same. This suggests that the increase in the motor map was due to higher excitability at the edges. Taub et al. [119] studied the effects of constraint-induced movement therapy (which forces a patient to use the affected limb). The findings were roughly the same as those of Liepert et al. [118] in the hemisphere contralateral to the affected limb. However, in addition they found a shrinking of the motor representation of the non-affected, constrained hand. Navigated TMS may offer a more objective monitor for these motor map fluctuations during functional recovery [120].

4.2 Facial Paralysis

When applied to the posterolateral section of the scalp, TMS may activate cranial nerves in the posterior fossa [121]. Therefore, the stimulus can access the facial nerve before it enters into the petrossal canal, where most lesions causing facial palsy take place. The facial nerve is highly excitable at the point of entrance to the internal acoustic meatus as its environment changes from low-resistance cerebrospinal fluid to high resistance petrous bone [122]. If conventional electrical stimulation of the facial nerve at the stylomastoid foramen is combined with TMS stimulation of cortical representations of the facial nerve, the experimenter can gather information on conduction across three valuable segments of the facial nerve projection: proximal (from cortex to posterior fossa), transpetrossal, and distal. Typically, cortical stimulation gives rise to bilateral MEPs due to the bilateral distribution of the corticonuclear tract [123–125] while facial nerve stimulation at the posterior fossa typically gives rise to only ipsilateral responses [126, 127].

TMS-induced MEPs and CMAP induced by peripheral nerve stimulation can be recorded over various facial muscles: the nasalis or buccinator being the most commonly used. Recordings are often performed bilaterally to allow for side-to-side comparison. Cortical MEPs often require facilitation through voluntary contraction of the target muscle [127]. MEPs in facial muscles are small and may sometimes be difficult to interpret due to contamination by uncrossed ipsilateral responses, the blink reflex, other facial reflex responses, peripheral stimulation of the ipsilateral facial nerve, or by responses of other neural structures such as muscles innervated by the trigeminal nerve. Posterior fossa stimulation of the facial nerve is best performed with the base of the coil over the mastoid process. Interestingly, the onset latency of the response does not vary much with changes in coil position as the selective stimulation point remains at the canalicular entrance [128].

It is relatively easy to calculate the transpetrossal conduction time of the facial nerve in both healthy subjects and patients if the responses are clear. In all idiopathic Bell's palsy, a greatly diminished (or absent) response to posterior fossa stimulation is observed starting from the time of symptom onset [129]. This reflects the hypoexcitability of the damaged nerve segment. Apparently, such hypoexcitability is not related to the clinical function in Bell's palsy and does not have a prognostic value [129]. However, it is useful in determining the site of nerve damage and it provides differential diagnosis with respect to varied sites of facial nerve lesión, such as intraaxial or distal. Absent responses to posterior fossa stimulation with preservation of responses to stylomastoid stimulus is highly suggestive of Bell's palsy. In contrast, a normal response to posterior fossa stimulation typically reflects dysfunction proximal to the internal acoustic meatus. Bilateral abnormalities are found in Guillain-Barré syndrome and other generalized neuropathies. These abnormalities are not restricted to the transpetrossal segment of the facial nerve but involve also the distal stretch of the facial nerve. Assessment of facial nerve conduction at transpetrossal and distal levels can be of help in the diagnosis of these and other syndromes, even in the absence of facial weakness. The inclusion of the cortico-proximal segment in an assessment may also aid in the diagnosis of central conduction time abnormalities in disorders involving the central nervous system, such as multiple sclerosis and ALS.

4.3 Spinal Cord Injury, Cervical Spondilotic Radiculo-Myelopathy, and Other Myelopathies

Trauma-related injuries are the most common source of lesions in the spinal cord. The level of the lesion and its severity determines the clinical picture. Impairment is commonly measured using the ASIA (American Spinal Injury Association) scale and a strong correlation has been reported between the ASIA and electrophysiological assessment using TMS—technique which may be of great help when assessing uncooperative or incomprehensive patients [130]. Complete lesions (ASIA-A) manifest as lack of movement caudal to the injury. Correspondingly, TMS will be unable to elicit MEPs in leg muscles. However, in patients with ASIA-B or ASIA-C (incomplete lesion), TMS may still not induce a well-defined MEP. In these instances, the technique of reflex modulation by TMS may help to ascertain if the lesion is indeed incomplete [91, 92]. Improvement in clinical functional scores for gait was correlated with an enlargement of MEP values and with TMS modulation of the H reflex [92].

A series of changes occur in the motor system at sites rostral to the spinal lesion. Macdonell and Donnan [131] described an early (6 h post-injury) increase in RMT, even in muscles that preserved innervation above the lesion level. These abnormal values returned to near normality 6 weeks post-injury. Cohen and colleagues [99, 132] described an increase in the area of cortical maps of muscles immediately rostral to the lesion level in patients with spinal cord injury. These findings, unfortunately, could not be replicated by Brouwer and Hopkins-Rosseel [133] who found higher rather than lower activation thresholds in gently contracting muscles innervated by nerve roots rostral to the lesion in patients with incomplete spinal cord lesions. However, in a more recent study, Saturno et al. [134] found decreased intracortical inhibition in spinal cord injury patients, indicating that plastic change does occur rostral to the lesion.

In patients with cervical spondilotic myelopathy, Lo et al. [135] reported a strong correlation between MRI and TMS features, with 98 %, sensitivity and specificity of the determination of CMCT if MRI is used as reference. However, images of anatomical compression do not always indicate the exact site of neural lesion. In many instances of anatomical compression, electrophysiological measures may indicate the location with more functional damage. TMS allows for measuring the conduction time to several muscles in the upper limbs, revealing the metameric level of impairment [136]. Care must be taken to explore whether concomitant root impairment [137] is distorting findings related to myelopathy. CMCT may be longer than normal in patients with cervical myelopathy but, when measured using spinal cord magnetic stimulation, root compression provides for further extending this time period [138]. Preoperative CMCT findings have no predictive value for the clinical outcome [139]. Topographic localization of focal myelopathy in cases of thoracic trauma, spondilotic lesions, or transverse myelitis due to inflammatory or demyelination processes has been attempted via the recording of MEPs along the paraspinal muscles [140, 141]. However, it is difficult to accurately assess conduction time due to the multiple segmental innervation. Ellaway and coworkers [141] demonstrated the change in RMT intensity is an important marker of the perilesional area.

CMCT may be abnormally lengthened in patients with myelopathies of various other etiologies, including degenerative diseases such as in idiopathic spastic paraparesis [90, 142], hereditary conditions such as spinocerebellar atrophies [143], infections such as HTLV-I [144, 145], or, less frequently, HIV [146] and metabolic disorders such as those related to chronic liver diseases [147] or in vitamin B12 deficiency [148]. In some of these latter conditions, early detection of involvement may lead to the appropriate treatment and limitation of disease progression [147, 149]. It is worth noting that CMCT may also be lengthened in diseases in which there is slowing of nerve conduction velocity at a radicular level as measured using spinal magnetic stimulation but not when using the F-wave method. However, in chronic inflammatory demyelinating neuropathy, some clinical, electrophysiological, and MRI studies have implicated CNS involvement [149].

4.4 Motoneuron Disease

Early stages of any form of motoneuron disease pose a diagnostic problem. The corticospinal tract may be involved from the very beginning in ALS and in primary lateral sclerosis (PLS). However, the degree of involvement is usually slight, causing only subtle abnormalities. ALS features both upper and lower motoneuron dysfunction. Upper motoneuron syndrome may lead to spasticity, weakness, brisk reflexes, and a Babinski sign while the involvement of the spinal motoneuron may lead to denervation signs. Therefore, depending on clinical presentation, there may be some difficulties in differentiating motoneuron disease and cervical myeloradiculopathy. Differential diagnosis can be aided by needle EMG, which typically reveals a widespread denervation in ALS and a limited involvement to the affected roots in cervical myeloradiculopathy. Also, the latter does not involve cranial nerve innervated muscles. TMS can also help in differential diagnosis. Corticospinal motor conduction time does not usually show very large abnormalities in ALS. Typically, there is only a slight delay or data in the upper

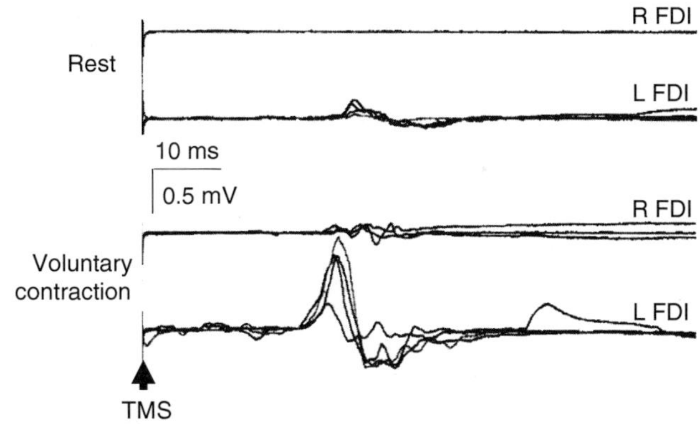

Fig. 4 Responses obtained in the first dorsal interosseous muscles (FDI) of a patient with multiple sclerosis showing motor deficit in the right arm. Note the absence of MEP in the *right side* and the low amplitude of the MEP in the *left side* when the stimuli were applied at rest (*upper two graphs*) and the presence of a small delayed MEP in the *right side* in comparison to a practically normal MEP on the *left side* during voluntary contraction (*lower two graphs*). Up to four traces are superimposed in each graph to show consistency of the results

range of normal values. This is related to the functional loss of large axons. On the other hand, CMCT is often found to be abnormal in cervical myeloradiculopathy.

It is possible that, in TMS studies of patients with motoneuron diseases, MEPs will unobtainable in hand muscles, even during facilitation by voluntary contraction [150]. In these cases, it may be worth attempting to distinguish a silent period. Also, one could, investigate alternate muscles, including proximal limb muscles. The MEP may be very small with slightly prolonged onset latency, leading to normal or slightly lengthened CMCT (Fig. 4). It's important to note that the slight CMCT delay will occur also in cranial nerve innervated muscles [151]. As such, this type of assessment may contribute largely to the electrophysiological diagnosis. A specific example is familial autosomal recessive ALS with the D90A CuZn-superoxide dismutase mutation [152]. These patients show extremely prolonged CMCT, probably due to selective degeneration of fastest-conducting cortico-motoneuronal cells with a concomitant preservation of slow-conducting cells [153].

Although most patients with ALS have an increased RMT (the MEP is unobtainable or requires a stimulus of high intensity), a number of researchers have described enhanced excitability at onset of the disorder with reduced RMT. This is most likely due to glutamate-induced excitatory neurotoxicity [154, 155]. Cortical inhibitory circuits are abnormal in ALS. This can be documented by a shortening of the silent period [154] and a decrease in short interval intracortical inhibition [156, 157]. The frequency of TMS measured abnormalities of patients with motoneuron disease varies in accordance with the severity and form of the disease and the thoroughness of the TMS exam. Obviously, the inclusion of slightly sophisticated techniques, such as TST [40, 42], can help improve the diagnostic yield. It may be worth building an index, based on data from TMS examinations to aid in follow-up assessments [22].

4.5 Multiple Sclerosis

Multiple sclerosis (MS) is a disease with variable presentations. As a consequence, not all patients with MS will exhibit the same clinical neurophysiological expression. Prolongation of CMCT is considered the most common single abnormality found in patients with multiple sclerosis [50, 158, 159]. The sensitivity of the measure increases if lower limb muscles are included in the test [160]. Abnormal CMCT is more pronounced in progressive than in relapsing-remitting forms [161] and, in the latter, there may be differences between the relapse and the remitting phases [162]. Focal central conduction block has been demonstrated using the TST [163], which may enhance detection of abnormalities in the MEP amplitude ratio [45].

TMS measures, including CMCT and RMT, are often integrated into a multimodal evoked potential score which combines the abnormalities across sensory and motor pathways. These scores may be used as a measure of disability [164], are usually well correlated with the Expanded Disability Status Scale (EDSS) score [165, 166], and may be helpful in assessing the disease evolution or the treatment effects [167, 168]. In a recent study of 131 patients with various degrees of disability, including patients with no apparent pyramidal tract dysfunction, Kale et al. [169] found that CMCT, latency, and MEP amplitude correlated significantly with EDSS scores ($p < 0.001$), suggesting that EPs predict disability even in subclinical presentations. On a similar note, Conte et al. [170] found that intracortical measures of excitability (RMT, SICI) were more likely to be abnormal in patients with high EDSS score and secondary progressive forms of MS than in patients with low EDSS score and relapsing-remitting forms of MS. Perretti et al. [171] reported abnormalities in the post-exercise depression of MEPs that could be related to the fatigue so often reported by patients with multiple sclerosis.

The corpus callosum is often used to quantify the atrophy and demyelination in multiple sclerosis [172] and its measurement correlates with disability [173]. Consequently, two neurophysiological measures thought to reflect conduction within the callosal fibers, the transcallosal conduction time and the ipsilateral silent period duration [174–176], have been used for the assessment of abnormal transcallosal conduction. An indirect way of evaluating interhemispheric connectivity is crossed reaction time tasks. Preliminary data recorded in our laboratory shows that patients with multiple sclerosis are slower than healthy controls in the execution of a simple reaction time task with one hand when somatosensory imperative stimuli are applied to the contralateral hand. This finding fits well with the recently published relationship between corpus callosum structure and abnormal intermanual tasks [177].

4.6 Movement Disorders

TMS has provided interesting data on the pathophysiological mechanisms underlying symptoms in diseases presenting movement disorders [87, 178–180], but currently it only has a limited role as a diagnostic test. There have been many reports on TMS abnormalities across various groups of patients with movement disorders. However, neither the specificity nor the sensitivity of the dysfunctions reported is enough to satisfy clinical requirements. Also, it is often difficult to attribute a specific finding to a certain pathophysiological mechanism as there are compensatory changes which accompany defective function in slowly progressive degenerative central nervous system diseases, such as those which present movement disorders. Also, a particular abnormality may actually be a trait of a disorder even if no clinical symptoms are yet evident. Therefore, the following is a comment on results in which there are some sort of consensus in the literature.

TMS may help in the differential diagnosis between diseases presenting parkinsonism. The fact that other areas of the central nervous system, apart from the substantia nigra, are involved in the degenerative quality of the so-called Parkinson-Plus syndromes may explain why many abnormal findings have been reported in multiple system atrophy, progressive supranuclear palsy, cortico-basal ganglionic degeneration, and Lewy-body disease [44, 181]. While no abnormalities have been reported in CMCT for patients with Parkinson's disease, CMCT may be lengthened in patients with multiple system atrophy [182, 183] and other forms of parkinsonism [180, 184]. Patients with multiple system atrophy also show defective modulation of MEPs generated by somatosensory stimuli presented to peripheral nerves [185]. The silent period induced in the contralateral hand to unilateral TMS is shorter in untreated Parkinson's disease patients than in healthy controls. This abnormality can be reduced with dopaminergic treatment [69, 78]. Testing of transcallosal inhibition with the ipsilateral silent period may also be used to distinguish between different parkinsonian syndromes [186]. In cortico-basal ganglionic degeneration, motor cortex excitability may be reduced. However, in a series of patients with prominent alien hand syndrome, Valls-Sole et al. [62] found signs compatible with hyperexcitability in the hemisphere contralateral to the alien hand and bilateral responses when unilateral TMS was applied to the contralateral hemisphere with the figure-of-eight coil (Fig. 5). Absence of SICI or unexpected facilitation may also be suggestive of cortico-basal ganglionic degeneration, but these observations need further confirmation.

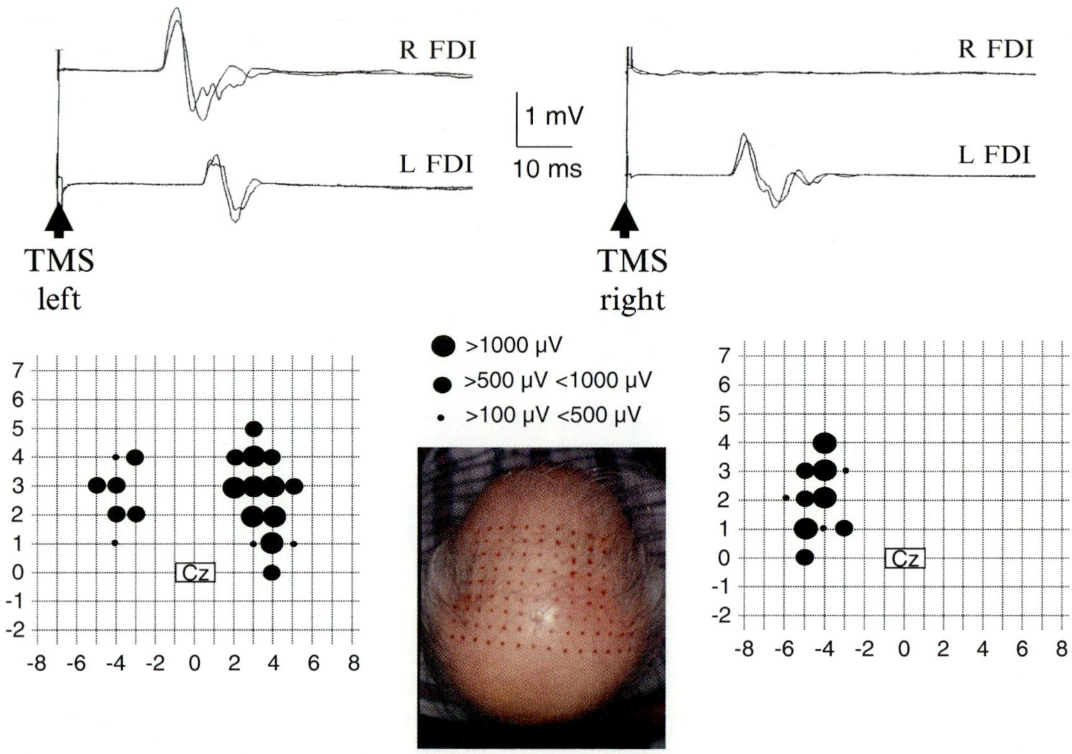

Fig. 5 MEPs obtained in the ipsilateral abductor pollicis brevis (APB) muscle to unilateral (focal) TMS in a patient with cortico-basal degeneration and alien hand syndrome in his left hand. The *upper graphs* show MEPs obtained simultaneously in the right and left APB to left hemisphere TMS (*left side graphs*) and right TMS (*right side graphs*). Note the presence of the ipsilateral MEP to stimulation of the right side only, with a latency of about 8 ms delayed with respect to the contralateral MEP. Two recordings are superimposed in each graph to show consistency of the findings. The *lower part* of the figure shows the cortical map of representation of the APB muscles in a grid of 1 cm squares. The *dots* represent the points at which focal TMS-induced MEPs of different sizes in left hand muscles (*left sided grid*) or right hand muscles (*right sided grid*)

In dystonia, the majority of conventional TMS measurements appear normal. However, the contralateral silent period is usually shortened [187]. There is also a decreased cortical inhibition (SICI) in patients with task-specific dystonia [188]. However, this is, again, a nonspecific finding—Espay et al. [189] found that this abnormality was also present in patients with psychogenic dystonia. Interestingly, Quartarone et al. [107] found that sensorimotor plasticity, examined with paired associative stimulation, was abnormal in patients with organic dystonia but not in patients with psychogenic dystonia.

5 Conclusion

TMS is helpful in the clinical assessment of many disorders. CMCT is undoubtedly the parameter most used in clinical studies, but care should be taken to use a methodology that allows for careful and consistent CMCT quantitation in individual assessment and across group comparisons. RMT, AMT, and MEP recruitment curves are used to test input–output characteristics of the motor system. Cortical maps of muscle representation may be useful in the assessment of plasticity changes after deafferentation, but they must be done carefully and with a consistent relation to RMT. The widespread use of neuronavigation systems will possibly bring more weight to the clinical use of cortical maps. The duration of the silent period in contralateral muscles indicates activity in inhibitory neurons at spinal and cortical levels, whereas the ipsilateral silent period may be used to assess conduction in callosal axons. Measures related to paired stimulation, specifically SICI, can be used to assess cortical excitability independently from the level of excitability in spinal motoneurons. Modulation of segmental reflexes and afferent modulation of motor cortical output are useful methods to explore the combined influence of activity in peripheral and central nervous system levels. Other parameters of clinical use stem from performing a basic assessment, which is described as a step by step summary procedure below:

1. Appropriateness of the exam
   The physician in charge of the exam has the ultimate responsibility for errors or mistakes derived from practicing TMS. Therefore, he or she must know if there are any contraindications for, or possible harm in, applying TMS. Check medication and clinical data.

2. Informing the patient
   The collaboration with the patient is required for many TMS tests. Attention, anxiety, nervousness, and other mood states may affect the results of the TMS exam. Therefore, the physician should prepare the patient for the exam and prepare the site for examination in the most convenient way. A preliminary demonstration of the effects of TMS may be needed in some cases.

3. Placing the recording electrodes
   It is important to gather as much information as possible from each TMS test. Therefore, machine settings and electrode placement must be considered thoroughly before applying TMS. Bilateral homonymous recording is typically a good practice. Skin preparation to reduce impedance to less than 5 kΩ helps prevent artifacts.

4. Defining the aim of the examination
   There should be a clear clinical question to be answered by TMS testing. The most basic tests involve recording the MEP in hand muscles; but this may not always be necessary.

5. Positioning the coil and locating the hot spot
   Navigation features are not presently common in most TMS laboratories. Therefore, positioning of the coil on the scalp must be done by trial and error, using anatomical knowledge.

Under these conditions, care has to be taken to maintain coil position during testing as even slight variation may influence the final results. In most settings, it is useful to use an assistant to either holding the coil or monitor the recording apparatus.

6. Determining RMT
   A compromise has to be taken for RMT determination as it can vary according to slight changes in coil position and the subject's attention. It is customary to ask for a slight muscle contraction first and for the subject to focus his or her attention on the muscle being tested. After locating the hot spot, stimulus intensity should be diminished until the appropriate data for RMT is obtained.
7. Recording the MEP and measuring latency and amplitude
   Once the hot spot and the RMT have been determined, the MEP can be recorded using a predetermined intensity as a percentage of RMT (usually 110 or 120 %). Reproducible MEPs to a number of subsequent stimuli should be used for measuring onset latency and peak amplitude. It may be useful to test contraction-induced facilitation and, if needed, the silent period. In these cases, it is convenient to use higher TMS intensity (i.e., 150 %) to avoid silent period duration variability.
8. Determining peripheral motor conduction time to the same muscle
   Once the latency of the MEP has been measured, it is useful to subtract from it the peripheral nerve conduction time to calculate CMCT. If done with the determination of the F wave, care must be taken to obtain the F wave of the shortest latency in the same muscle used for recording the MEP. If done with magnetic stimulation at the spinal cord, the coil has to be positioned in order to stimulate each side of the root canal. In either case, the size of the CMAP in the target muscle should be obtained to calculate the amplitude ratio.
9. Carry on further exams as needed
   The recording of the MEP in non-hand muscles should be performed as needed, following similar methods as described above. Similarly, other tests of cortical excitability, modulation of segmental reflexes, or afferent modulation of the MEPs should be performed under strict methodological criteria (intensity, coil position, muscle contraction, etc.).
10. Observe reactions and record side effects
    Slight headache can be generated by any TMS exam and is more likely after long sessions. Usually they fade away in a few minutes. It is necessary to report in the patient's file the occurrence of headache, nausea, tinnitus, sweating, or any other light adverse effects.

References

1. Barker AT, Jalinous R, Freeston IL (1985) Non-invasive magnetic stimulation of human motor cortex. Lancet 1:1106–1107
2. Day BL, Thompson PD, Dick JP, Nakashima K, Marsden CD (1987) Different sites of action of electrical and magnetic stimulation of the human brain. Neurosci Lett 75:101–106
3. Rothwell JC, Thompson PD, Day BL, Boyd S, Marsden CD (1991) Stimulation of the human motor cortex through the scalp. Exp Physiol 76:159–200
4. Rossini PM, Desiato MT, Caramia MD (1992) Age-related changes of motor evoked potentials in healthy humans: non-invasive evaluation of central and peripheral motor tracts excitability and conductivity. Brain Res 593: 14–19
5. Jalinous R (1991) Technical and practical aspects of magnetic nerve stimulation. J Clin Neurophysiol 8:10–25
6. Maccabee PJ, Amassian VE, Eberle LP, Cracco RQ (1993) Magnetic coil stimulation of straight and bent amphibian and mammalian peripheral nerve in vitro: locus of excitation. J Physiol 460:201–219
7. Brasil-Neto JP, Cohen LG, Panizza M, Nilsson J, Roth BJ, Hallett M (1991) Optimal focal transcranial magnetic activation of the human motor cortex: effects of coil orientation, shape of the induced current pulse, and stimulus intensity. J Clin Neurophysiol 9:132–136
8. Kammer T, Beck S, Erb M, Grodd W (2001) The influence of current direction on phosphene thresholds evoked by transcranial magnetic stimulation. Clin Neurophysiol 112:2015–2021
9. Kammer T, Beck S, Thielscher A, Laubis-Herrmann U, Topka H (2001) Motor thresholds in humans: a transcranial magnetic stimulation study comparing different pulse waveforms, current directions and stimulator types. Clin Neurophysiol 112:250–258
10. Wolf SL, Butler AJ, Campana GI, Parris TA, Struys DM, Weinstein SR, Weiss P (2004) Intra-subject reliability of parameters contributing to maps generated by transcranial magnetic stimulation in able bodied adults. Clin Neurophysiol 115:1740–1747
11. Rossini PM, Barker AT, Berardelli A, Caramia MD, Caruso G, Cracco RQ et al (1994) Non-invasive electrical and magnetic stimulation of the brain, spinal cord and roots: basic principles and procedures for routine clinical application. Report of an IFCN committee. Electroencephalogr Clin Neurophysiol 91:79–92
12. Wassermann EM (2002) Variation in the response to transcranial magnetic brain stimulation in the general population. Clin Neurophysiol 113:1165–1171
13. Pitcher JB, Ogston KM, Miles TS (2003) Age and sex differences in human motor cortex input–output characteristics. J Physiol 546: 605–613
14. Koski L, Schrader LM, Wu AD, Stern JM (2005) Normative data on changes in transcranial magnetic stimulation measures over a ten hour period. Clin Neurophysiol 116: 2099–2109
15. McConnell KA, Nahas Z, Shastri A, Lorberbaum JP, Kozel FA, Bohning DE, George MS (2001) The transcranial magnetic stimulation motor threshold depends on the distance from coil to underlying cortex: a replication in healthy adults comparing two methods of assessing the distance to cortex. Biol Psychiatry 49:454–459
16. Chen R, Cros D, Curra A, Di Lazzaro V, Lefaucheur JP, Magistris MR, Mills K, Rösler KM, Triggs WJ, Ugawa Y, Ziemann U (2008) The clinical diagnostic utility of transcranial magnetic stimulation: report of an IFCN committee. Clin Neurophysiol 119:504–532
17. Siniatchkin M, Groppa S, Siebner H, Stephani U (2006) A single dose of sulthiame induces a selective increase in resting motor threshold in human motor cortex: a transcranial magnetic stimulation study. Epilepsy Res 72:18–24
18. Ziemann U (2004) TMS and drugs. Clin Neurophysiol 115:1717–1729
19. Sparing R, Dafotakis M, Buelte D, Meister IG, Noth J (2007) Excitability of human motor and visual cortex before, during, and after hyperventilation. J Appl Physiol 102:406–411
20. Nardone R, Ausserer H, Bratti A, Covi M, Lochner P, Marth R, Tezzon F (2006) Cabergoline reverses cortical hyperexcitability in patients with restless legs syndrome. Acta Neurol Scand 114:244–249
21. Franca M, Koch G, Mochizuki H, Huang YZ, Rothwell JC (2006) Effects of theta burst stimulation protocols on phosphene threshold. Clin Neurophysiol 117:1808–1813
22. De Carvalho M, Turkman A, Swash M (2003) Motor responses evoked by transcranial magnetic stimulation and peripheral nerve stimulation in the ulnar innervation in amyotrophic lateral sclerosis: the effect of upper and lower motor neuron lesion. J Neurol Sci 210: 83–90
23. Groppa S, Siebner HR, Kurth C, Stephani U, Siniatchkin M (2008) Abnormal response of motor cortex to photic stimulation in

idiopathic generalized epilepsy. Epilepsia 49: 2022–2029
24. Reutens DC, Berkovic SF, Macdonell RAL, Baldin PF (1993) Magnetic stimulation of the brain in generalized epilepsy: reversal of cortical hyperexcitability by anticonvulsants. Ann Neurol 34:351–355
25. Kujirai T, Caramia MD, Rothwell JC, Day BL, Thompson PD, Ferbert A, Wroe S, Asselman P, Marsden CD (1993) Corticocortical inhibition in human motor cortex. J Physiol 471: 501–519
26. Di Lazzaro V, Oliviero A, Pilato F, Mazzone P, Insola A, Ranieri F et al (2003) Corticospinal volleys evoked by transcranial stimulation of the brain in conscious humans. Neurol Res 25:143–150
27. Di Lazzaro V, Restuccia D, Oliviero A, Profice P, Ferrara L, Insola A, Mazzone P, Tonali P, Rothwell JC (1998) Magnetic transcranial stimulation at intensities below active motor threshold activates intracortical inhibitory circuits. Exp Brain Res 119:265–268
28. Reynolds C, Ashby P (1999) Inhibition in the human motor cortex is reduced just before a voluntary contraction. Neurology 53: 730–735
29. Soto O, Valls-Sole J, Shanahan P, Rothwell J (2006) Reduction of intracortical inhibition in soleus muscle during postural activity. J Neurophysiol 96:1711–1717
30. Floeter MK, Rothwell JC (1999) Releasing the brakes before pressing the gas pedal. Neurology 53:664–665
31. Ridding MC, Taylor JL, Rothwell JC (1995) The effect of voluntary contraction on cortico-cortical inhibition in human motor cortex. J Physiol 487:541–548
32. Kumru H, Soto O, Casanova J, Valls-Sole J (2008) Motor cortex excitability changes during imagery of simple reaction time. Exp Brain Res 30:1433–1436
33. Ridding MC, Sheean G, Rothwell JC, Inzelberg R, Kujirai T (1995) Changes in the balance between motor cortical excitation and inhibition in focal, task specific dystonia. J Neurol Neurosurg Psychiatry 59:493–498
34. Ridding MC, Inzelberg R, Rothwell JC (1995) Changes in excitability of motor cortical circuitry in patients with Parkinson's disease. Ann Neurol 37:181–188
35. Berardelli A, Rona S, Inghilleri M, Manfredi M (1996) Cortical inhibition in Parkinson's disease. A study with paired magnetic stimulation. Brain 119:71–77
36. Rösler KM, Petrow E, Mathis J, Arányi Z, Hess CW, Magistris MR (2002) Effect of discharge desynchronization on the size of motor evoked potentials: an analysis. Clin Neurophysiol 113:1680–1687
37. Valls-Solé J, Alvarez R, Tolosa ES (1994) Vibration-induced presynaptic inhibition of the soleus H reflex is temporarily reduced by cortical magnetic stimulation in human subjects. Neurosci Lett 170:149–152
38. Ray JL, McNamara B, Priest A, Boniface SJ (2002) Measuring TMS stimulus/response characteristics from both hemispheres simultaneously for proximal and distal upper limb muscles. J Clin Neurophysiol 19:371–375
39. Möller C, Arai N, Lücke J, Ziemann U (2009) Hysteresis effects on the input–output curve of motor evoked potentials. Clin Neurophysiol 120:1003–1008
40. Magistris MR, Rösler KM, Truffert A, Landis T, Hess CW (1999) A clinical study of motor evoked potentials using a triple stimulation technique. Brain 122:265–279
41. Roth G, Magistris MR (1989) Identification of motor conduction block despite desynchronisation. A method. Electromyogr Clin Neurophysiol 29:305–313
42. Komissarow L, Rollnik JD, Bogdanova D, Krampfl K, Khabirov FA, Kossev A, Dengler R, Bufler J (2004) Triple stimulation technique (TST) in amyotrophic lateral sclerosis. Clin Neurophysiol 115:356–360
43. Sakuma K, Adachi Y, Fukuda H, Kai T, Nakashima K (2005) Triple stimulation technique in patients with spinocerebellar ataxia type 6. Clin Neurophysiol 116:2586–2591
44. Eusebio A, Azulay JP, Witjas T, Rico A, Attarian S (2007) Assessment of corticospinal tract impairment in multiple system atrophy using transcranial magnetic stimulation. Clin Neurophysiol 118:815–823
45. Rico A, Audoin B, Franques J, Eusebio A, Reuter F, Malikova I, Ali Cherif A, Pouget J, Pelletier J, Attarian S (2009) Motor evoked potentials in clinically isolated syndrome suggestive of multiple sclerosis. Mult Scler 15: 355–362
46. Ugawa Y, Uesaka Y, Terao Y, Hanajima R, Kanazawa I (1994) Magnetic stimulation of corticospinal pathways at the foramen magnum level in humans. Ann Neurol 36:618–624
47. Taylor JL, Gandevia SC (2004) Noninvasive stimulation of the human corticospinal tract. J Appl Physiol 96:1496–1503
48. Molinuevo JL, Cruz-Martínez A, Graus F, Serra J, Ribalta T, Valls-Solé J (1999) Central motor conduction time in patients with multifocal motor conduction block. Muscle Nerve 22:926–932
49. Cicinelli P, Pasqualetti P, Zaccagnini M, Traversa R, Oliveri M, Rossini PM (2003) Interhemispheric asymmetries of motor cortex excitability in the postacute stroke stage: a paired-pulse transcranial magnetic stimulation study. Stroke 34:2653–2658

50. Hess CW, Mills KR, Murray NM (1986) Measurement of central motor conduction in multiple sclerosis by magnetic brain stimulation. Lancet 2:355–358
51. Chen R, Corwell B, Yaseen Z, Hallett M, Cohen LG (1998) Mechanisms of cortical reorganization in lower-limb amputees. J Neurosci 18:3443–3450
52. Penfield W, Boldrey E (1937) Somatotopic motor and sensory representation in the cerebral cortex of man as studied by electrical stimulation. Brain 60:389–443
53. Cohen LG, Hallett M (1988) Noninvasive mapping of human motor cortex. Neurology 38:904–909
54. Wassermann EM, McShane LM, Hallett M, Cohen LG (1992) Noninvasive mapping of muscle representations in human motor cortex. Electroencephalogr Clin Neurophysiol 85: 1–8
55. Thickbroom GW, Byrnes ML, Mastaglia FL (1999) Methodology and application of TMS mapping. Electroencephalogr Clin Neurophysiol Suppl 51:48–54
56. Thickbroom G, Sammut F, Mastaglia R (1998) Magnetic stimulation mapping of motor cortex: factors contributing to map area. Electroencephalogr Clin Neurophysiol 109:79–84
57. Brasil-Neto JP, Cohen LG, Pascual-Leone A, Jabir FK, Wall RT, Hallett M (1992) Rapid reversible modulation of human motor outputs after transient deafferentation of the forearm: a study with transcranial magnetic stimulation. Neurology 42:1302–1306
58. Darling WG, Wolf SL, Butler AJ (2006) Variability of motor potentials evoked by transcranial magnetic stimulation depends on muscle activation. Exp Brain Res 174: 376–385
59. Cincotta M, Borgheresi A, Balestrieri F, Giovannelli F, Ragazzoni A, Vanni P, Benvenuti F, Zaccara G, Ziemann U (2006) Mechanisms underlying mirror movements in Parkinson's disease: a transcranial magnetic stimulation study. Mov Disord 21: 1019–1025
60. Cincotta M, Ziemann U (2008) Neurophysiology of unimanual motor control and mirror movements. Clin Neurophysiol 119: 744–762
61. Civardi C, Vicentini R, Collini A, Boccagni C, Cantello R, Monaco F (2009) Motor cortical organization in an adult with hemimegalencephaly and late onset epilepsy. Neurosci Lett 460:126–129
62. Valls-Solé J, Tolosa E, Marti MJ, Valldeoriola F, Revilla M, Pastor P, Blesa R (2001) Examination of motor output pathways in patients with corticobasal ganglionic degeneration using transcranial magnetic stimulation. Brain 124:1131–1137
63. Martinez M, Brezun JM, Zennou-Azogui Y, Baril N, Xerri C (2009) Sensorimotor training promotes functional recovery and somatosensory cortical map reactivation following cervical spinal cord injury. Eur J Neurosci 30: 2356–2367
64. Karl A, Birbaumer N, Lutzenberger W, Cohen LG, Flor H (2001) Reorganization of motor and somatosensory cortex in upper extremity amputees with phantom limb pain. J Neurosci 21:3609–3618
65. Picht T, Mularski S, Kuehn B, Vajkoczy P, Kombos T, Suess O (2009) Navigated transcranial magnetic stimulation for preoperative functional diagnostics in brain tumor surgery. Neurosurgery 65(6 Suppl):93–99
66. Forster MT, Hattingen E, Senft C, Gasser T, Seifert V, Szelényi A (2011) Navigated transcranial magnetic stimulation and functional magnetic resonance imaging: advanced adjuncts in preoperative planning for central region tumors. Neurosurgery 68:1317–1325
67. Picht T, Schmidt S, Brandt S, Frey D, Hannula H, Neuvonen T, Karhu J, Vajkoczy P, Suess O (2011) Preoperative functional mapping for rolandic brain tumor surgery: comparison of navigated transcranial magnetic stimulation to direct cortical stimulation. Neurosurgery 69: 581–588
68. Fuhr P, Agostino R, Hallett M (1991) Spinal motor neuron excitability during the silent period after cortical stimulation. Electroencephalogr Clin Neurophysiol 81:257–262
69. Cantello R, Gianelli M, Civardi C, Mutani R (1992) Magnetic brain stimulation: the silent period after the motor evoked potential. Neurology 42:1951–1959
70. Triggs WJ, Macdonell RA, Cros D, Chiappa KH, Shahani BT, Day BJ (1992) Motor inhibition and excitation are independent effects of magnetic cortical stimulation. Ann Neurol 32:345–351
71. Roick H, von Giesen HJ, Benecke R (1993) On the origin of the postexcitatory inhibition seen after transcranial magnetic brain stimulation in awake human subjects. Exp Brain Res 94:489–498
72. Inghilleri M, Berardelli A, Cruccu G, Manfredi M (1993) Silent period evoked by transcranial stimulation of the human cortex and cervicomedullary junction. J Physiol 466:521–534
73. Wassermann EM, Pascual-Leone A, Valls-Solé J, Toro C, Cohen LG, Hallett M (1993) Topography of the inhibitory and excitatory responses to transcranial magnetic stimulation in a hand muscle. Electroencephalogr Clin Neurophysiol 89:424–433

74. Brasil-Neto JP, Cammarota A, Valls-Solé J, Pascual-Leone A, Hallett M, Cohen LG (1995) Role of intracortical mechanisms in the late part of the silent period to transcranial stimulation of the human motor cortex. Acta Neurol Scand 92:383–386
75. Werhahn KJ, Kunesch E, Noachtar S, Benecke R, Classen J (1999) Differential effects on motorcortical inhibition induced by blockade of GABA uptake in humans. J Physiol 517(2): 591–597
76. Nakashima K, Wang Y, Shimoda M, Sakuma K, Takahashi K (1995) Shortened silent period produced by magnetic cortical stimulation in patients with Parkinson's disease. J Neurol Sci 130:209–214
77. Ziemann U, Bruns D, Paulus W (1996) Enhancement of human motor cortex inhibition by the dopamine receptor agonist pergolide: evidence from transcranial magnetic stimulation. Neurosci Lett 208:187–190
78. Cantello R, Tarletti R, Civardi C (2002) Transcranial magnetic stimulation and Parkinson's disease. Brain Res Brain Res Rev 38:309–327
79. Ferbert A, Priori A, Rothwell JC, Day BL, Colebatch JG, Marsden CD (1992) Interhemispheric inhibition of the human motor cortex. J Physiol 453:525–546
80. Meyer BU, Roricht S, Grafin von Einsiedel H, Kruggel F, Weindl A (1995) Inhibitory and excitatory interhemispheric transfers between motor cortical areas in normal humans and patients with abnormalities of the corpus callosum. Brain 118:429–440
81. Taylor JL, Fogel W, Day BL, Rothwell JC (1995) Ipsilateral cortical stimulation inhibited the long-latency response to stretch in the long finger flexors in humans. J Physiol 488: 821–831
82. Kato T, Kasai T, Maehara T (2002) Effects of transcranial magnetic stimulation to the reciprocal Ia inhibitory interneurones in the human wrist. Percept Mot Skills 94:575–594
83. Rothwell JC, Gandevia SC, Burke D (1990) Activation of fusimotor neurones by motor cortical stimulation in human subjects. J Physiol 431:743–756
84. Valls-Solé J, Pascual-Leone A, Brasil-Neto JP, Cammarota A, McShane L, Hallett M (1994) Abnormal facilitation of the response to transcranial magnetic stimulation in patients with Parkinson's disease. Neurology 44:735–741
85. Cowan JM, Day BL, Marsden C, Rothwell JC (1986) The effect of percutaneous motor cortex stimulation on H reflexes in muscles of the arm and leg in intact man. J Physiol 377(1): 333–347
86. Poon DE, Roy FD, Gorassini MA, Stein RB (2008) Interaction of paired cortical and peripheral nerve stimulation on human motor neurons. Exp Brain Res 188:13–21
87. Valls-Solé J, Valldeoriola F (2002) Neurophysiological correlate of clinical signs in Parkinson's disease. Clin Neurophysiol 113: 792–805
88. Petersen N, Christensen LO, Nielsen J (1998) The effect of transcranial magnetic stimulation on the soleus H reflex during human walking. J Physiol 513:599–610
89. Inghilleri M, Lorenzano C, Conte A, Frasca V, Manfredi M, Berardelli A (2003) Effects of transcranial magnetic stimulation on the H reflex and F wave in the hand muscles. Clin Neurophysiol 114:1096–1101
90. Serranová T, Valls-Solé J, Muñoz E, Genís D, Jech R, Seeman P (2008) Abnormal corticospinal tract modulation of the soleus H reflex in patients with pure spastic paraparesis. Neurosci Lett 437:15–19
91. Wolfe DL, Hayes KC, Potter PJ, Delaney GA (1996) Conditioning lower limb H-reflexes by transcranial magnetic stimulation of motor cortex reveals preserved innervation in SCI patients. J Neurotrauma 13:281–291
92. Benito Penalva J, Opisso E, Medina J, Corrons M, Kumru H, Vidal J, Valls-Solé J (2010) H reflex modulation by transcranial magnetic stimulation in spinal cord injury subjects after gait training with electromechanical systems. Spinal Cord 48:400–406
93. Mariorenzi R, Zarola F, Caramia MD, Paradiso C, Rossini PM (1991) Non-invasive evaluation of central motor tract excitability changes following peripheral nerve stimulation in healthy humans. Electroencephalogr Clin Neurophysiol 81:90–101
94. Kasai T, Hayes KC, Wolfe DL, Allatt RD (1992) Afferent conditioning of motor evoked potentials following transcranial magnetic stimulation of motor cortex in normal subjects. Electroencephalogr Clin Neurophysiol 85: 95–101
95. Kossev A, Siggelkow S, Schubert M, Wohlfarth K, Dengler R (1999) Muscle vibration:different effects on transcranial magnetic and electrical stimulation. Muscle Nerve 22:946–948
96. Kofler M, Glocker FX, Leis AA, Seifert C, Wissel J, Kronenberg MF, Fuhr P (1998) Modulation of upper extremity motoneurone excitability following noxious finger tip stimulation in man: a study with transcranial magnetic stimulation. Neurosci Lett 246:97–100
97. Valeriani M, Restuccia D, Di Lazzaro V, Oliviero A, Le Pera D, Profice P, Saturno E, Tonali P (2001) Inhibition of biceps brachii muscle motor area by painful heat stimulation of the skin. Exp Brain Res 139:168–172
98. Ziemann U, Corwell B, Cohen LG (1998) Modulation of plasticity in human motor cortex after forearm ischemic nerve block. J Neurosci 18:1115–1123

99. Cohen LG, Bandinelli S, Topka HR, Fuhr P, Roth BJ, Hallett M (1991) Topographic maps of human motor cortex in normal and pathological conditions: mirror movements, amputations and spinal cord injuries. Electroencephalogr Clin Neurophysiol Suppl 43: 36–50
100. Kew JJ, Ridding MC, Rothwell JC, Passingham RE, Leigh PN, Sooriakumaran S, Frackowiak RS, Brooks DJ (1994) Reorganization of cortical blood flow and transcranial magnetic stimulation maps in human subjects after upper limb amputation. J Neurophysiol 72:2517–2524
101. Ziemann U, Hallett M, Cohen LG (1998) Mechanisms of deafferentation-induced plasticity in human motor cortex. J Neurosci 18: 7000–7007
102. Stefan K, Kunesch E, Cohen LG, Benecke R, Classen J (2000) Induction of plasticity in the human motor cortex by paired associative stimulation. Brain 123:572–584
103. Classen J, Wolters A, Stefan K, Wycislo M, Sandbrink F, Schmidt A, Kunesch E (2004) Paired associative stimulation. Suppl Clin Neurophysiol 57:563–569
104. Lourenço G, Meunier S, Vidailhet M, Simonetta-Moreau M (2007) Impaired modulation of motor cortex excitability by homonymous and heteronymous muscle afferents in focal hand dystonia. Mov Disord 22:523–527
105. Tamburin S, Manganotti P, Marzi CA, Fiaschi A, Zanette G (2002) Abnormal somatotopic arrangement of sensorimotor interactions in dystonic patients. Brain 125:2719–2730
106. Bagnato S, Agostino R, Modugno N, Quartarone A, Berardelli A (2006) Plasticity of the motor cortex in Parkinson's disease patients on and off therapy. Mov Disord 21: 639–645
107. Quartarone A, Rizzo V, Terranova C, Morgante F, Schneider S, Ibrahim N, Girlanda P, Bhatia KP, Rothwell JC (2009) Abnormal sensorimotor plasticity in organic but not in psychogenic dystonia. Brain 132:2871–2877
108. Rapisarda G, Bastings E, de Noordhout AM et al (1996) Can motor recovery in stroke patients be predicted by early transcranial magnetic stimulation? Stroke 27:2191–2196
109. Trompetto C, Assini A, Buccolieri A, Marchese R, Abbruzzese G (2000) Motor recovery following stroke: a transcranial magnetic stimulation study. Clin Neurophysiol 111:1860–1867
110. Escudero JV, Sancho J, Bautista D, Escudero M, López-Trigo J (1998) Prognostic value of motor evoked potential obtained by transcranial magnetic brain stimulation in motor function recovery in patients with acute ischemic stroke. Stroke 29:1854–1859
111. Di Lazzaro V, Oliviero A, Profice P, Saturno E, Pilato F, Tonali P (1999) Motor cortex excitability changes within 8 hours after ischaemic stroke may predict the functional outcome. Eur J Emerg Med 6:119–121
112. Heald A, Bates D, Cartlidge NE et al (1993) Longitudinal study of central motor conduction time following stroke. 2. Central motor conduction measured within 72 h after stroke as a predictor of functional outcome at 12 months. Brain 116:1371–1385
113. Turton A, Wroe S, Trepte N et al (1996) Contralateral and ipsilateral EMG responses to transcranial magnetic stimulation during recovery of arm and hand function after stroke. Electroencephalogr Clin Neurophysiol 101:316–328
114. Butler AJ, Wolf SL (2007) Putting the brain on the map: use of transcranial magnetic stimulation to assess and induce cortical plasticity of upper-extremity movement. Phys Ther 87: 719–736
115. Alagona G, Delvaux V, Gérard P, De Pasqua V, Pennisi G, Delwaide PJ, Nicoletti F, Maertens de Noordhout A (2001) Ipsilateral motor responses to focal transcranial magnetic stimulation in healthy subjects and acute-stroke patients. Stroke 32:1304–1309
116. Werhahn KJ, Conforto AB, Kadom N, Hallett M, Cohen LG (2003) Contribution of the ipsilateral motor cortex to recovery after chronic stroke. Ann Neurol 54:464–472
117. Muellbacher W, Artner C, Mamoli B (1998) Motor evoked potentials in unilateral lingual paralysis after monohemispheric ischaemia. J Neurol Neurosurg Psychiatry 65:755–761
118. Liepert J, Graef S, Uhde I et al (2000) Training induced changes of motor cortex representations in stroke patients. Acta Neurol Scand 101:321–326
119. Taub E, Uswatte G, Pidikiti R (1999) Constraint-induced movement therapy: a new family of techniques with broad application to physical rehabilitation—a clinical review. J Rehabil Res Dev 36:237–251
120. Ruohonen J, Karhu J (2010) Navigated transcranial magnetic stimulation. Neurophysiol Clin 40:7–17
121. Benecke R, Meyer BU, Schönle P, Conrad B (1988) Transcranial magnetic stimulation of the human brain: responses in muscles supplied by cranial nerves. Exp Brain Res 71: 623–632
122. Schmid UD, Moller AR, Schmid J (1991) Transcranial magnetic stimulation excites the labyrinthine segment of the facial nerve: an intraoperative electrophysiological study in man. Neurosci Lett 124:273–276
123. Benecke R, Meyer BU (1991) Magnetic stimulation of corticonuclear systems and of cranial nerves in man: physiological basis and

clinical application. Electroencephalogr Clin Neurophysiol Suppl 43:333–343
124. Costa J, Valls-Sole J, Valldeoriola F, Rumia J, Tolosa E (2007) Motor responses of muscles supplied by cranial nerves to subthalamic nucleus deep brain stimuli. Brain 130: 245–255
125. Paradiso GO, Cunic DI, Gunraj CA, Chen R (2005) Representation of facial muscles in human motor cortex. J Physiol 567:323–336
126. Schriefer TN, Mills KR, Murray NM, Hess CW (1988) Evaluation of proximal facial nerve conduction by transcranial magnetic stimulation. J Neurol Neurosurg Psychiatry 51:60–66
127. Rösler KM, Hess CW, Schmid UD (1989) Investigation of facial motor pathways by electrical and magnetic stimulation: sites and mechanisms of excitation. J Neurol Neurosurg Psychiatry 52:1149–1156
128. Schmid UD, Moller AR, Schmid J (1992) Transcranial magnetic stimulation of the facial nerve: intraoperative study on the effect of stimulus parameters on the excitation site in man. Muscle Nerve 15:829–836
129. Glocker FX, Magistris MR, Rosler KM, Hess CW (1994) Magnetic transcranial and electrical stylomastoidal stimulation of the facial motor pathways in Bell's palsy: time course and relevance of electrophysiological parameters. Electroencephalogr Clin Neurophysiol 93:113–120
130. Curt A, Keck ME, Dietz V (1998) Functional outcome following spinal cord injury: significance of motor-evoked potentials and ASIA scores. Arch Phys Med Rehabil 79:81–86
131. Macdonell RA, Donnan GA (1995) Magnetic cortical stimulation in acute spinal cord injury. Neurology 45:303–306
132. Topka H, Cohen LG, Cole RA, Hallett M (1991) Reorganization of corticospinal pathways following spinal cord injury. Neurology 41:1276–1283
133. Brouwer B, Hopkins-Rosseel DH (1997) Motor cortical mapping of proximal upper extremity muscles following spinal cord injury. Spinal Cord 35:205–212
134. Saturno E, Bonato C, Miniussi C, Lazzaro V, Callea L (2008) Motor cortex changes in spinal cord injury: a TMS study. Neurol Res 30: 1084–1085
135. Lo YL, Chan LL, Lim W, Tan SB, Tan CT, Chen JL, Fook-Chong S, Ratnagopal P (2004) Systematic correlation of transcranial magnetic stimulation and magnetic resonance imaging in cervical spondylotic myelopathy. Spine 29:1137–1145
136. Chan KM, Nasathurai S, Chavin JM, Brown WF (1998) The usefulness of central motor conduction studies in the localization of cord involvement in cervical spondylytic myelopathy. Muscle Nerve 21:1220–1223
137. Abbruzzese G, Dall'Agata D, Morena M, Simonetti S, Spadavecchia L, Severi P, Andrioli GC, Favale E (1988) Electrical stimulation of the motor tracts in cervical spondylosis. J Neurol Neurosurg Psychiatry 51:796–802
138. Maertens de Noordhout A, Remacle JM, Pepin JL, Born JD, Delwaide PJ (1991) Magnetic stimulation of the motor cortex in cervical spondylosis. Neurology 41:75–80
139. Jaskolski DJ, Laing RJ, Jarratt JA, Jukubowski J (1990) Pre- and postoperative motor conduction times, measured using magnetic stimulation, in patients with cervical spondylosis. Br J Neurosurg 4:187–192
140. Hashimoto T, Uozumi T, Tsuji S (2000) Paraspinal motor evoked potentials by magnetic stimulation of the motor cortex. Neurology 55:885–888
141. Ellaway PH, Catley M, Davey NJ, Kuppuswamy A, Strutton P, Frankel HL, Jamous A, Savic G (2007) Review of physiological motor outcome measures in spinal cord injury using transcranial magnetic stimulation and spinal reflexes. J Rehabil Res Dev 44:69–76
142. Nardone R, Tezzon F (2003) Transcranial magnetic stimulation study in hereditary spastic paraparesis. Eur Neurol 49:234–237
143. Schwenkreis P, Tegenthoff M, Witscher K, Börnke C, Przuntek H, Malin JP, Schöls L (2002) Motor cortex activation by transcranial magnetic stimulation in ataxia patients depends on the genetic defect. Brain 125: 301–309
144. Young RE, Morgan OS, Forster A (1998) Motor pathway analysis in HAM/TSP using magnetic stimulation and F-waves. Can J Neurol Sci 25:48–54
145. Shimizu H, Shiga Y, Fujihara K, Ohnuma A, Itoyama Y (2001) Clinical and physiological significance of abnormally prolonged central motor conduction time in HAM/TSP. J Neurol Sci 185:39–42
146. Parisi L, Calandriello E, Terracciano M, Valente G, Amabile GA (1992) Analysis of central motor pathways with CCT study in HIV infections. Electroencephalogr Clin Neurophysiol 82:27–28
147. Nardone R, Buratti T, Oliviero A, Lochmann A, Tezzon F (2006) Corticospinal involvement in patients with a portosystemic shunt due to liver cirrhosis: a MEP study. J Neurol 253:81–85
148. Caramia MD, Bernardi G, Zarola F, Rossini PM (1988) Neurophysiological evaluation of the central nervous impulse propagation in patients with sensorimotor disturbances.

Electroencephalogr Clin Neurophysiol 70: 16–25
149. Pineda AA, Ogata K, Osoegawa M, Murai H, Shigeto H, Yoshiura T, Tobimatsu S, Kira J (2007) A distinct subgroup of chronic inflammatory demyelinating polyneuropathy with CNS demyelination and a favorable response to immunotherapy. J Neurol Sci 255:1–6
150. Eisen A (2004) Clinical neurophysiology of motor neuron diseases. In: Daube JR, Mauguière F (eds) Handbook of clinical neurophysiology, vol 4. Elsevier, Amsterdam
151. Urban PP, Wicht S, Hopf HC (2001) Sensitivity of transcranial magnetic stimulation of cortico-bulbar vs. cortico-spinal tract involvement in Amyotrophic Lateral Sclerosis (ALS). J Neurol 248:850–855
152. Rosen DR, Siddique T, Patterson D, Figlewicz DA, Sapp P, Hentati A, Donaldson D, Goto J, O'Regan JP, Deng HX et al (1993) Mutations in Cu/Zn superoxide dismutase gene are associated with familial amyotrophic lateral sclerosis. Nature 362:59–62
153. Weber M, Eisen A, Stewart HG, Andersen PM (2000) Preserved slow conducting corticomotoneuronal projections in amyotrophic lateral sclerosis with autosomal recessive D90A CuZn-superoxide dismutase mutation. Brain 123:1505–1515
154. Mills KR (2003) The natural history of central motor abnormalities in amyotrophic lateral sclerosis. Brain 126:2558–2566
155. Caramia MD, Cicinelli P, Paradiso C, Mariorenzi R, Zarola F, Bernardi G, Rossini PM (1991) Excitability changes of muscular responses to magnetic brain stimulation in patients with central motor disorders. Electroencephalogr Clin Neurophysiol 81:243–250
156. Yokota T, Yoshino A, Inaba A, Saito Y (1996) Double cortical stimulation in amyotrophic lateral sclerosis. J Neurol Neurosurg Psychiatry 61:596–600
157. Zanette G, Tamburin S, Manganotti P, Refatti N, Forgione A, Rizzuto N (2002) Changes in motor cortex inhibition over time in patients with amyotrophic lateral sclerosis. J Neurol 249:1723–1728
158. Humm AM, Z'Graggen WJ, Bühler R, Magistris MR, Rösler KM (2006) Quantification of central motor conduction deficits in multiple sclerosis patients before and after treatment of acute exacerbation by methylprednisolone. J Neurol Neurosurg Psychiatry 77:345–350
159. Ravnborg M, Christiansen P, Larsson H, Sorensen PS (1992) The diagnostic reliability of magnetically evoked motor potentials in multiple-sclerosis. Neurology 42:1296–1301
160. Mayr N, Baumgartner C, Zeitlhofer J, Deecke L (1991) The sensitivity of transcranial cortical magnetic stimulation in detecting pyramidal tract lesions in clinically definite multiple sclerosis. Neurology 41:566–569
161. Humm AM, Magistris MR, Truffert A, Hess CW, Rosler KM (2003) Central motor conduction differs between acute relapsing-remitting and chronic progressive multiple sclerosis. Clin Neurophysiol 114:2196–2203
162. Caramia MD, Palmieri MG, Desiato MT, Boffa L, Galizia P, Rossini PM, Centonze D, Bernardi G (2004) Brain excitability changes in the relapsing and remitting phases of multiple sclerosis: a study with transcranial magnetic stimulation. Clin Neurophysiol 115:956–965
163. Humm AM, Z'Graggen WJ, von Hornstein NE, Magistris MR, Rosler KM (2004) Assessment of central motor conduction to intrinsic hand muscles using the triple stimulation technique: normal values and repeatability. Clin Neurophysiol 115:2558–2566
164. Schmierer K, Irlbacher K, Grosse P, Röricht S, Meyer BU (2002) Correlates of disability in multiple sclerosis detected by transcranial magnetic stimulation. Neurology 59:1218–1224
165. Bednarik J, Kadanka Z (1992) Multimodal sensory and motor evoked-potentials in a 2-year follow-up-study of MS patients with relapsing course. Acta Neurol Scand 86: 15–18
166. Leocani L, Rovaris M, Boneschi FM, Medaglini S, Rossi P, Martinelli V, Amadio S, Comi G (2006) Multimodal evoked potentials to assess the evolution of multiple sclerosis: a longitudinal study. J Neurol Neurosurg Psychiatry 77:1030–1035
167. Feuillet L, Pelletier J, Suchet L, Rico A, Ali Cherif A, Pouget J, Attarian S (2007) Prospective clinical and electrophysiological follow-up on a multiple sclerosis population treated with interferon beta-1 a: a pilot study. Mult Scler 13:348–356
168. Comi G, Rovaris M, Leocani L, Martinelli V, Filippi M (2001) Clinical and MRI assessment of brain damage in MS. Neurol Sci 22(Suppl 2):S123–S127
169. Kale N, Agaoglu J, Onder G, Tanik O (2009) Correlation between disability and transcranial magnetic stimulation abnormalities in patients with multiple sclerosis. J Clin Neurosci 16: 1439–1442
170. Conte A, Lenzi D, Frasca V, Gilio F, Giacomelli E, Gabriele M, Bettolo CM, Iacovelli E, Pantano P, Pozzilli C, Inghilleri M (2009) Intracortical excitability in patients with relapsing-remitting and secondary progressive multiple sclerosis. J Neurol 256:933–938
171. Perretti A, Balbi P, Orefice G, Trojano L, Marcantonio L, Brescia-Morra V, Ascione S, Manganelli F, Conte G, Santoro L (2004) Post-exercise facilitation and depression of

motor evoked potentials to transcranial magnetic stimulation: a study in multiple sclerosis. Clin Neurophysiol 115:2128–2133
172. Ozturk A, Smith SA, Gordon-Lipkin EM, Harrison DM, Shiee N, Pham DL, Caffo BS, Calabresi PA, Reich DS (2010) MRI of the corpus callosum in multiple sclerosis: association with disability. Mult Scler 16:166–177
173. Yaldizli O, Atefy R, Gass A, Sturm D, Glassl S, Tettenborn B, Putzki N (2010) Corpus callosum index and long-term disability in multiple sclerosis patients. J Neurol 257:1256–1264
174. Boroojerdi B, Diefenbach K, Ferbert A (1996) Transcallosal inhibition in cortical and subcortical cerebral vascular lesions. J Neurol Sci 144:160–170
175. Hoppner J, Kunesch E, Buchmann J, Hess A, Grossmann A, Benecke R (1999) Demyelination and axonal degeneration in corpus callosum assessed by analysis of transcallosally mediated inhibition in multiple sclerosis. Clin Neurophysiol 110:748–756
176. Jung P, Beyerle A, Humpich M, Neumann-Haefelin T, Lanfermann H, Ziemann U (2006) Ipsilateral silent period: a marker of callosal conduction abnormality in early relapsing-remitting multiple sclerosis? J Neurol Sci 250:133–139
177. Bonzano L, Tacchino A, Roccatagliata L, Mancardi GL, Abbruzzese G, Bove M (2011) Structural integrity of callosal midbody influences intermanual transfer in a motor reaction-time task. Hum Brain Mapp 32(2): 218–228
178. Currà A, Modugno N, Inghilleri M, Manfredi M, Hallett M, Berardelli A (2002) Transcranial magnetic stimulation techniques in clinical investigation. Neurology 59:1851–1859
179. Kühn AA, Grosse P, Holtz K, Brown P, Meyer BU, Kupsch A (2004) Patterns of abnormal motor cortex excitability in atypical parkinsonian syndromes. Clin Neurophysiol 115: 1786–1795
180. Valls-Solé J (2007) Neurophysiology of motor control and movement disorders. In: Jankovic J, Tolosa E (eds) Parkinson's disease and movement disorders. Lippincott Williams and Wilkins, Philadelphia, pp 7–22
181. Valls-Solé J (2005) Role of electrophysiology in diagnosis and research in atypical parkinsonian disorders. In: Litvan I (ed) Atypical parkinsonian disorders, Clinical and research aspects. Humana Press, New Jersey, pp 409–429
182. Cruz Martínez A, Arpa J, Alonso M, Palomo F, Villoslada C (1995) Transcranial magnetic stimulation in multiple system and late onset cerebellar atrophies. Acta Neurol Scand 92: 218–224
183. Abbruzzese G, Marchese R, Trompetto C (1997) Sensory and motor evoked potentials in multiple system atrophy: a comparative study with Parkinson's disease. Mov Disord 12:315–321
184. De Rosa A, Volpe G, Marcantonio L, Santoro L, Brice A, Filla A, Perretti A, De Michele G (2006) Neurophysiological evidence of corticospinal tract abnormality in patients with Parkin mutations. J Neurol 253:275–279
185. Mascia MM, Valls-Sole J, Marti MJ, Salazar G (2005) Sensorimotor integration in patients with parkinsonian type multisystem atrophy. J Neurol 252:473–481
186. Wolters A, Classen J, Kunesch E, Grossmann A, Benecke R (2004) Measurements of transcallosally mediated cortical inhibition for differentiating parkinsonian syndromes. Mov Disord 19:518–528
187. Filipovic SR, Ljubisavljevic M, Svetel M, Milanovic S, Kacar A, Kostic VS (1997) Impairment of cortical inhibition in writer's cramp as revealed by changes in electromyographic silent period after transcranial magnetic stimulation. Neurosci Lett 222: 167–170
188. Butefisch CM, Boroojerdi B, Chen R, Battaglia F, Hallett M (2005) Task-dependent intracortical inhibition is impaired in focal hand dystonia. Mov Disord 20:545–551
189. Espay AJ, Morgante F, Purzner J, Gunraj CA, Lang AE, Chen R (2006) Cortical and spinal abnormalities in psychogenic dystonia. Ann Neurol 59:825–834

Chapter 14

A Review of Current Clinical Practice in the Treatment of Major Depression

Mark A. Demitrack and David G. Brock

Abstract

Transcranial magnetic stimulation (TMS) is a method of using high-strength pulsed magnetic fields to induce a depolarizing electric current in a localized area of the cerebral cortex. A large, replicated body of multisite, randomized sham-controlled studies has established the safety and effectiveness of a specific TMS device for the treatment of patients with pharmacoresistant major depressive disorder. This chapter reviews the current understanding of the physiologic effects and presumed mechanism of effect of TMS as an antidepressant. In addition, the scientific evidence base for its safety and efficacy are reviewed, along with a summary of its current acceptance by professional societies. The introduction of TMS as an office-based treatment in clinical psychiatry has introduced the need to understand the basic elements of organization and implementation of a TMS treatment option within a psychiatric clinical practice setting. This chapter provides a broad overview and specific recommendations for the organization of a TMS treatment team, and presents a basic framework and clinical approach to the implementation of TMS as a therapeutic technique. Aspects of patient diagnostic assessment, treatment planning, treatment team roles and responsibilities, and patient safety are reviewed.

Key words Transcranial magnetic stimulation, Antidepressant, Clinical trials, Clinical practice

1 Introduction

Transcranial magnetic stimulation (TMS) is a method of using high-strength pulsed magnetic fields to induce an electrical current in a localized area of the cerebral cortex. When applied at a magnetic field strength sufficient to induce neuronal depolarization, different standardized simple patterns of pulsed activity will result in varied but predictable and enduring changes in the functional activity of cortical neurons. For example, slower pulse repetition rates (i.e., at periods of 1 pulse per second or less) have been observed to result in inhibition of neuronal function, while more rapid patterns of activity (i.e., at periods of more than 1 pulse per second) have been observed to result in excitation of neuronal activity [1]. More complex compound patterns of stimulation that

Alexander Rotenberg et al. (eds.), *Transcranial Magnetic Stimulation*, Neuromethods, vol. 89,
DOI 10.1007/978-1-4939-0879-0_14, © Springer Science+Business Media New York 2014

combine both fast and slow pulse sequences, such as theta burst TMS, are currently being explored, but their functional action and potential clinical benefits are less well understood at this time [2].

In general, the application of rapidly pulsed TMS has been an active area of investigation for some time as a therapeutic option for patients who have failed to receive acceptable levels of clinical improvement from initial treatment attempts using conventional antidepressant medications. Why should TMS even be considered as a plausible method of antidepressant treatment? There are several compelling circumstantial reasons that address this question. First, there is now ample evidence from functional neuroimaging studies of depressed patients that implicates reduction of metabolic activity in specific areas of the brain, most notably the left prefrontal cortex [3–5]. Second, in human experimental models, TMS has been shown to generate effects on several surrogate physiological endpoints that are consistent with potential antidepressant action, including increasing brain metabolic activity in targeted regions of the cortex [6], delaying the onset of REM sleep [7], and normalizing hypothalamic pituitary adrenal axis function [8]. Third, in animal models of depression, TMS has been shown to produce behavioral effects consistent with those known to be predictive of antidepressant action in humans [9]. Fourth, TMS has been shown in both animal and human models to produce changes in a variety of neurochemical and cellular targets that are implicated in antidepressant action [10]. In sum, TMS has a substantial preclinical and clinical evidence base that would strongly implicate it as a candidate human antidepressant.

Based on this evidence, it is not surprising that TMS has been shown to result in a statistically significant and clinically meaningful antidepressant effect in patients with major depression. In fact, the evidence for the clinical efficacy of TMS in the treatment of depression is considerable, spanning more than 30 controlled clinical studies in over 2,000 patients. To date, this aggregate data has been examined in more than ten meta-analyses and two qualitative reviews published in the peer-reviewed literature from 2001 to 2011 [11–23]. Overall, these reviews provide a consistent, comprehensive, and replicated literature base that supports the specific use of left-prefrontal, high frequency TMS in patients with treatment-resistant depression. Among the most recent published meta-analyses, the report by Schutter et al. [21] incorporates data from 30 studies involving 1,164 patients. The author reported an estimated standardized effect size of 0.39 (95 % confidence interval: 0.25–0.54) and concluded, "…high frequency TMS over the left dorsolateral prefrontal cortex (DLPFC) is superior to sham in the treatment of depression. The effect size is robust and comparable to at least a subset of commercially available antidepressant drug agents…." Another meta-analysis, and among the largest to date [22], extends these observations. This latter report examined data from 34 studies involving 1,383 patients. These authors reported an effect size of 0.55 (95 % confidence interval 0.38–0.72) for the use of TMS in the treatment of depression, and concluded, "...rTMS deserves a place in the standard toolbox of psychiatric treatment methods, as it is effective for depression and has a mild side effect profile... Although rTMS cannot replace ECT in depressive patients, there may be subgroups in which rTMS can replace antidepressant medication...."

Comprehensive medical technology reviews also provide support that TMS is an evidence-based treatment option for patients who have failed to benefit from initial acute phase treatment of major depression. In 2011, the federally funded Effective Health Care Program of the Agency for Healthcare Research and Quality (AHRQ), published a Comparative Effectiveness Review (Number 33), entitled, "Nonpharmacologic Interventions for Treatment-Resistant Depression in Adults" [24]. Overall, the AHRQ Panel concluded that there is a substantial and well-replicated body of evidence from randomized, sham-controlled clinical trials that provide a "high strength of evidence" that TMS produces significantly greater decreases in depression severity, response rate, and remission rate when compared to a sham treatment condition in the majority of peer-reviewed published clinical trials. Specifically, they noted, "...rTMS averaged a decrease in depressive severity measured by the Hamilton Rating Scale for Depression (HAM-D) of more than five points relative to sham control, and this change meets the minimum threshold of the 3-point HAM-D difference that is considered clinically meaningful. Response rates were greater with rTMS than sham (also high strength of evidence); those receiving rTMS were more than three times as likely to achieve a depressive response as patients receiving sham procedure. Finally, rTMS was also more likely to produce remission than the control procedure (moderate strength of evidence); patients receiving rTMS were more than six times as likely to achieve remission as those receiving the sham." The AHRQ Comparative Effectiveness report is particularly important because their findings represent a rigorously conducted, unbiased assessment of the available scientific evidence. They stand in a unique and authoritative position as a statement on the favorable scientific and clinical conclusions that can be drawn from the peer-reviewed, published literature on the use of TMS in depression. Finally, they are consistent with the prevailing conclusions in the broader scientific literature regarding the safety and efficacy of the use of TMS in pharmacoresistant major depression.

The clinical utility of a specific TMS treatment protocol as delivered by a specific medical device (NeuroStar TMS Therapy System, Neuronetics, Inc., Malvern, PA, USA) was established in a large (N = 301 patients), multisite (N = 23 study sites), randomized sham-controlled trial. This device was the first therapeutic TMS device cleared for clinical use in the United States by the US Food and Drug Administration (FDA) in October of 2008 [25–27]. Recently, a second, large (N= 190 patients), multisite (N= 4 study sites), randomized sham-controlled trial has been reported and provides confirmation of the efficacy of this specific TMS device in pharmacoresistant major depression [28]. This second study is particularly important as it was an NIH-sponsored study and was, therefore, conducted independent of industry support. In addition, this study incorporated several methodological improvements on prior work, such as the use of an active sham TMS control with rigorous attention to the integrity of study blinding. These investigators also focused on the primary efficacy endpoint of remission, as measured using the 24 item Hamilton Depression Rating Scale (HAMD24), in order to clearly determine the clinical significance of the study outcome. Using the same device cleared for commercial use by the FDA and the same treatment parameters as reported in prior studies [25], these authors reported that there was a significant effect of active treatment on the proportion of remitters (15 % active TMS vs. 4 % sham control group, P= 0.015), representing 4.2 greater odds of reaching remission with active TMS compared to sham control. They concluded, "…daily left prefrontal rTMS as monotherapy produced significant and clinically meaningful antidepressant therapeutic effects greater than sham…" [28].

In addition to these two pivotal trials, the NeuroStar TMS system has been studied in two open-label extension trials in patients who did not respond to the initial TMS or sham treatment in the above described studies [29, 30] and in a large, multisite, open-label post-market study evaluating the efficacy in naturalistic use in patients across a range of antidepressant treatment resistance [31]. Clinical outcomes in this latter observational study showed response and adherence rates under conditions of general clinical use that were similar to those reported in open-label clinical trials in the research study populations. These real world data validate the efficacy of NeuroStar TMS Therapy reported in published controlled trials, and further support TMS as an effective and well-tolerated treatment for those patients who have failed to benefit from antidepressant medication.

The NeuroStar TMS Therapy System has been cleared by the US FDA and is indicated for the treatment of Major Depressive Disorder in adult patients who have failed to receive satisfactory improvement from prior antidepressant medication in the current episode. The NeuroStar System is used for patient treatment by prescription only under the supervision of a licensed physician. The purpose of this chapter is to provide a detailed description of the standard treatment protocol used with the NeuroStar System and to review basic issues concerning patient selection, treatment planning, and follow-up within the context of the evidence-based treatment protocol. A second TMS device has recently been cleared by the FDA (Brainsway, Inc., Tel Aviv, Israel). That device uses a different coil design than used in the Neuronetics NeuroStar TMS device, and therefore differs in certain pulse characteristics. At the time of this chapter's writing, the safety and efficacy data from that device's registration trial has not been reported in the peer-reviewed literature, nor are there any reports of clinical use of the device in routine practice. Therefore, the remainder of this chapter will consider the specific design and treatment considerations for the use of the NeuroStar TMS Therapy device.

2 Overview of TMS Practice Setting and Introducing TMS to the Patient

In this section we will review several aspects of the clinical delivery of TMS, including a discussion of the basic requirements of an office-based practice and several important aspects involved in the preparation and introduction of the TMS procedure to the patient.

2.1 The TMS Suite: Key Components

TMS has been designed for use primarily as a convenient, outpatient, office-based medical procedure. This is possible in large part because of its ease of use and the tolerability and safety of the procedure when delivered in a manner consistent with the labeled treatment parameters. The standard TMS office suite should have a dedicated space for the TMS equipment separate from other areas of the office for patient privacy. Optimal floor dimensions of approximately 12 ft by 15 ft should be anticipated (Fig. 1). TMS utilizes the same type of pulsed magnetic field technology employed in magnetic resonance imaging devices. However, because the size of the magnet and the resulting magnetic field generated by a TMS device is confined to a much smaller spatial volume than an MR imaging device, there is no requirement of special magnetic field shielding for the room that houses the device. It is recommended that the room be arranged in a manner oriented toward optimizing patient comfort, since the 40–55 min procedure is performed with the patient awake and alert throughout. Therefore, comfortable surroundings likely enhance patient acceptance of the treatment itself. Access to seating and a workspace for the clinical staff in attendance during the procedure is also recommended, since a TMS session should be continuously observed by either the physician or their clinical staff designee to manage any during treatment emergent clinical issues.

2.2 Roles and Responsibilities of the Physician and the TMS Treater

Current labeling limits the circumstances under which TMS can be prescribed for therapeutic management of patients with major depression. Specifically, prescription of the treatment must be under the authority of a licensed physician, who is ultimately responsible for the overall day-to-day management of the TMS treatment course. It is recommended that the prescribing physician perform the initial motor threshold determination, establish the anticipated clinical treatment plan based on assessment of the

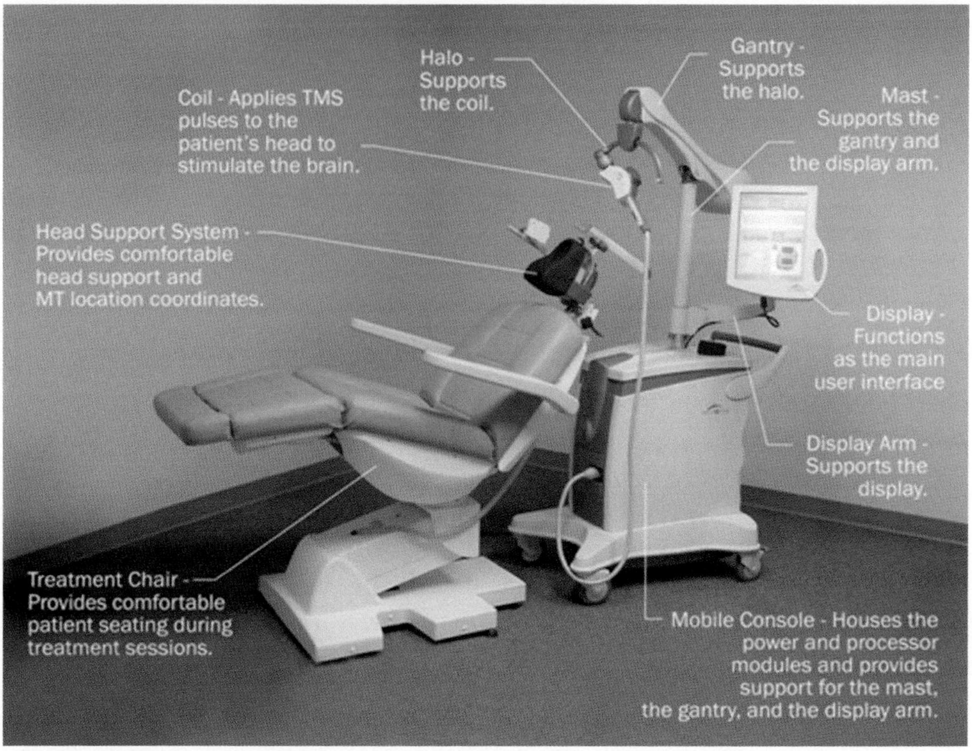

Fig. 1 NeuroStar TMS therapy system: key components

patient's clinical history, and review this treatment plan with the patient prior to beginning the course of treatment. The physician must also review the clinical course of each daily treatment session in order to determine whether any modifications to the subsequent daily treatment should occur. For example, the physician should evaluate whether a redetermination of motor threshold is required prior to the following day's session and respond to adverse events as they occur. The attendance and management of daily treatment sessions may sometimes be delegated by the treating physician to another member of the clinical staff. Currently, there are no published or publically defined consensus accepted minimum standards for the basic skills needed by TMS treater. Neuronetics has established internal procedures and standards for TMS clinical training and recommends that TMS clinical staff be clinical professionals and maintain appropriate training to support their role as first responders to potential medical emergencies (see sections below). However, because TMS is a medically complex treatment, the physician must remain accessible at all times to the TMS treater in order to respond in the event of a medical emergency.

In order to understand the importance of the role of the attending physician in the overall management and clinical coordination of the TMS treatment team, it is important to consider the potential medical risks associated with the use of TMS. The most important safety risk associated with TMS is the inadvertent induction of a seizure. Therefore, it is essential that both the supervising physician and the TMS treatment staff are familiar with proper first responder capability for such an event. The incidence of seizure with TMS is small and appears slightly lower than the incident risk reported for the use of current antidepressant medications [26]. It is of interest to note that the typical characteristics of an inadvertently induced TMS seizure are remarkable in their pattern and predictability. Specifically, seizure events that have been reported in the scientific literature with TMS have all been generalized, tonic-clonic seizure episodes which terminated with conservative intervention only [26]. Further, they have only been reported to occur during the actual treatment session itself while the TMS device is actively pulsing. Finally, there have been no reports of an inadvertently induced seizure progressing to status epilepticus, or to the later development of a spontaneous seizure disorder, or to persistent abnormalities in the post-ictal EEG, when obtained. It has been reported [26] that the two most critical TMS treatment parameters that may contribute to an increased seizure risk are: (1) the duration of stimulation at a given frequency and magnetic field intensity [32], and (2) the duration of time between stimulation periods [33]. In clinical practice, the use of an appropriately worded informed consent procedure (discussed in the following section) is recommended, as are adequate methods for pretreatment clinical screening of potential seizure risk and continuous clinical monitoring of the TMS treatment session itself. All clinical personnel involved in the delivery of TMS care must be trained as "first responders" in order to provide appropriate initial management for a seizure or other medical event. Additional attention to existing recommendations for TMS treatment parameter limits also appears to minimize the risk of this medical event. There were no seizures during the 10,000 treatments conducted in the clinical trials of the NeuroStar TMS System prior to FDA clearance. Since commercial introduction of TMS in the United States in October, 2008, there have been eight reported seizures. Based on this, the overall risk of seizure is estimated to be less than 1 in 30,000 treatment sessions (<0.003 %) or less than 1 in 1,000 patient exposures (<0.1 %) (Neuronetics, Inc., data on file). Finally, it is worth noting that although access to nearby medical support is common practice, there are no specific labeling requirements for advanced resuscitative equipment to be present in the TMS treatment room.

To remain consistent with the approved product labeling, standards of practice and regulatory requirements for device training have been established by the TMS medical device manufacturer

(Neuronetics, Inc., Malvern, PA, USA). In some settings, additional local institutional requirements have been developed to document credentialing expectations prior to use or to define local standard operating procedures and informed consent processes. Other professional organizations are beginning to offer additional continuing medical education courses on the theory and practice of TMS thereby providing important supplemental information.

2.3 Evaluation and Presentation of TMS to the Patient

Because the TMS device produces a strong magnetic field in the immediate vicinity of the treatment coil, careful screening, and evaluation of the patient's medical history should be performed to determine whether any ferromagnetically active metal is present. The use of TMS is contraindicated in the presence of conductive, ferromagnetic, or other magnetically active metals that are implanted in the head or are otherwise nonremovable and are present within 30 cm of the face of the coil. The reason for this is that, within 30 cm of the face of the treatment coil, the peak magnetic field can be greater than 5 G, which is the recommended static magnetic field exclusion level for many electronic devices.

The clinical circumstances described above are the only absolute contraindication to the use of TMS. TMS has been administered in the presence of other types of implanted metallic objects, including implanted medical devices that were located outside of the 30 cm exclusion limit [32]. However, caution should be taken in the decision to use TMS in such situations and a careful assessment of the potential benefits of TMS should be weighed against the risks of inadvertently causing a malfunction of such implanted devices. Screening checklists for metallic objects (specifically highlighting those that are unsafe to use with TMS and those that require clinical caution) are provided with product documentation and should be consulted during safety screening prior to treatment.

2.4 Diagnostic Assessment and Evaluation of Prior Treatment History

In the clinical studies that led to the FDA clearance of the NeuroStar TMS Therapy System, patients were carefully characterized in terms of both their illness history and their prior treatment history. Under the conditions of use defined in the FDA-cleared evidence base, the patient population for whom TMS has been shown to be effective is clinically well defined. In addition to meeting the DSM-IV disease diagnostic criteria for a major depressive disorder, with a nonpsychotic, unipolar course of illness, patients for whom TMS is a clinically appropriate treatment intervention also demonstrated the following demographic and clinical features [25]:

- *Moderate to severe treatment resistance in the current treatment episode*
  - In registration clinical studies, the average patient had already received at least 4 antidepressant treatment attempts in the current episode, with a range of 1 to as many as 23 antidepressant treatment attempts reported.

  - Among these treatment attempts, the patient had been able to tolerate at least one antidepressant treatment at an acceptable dose and duration of exposure to validate the presence of pharmacoresistance in the current illness episode.
  - Lifetime treatment attempts were not limited in these clinical studies.
- *A recurrent course of illness*
  - In registration clinical studies for the NeuroStar TMS System, over 95 % of patients had experienced prior illness episodes.
  - The average patient age was approximately 49 years, which reflected an average age about a decade older than a first episode patient population.
- *Moderate to severe illness severity upon initial clinical evaluation, and*
- *Work productivity reflecting significant functional morbidity*
  - Nearly 50 % of patients were currently unemployed due to their illness.
  - About 30 % of patients were receiving disability due to their current illness.

These clinical characteristics describe a patient population who have failed to receive satisfactory clinical improvement from initial acute phase antidepressant medication treatment and are faced with the prospect of progressing to more complex, difficult-to-tolerate antidepressant medication regimens as the traditional next step in their clinical care. These regimens, many of which have not been specifically studied or approved for use by the FDA in a treatment-resistant population, include the use of augmentation or combination treatments such as lithium, thyroid hormone, or second-generation antipsychotic medications added to a primary antidepressant. The use of older medications, such as tricyclic antidepressants or monoamine oxidase inhibitors, may also be considered at this stage of the patient's care. For many patients, the risk of treatment discontinuation due to intolerance of these more cumbersome treatments increases dramatically. This latter point has been demonstrated vividly in the recently completed Sequenced Treatment Alternatives to Relieve Depression (STAR*D) Study [34]. In this series of studies, the likelihood of treatment discontinuation due to adverse events to the next offered treatment option nearly tripled after one treatment failure and nearly quintupled after three treatment failures [34]. Therefore, the option of using a better tolerated antidepressant treatment such as TMS offers an attractive alternative for the patient who has failed initial acute phase pharmacotherapy. The importance of TMS as a treatment alternative for patients at this stage of their illness is now recognized in the recently reported 3rd Edition of the Practice Guidelines for the Treatment of Major Depressive Disorder from the

American Psychiatric Association. Other published international treatment guidelines also acknowledge the importance of TMS for this patient population [35, 36].

From the evidence obtained in the clinical studies of the NeuroStar TMS system [25–27], and consistent with other observations in the scientific literature [28], it is clear that an accurate assessment of the patient's prior antidepressant treatment history is a critical aspect of the initial assessment process. This is because the level of prior treatment resistance is one of the most powerful predictors of future treatment outcome with TMS or any other antidepressant treatment. Several validated tools are available to assist in defining this clinical characteristic. The Antidepressant Treatment History Form (ATHF) is the best studied tool and has been shown to have prospective validity in determining treatment outcome [37]. A simplified, clinician-adapted and validated version of this form, the Antidepressant Treatment Record (ATR), was developed by Neuronetics during the registration clinical studies and is available upon author request for routine use in clinical practice.

2.5 Concurrent Medical and Psychiatric Illnesses

Major depression rarely occurs in isolation and usually presents in clinical practice concurrent with other medical or psychiatric illnesses. For example, nearly one third of patients in the clinical studies leading to the clearance of the NeuroStar TMS system had a concurrent anxiety disorder. Moreover, medical comorbidities, including cardiovascular illness and diabetes, were common. There was no evidence that the concurrent existence of a medical or psychiatric illness had any effect on the response to TMS.

Concurrent neurologic illnesses represent an area of special interest. It is important to determine whether any accompanying neurologic illness may be expected to alter seizure threshold. While not a contraindication to the use of TMS, the existence of an alteration in seizure threshold due to a concurrent neurologic condition should be included in the risk-benefit considerations. It is worth noting that this clinical risk-benefit consideration is an intrinsic part of the management of the depressed patient with an altered seizure threshold as virtually every known antidepressant treatment option confers a slight increased risk of seizure.

2.6 Presenting TMS to the Patient and Obtaining Informed Consent

Once a decision has been made to prescribe the use of TMS as a treatment option, it is crucial that the patient has a thorough, accurate, and informative presentation of what treatment with TMS entails. It is generally helpful to escort the patient to the treatment room and allow them to see, touch, hear, and understand the basic purpose of all parts of the TMS device. Remember that during the treatment session, since the patient is reclined in the treatment chair, the majority of the operating aspects of the device are out of the patient's field of view. As such, reducing anxiety regarding

Table 1
NeuroStar TMS therapy system standard TMS treatment parameter set

Treatment parameter	NeuroStar TMS therapy characteristic
Motor threshold (MT)	Lowest intensity over primary motor cortex to produce visual contraction of the first dorsal (thumb) interosseous or abductor pollicis brevis muscle
Treatment coil location	Left dorsolateral prefrontal cortex (DLPFC)—5 cm anterior from the MT location along a left superior oblique plane on the surface of the head
Number of pulses per session	3,000
Magnetic field intensity	120 % of MT (typical setting)
Pulse repetition rate	10 pulses per second
Stimulation time (i.e., pulse train duration)	4 s
Stimulation interval (i.e., between pulse trains)	26 s
Length of standard treatment session	37.5 min

the nature of the treatment process is essential prior to treatment commencement. A variety of visual aids are provided with the product documentation, including treatment brochures and videos, which can also be used to instruct the patient on the treatment process. It is generally encouraged to invite family members into the consultation room to address any questions they may have. Only when the procedure is well understood and questions have been answered should written informed consent be obtained and documented in the medical record.

3 The Treatment Session: Dose Estimation, Treatment Localization, and Treatment Parameters

3.1 Summary of Standard Protocol Parameters

For acute phase treatment of patients with major depression who have not benefited from initial treatment, the NeuroStar TMS System has been demonstrated to be safe and effective when used with a specific treatment regimen, described in the FDA-cleared labeling for the device. This regimen has the most replicated body of evidence for efficacy of any specific treatment parameter set. The regimen consists of daily (five times per week) treatment sessions administered over the left dorsolateral prefrontal cortex using the specific treatment parameter set shown in Table 1. Based on the studies conducted for FDA clearance, the recommended minimum duration of treatment is 4 weeks, or 20 sessions, and most patients should receive benefit by 6 weeks or 30 sessions. The convention of daily sessions is largely based on the substantial safety profile of TMS which permits apparent unlimited lifetime exposure and on clinical anecdote in research settings which suggests that treatment interruption for extended periods during the period of acute phase treatment degrades the potential maximum benefit. During the NeuroStar clinical development, periods of treatment interruption longer than 3 consecutive days were not permitted.

There is no systematic controlled evidence at the present time that a greater duration of the acute treatment session length (i.e., more than 3,000 pulses per session) confers any increased efficacy. On the other hand, there is no evidence to suggest that there is any new safety concern that emerges with extended exposure [38]. Alternative treatment regimens for TMS that may increase the potential for efficacy are an area of active research scrutiny. Probably the best studied of these alternative regimens is the use of bilateral TMS treatment, usually conducted as sequential left high frequency treatment followed by low frequency treatment on the symmetric location on the right hemisphere [39–41]. No standard right hemisphere treatment regimen is established at the present time. The interested reader is referred to several excellent research investigations of this technique for more details [39–41].

3.2 Methods of Treatment Localization and Dose Estimation

Early in the scientific investigation of the antidepressant efficacy of TMS, several research investigators adopted a simple, indirect method of locating the left dorsolateral prefrontal cortex in humans based on surface landmarks. This method exploits the fact that stimulation over the motor strip serves as a convenient functional landmark for TMS action in that depolarization of motor neurons results in an observable movement in the contralateral limb. The location on the motor cortex of the thumb or fingers in most humans is oriented at a point approximately 30° from the midline. Following along an oblique plane with a rotation point about the patient's nose, and moving anterior from the motor strip approximately 5 cm locates the coil at a point on the surface of the head that is nearly coincident with the dorsolateral prefrontal cortex [52]. It is obvious that, although generally accurate most of the time, this technique is vulnerable to anatomic variation of the intended location. For example, variations in head circumference or variation in motor strip organization may influence the starting location and will result in subsequent variability in location of the site of treatment. The error associated with this method has been demonstrated formally in the research literature [42]. Despite the degree of anatomic variability, this method has been robust with regard to demonstration of efficacy in replicated studies of pharmacoresistant depression. This is likely due to the fact that the specific depression "locus" in the left frontal lobe is not small, but probably involves large areas of cortical tissue that are greater even in size than the intrinsic variation of this method. Nevertheless, investigators have begun to explore whether alternative methods of treatment localization that account for head size or involve imaging of the patient's brain with neuronavigation to the site of treatment may confer greater degrees of efficacy than the external landmark method described above [43, 44]. Whether these methods will result in improved outcome is not yet clear, however, it is unlikely that these modifications of treatment localization will affect the safety of the treatment in any meaningful way.

Dose estimation for a TMS session is based upon the relative degree of cortical excitability of a patient. This is done because it is known that increasing the magnetic field intensity to a high enough level increases the risk of inadvertently inducing seizure. Rather than measure seizure threshold directly, estimation of cortical excitability is used as an indirect proxy. By convention, this is assessed by determining the patient's motor threshold, which is then used as a base reference to calculate the strength of the applied magnetic field used over the treatment location. The motor threshold is defined as the magnetic field strength required to produce a visible movement of the thumb (or a nearby finger if the thumb is not localizable) one out of every two times a magnetic pulse is applied. The NeuroStar System software uses a proprietary mathematical algorithm based on a maximum likelihood estimation method to assist the operator in performing this estimation method (MT Assist, Neuronetics, Inc., Malvern, PA, USA). The device automatically adjusts the intensity of the magnetic field after each pulse in response to a query of the TMS operator as to whether a visible movement of the finger was observed. Once sufficient mathematical information is obtained, the motor threshold value can be estimated and displayed. Some investigators alternatively use surface electromyography of the hand to validate muscle contraction. However, there is little evidence that use of an EMG results in a substantially different result than that based on visible movement alone [45]. The treatment dose in the NeuroStar TMS standard protocol is fixed at 120 % of the patient's observed motor threshold.

3.3 Special Topics: Seizure Management, Concomitant Medication Use, and Auditory Precautions

As noted in previous sections, the most significant medical risk with the use of TMS is the inadvertent induction of a seizure. Though the risk is exceedingly low, the TMS team members in the standard office set up should be ready to anticipate such an event. Written standard operating procedures for seizure response should be posted and all staff trained on these steps. It is recommended that all clinical staff be formally trained as first responders. Familiarity and up-to-date training on basic life support (BLS) would accomplish this goal. There has been no incidence in the reported literature of the use of any advanced resuscitative equipment when a seizure has occurred during a TMS session. As such, the labeling does not require the presence of such advanced equipment as this is unlikely to be of any relevance to immediate clinical management of the patient. On occasion, local requirements may dictate access to such equipment as a precaution.

The NeuroStar TMS System was studied in controlled trials as a monotherapy for the treatment of pharmacoresistant major depression. During the taper phase of that study, during the 6 month longer term follow-up study [46], and in the large, multisite naturalistic outcomes study [31] TMS was administered in conjunction with antidepressant medication. The majority of the prior clinical trial literature concerning the use of TMS in the treatment of depression has also been performed using TMS as an add-on treatment to complex antidepressant pharmacotherapy. Therefore, the labeling for the NeuroStar System permits the clinician to determine whether concurrent administration of TMS with other psychotropic medication is in the patient's best interest under the circumstances of that specific treatment setting. Most clinicians will make this determination and revise any ongoing medications to a stable regimen prior to initiating treatment. If medication changes occur during the course of treatment, it is recommended that the motor threshold be redetermined prior to the next treatment session, since a majority of the psychotropic medications commonly used in the setting of major depression are known to alter seizure threshold.

During a TMS session, the magnetic pulse produces an audible high frequency clicking sound. It has been shown that without routine ear protection, short-term decreases in air conduction auditory threshold may occur [47, 48]. Therefore, an additional safety precaution that is now standard in all TMS treatments is the use of ear protection with ear plugs capable of at least 30 dB sound reduction. Such a precaution eliminates the risk of changes in auditory threshold with treatment.

4 The TMS Treatment Plan

4.1 Planning an Acute Treatment Course

As described above, the standard treatment regimen studied in the NeuroStar clinical development plan showed gradual and continued benefit through the 4th week of treatment, typically plateauing by the end of the 6th week. Therefore, patients should be advised of this prior to initiating treatment in order to set appropriate expectations of the time course of benefit and when and how assessment of efficacy should reasonably be estimated. As with antidepressant medications, unusually effective first treatments should be appraised in context and not considered to represent or define whether TMS will or will not ultimately be shown to have long-term benefit for a patient.

Objective documentation of clinical benefit during a treatment course is an important component of clinical care. Although a general mental status examination is an essential foundational source of information, it is advised that a disease-specific, validated measurement tool be used to specifically characterize the course and outcome of treatment. Such tools are useful for several purposes. Most importantly, they ensure a thorough review of all relevant disease symptoms in the conversation between the physician and the patient and thereby serve as an important source of information allowing the patient to increase their awareness of specific disease features and participate more actively in their own recovery. Measurement of disease progress also provides an objective medical record documenting the progress of clinical improvement for the specific disease state, and can be useful in furnishing supporting medical information to a third party payor when reimbursement claims are filed. There are several validated patient-reported outcome measures of depression symptoms that are available in the public domain, along with their methods of scoring. Examples include the Patient Health Questionnaire, 9-Item scale (PHQ-9; [49]; http://www.depression-primarycare.org/clinicians/toolkits/materials/forms/phq9/) and the Quick Inventory of Depressive Symptoms—Self Report (QIDS-SR; 50; http://www.ids-qids.org/). These scales have been shown to provide valid and reliable measures of depression symptom change, are easy to use, and correspond closely to change measured with more sophisticated research-based clinician-administered rating scales [50].

Lack of meaningful clinical benefit by 6 weeks (30 sessions) under the standard protocol raises several questions. First, throughout the treatment course, attention and care should be given to the estimation of motor threshold. If the initial motor threshold determination was unusually difficult (raising the possibility that the initial treatment dose estimation may have been inappropriately low or the treatment location may not have been accurately estimated), the clinician should consider early in the course of treatment whether a redetermination of the motor threshold should be performed. If this is deemed necessary, the entire motor threshold procedure, including localization of the motor threshold site itself, should be performed again. Second, changes in concomitant medication during treatment should be avoided. If this has occurred, however, the clinician should consider whether lack of response to TMS is an artifact of medication change or representative of an effect of medication withdrawal. Third, in the determination of treatment resistance it is well known that late response may occur; therefore, some patients may benefit from an extended duration of acute treatment if possible. During the NeuroStar clinical trials, some patients were treated for acute phase durations as long as 12 weeks when they proceeded from the randomized controlled trial into the open-label extension study [29]. In that setting, some patients who had not benefited by the end of 4–6 weeks in the randomized study eventually achieved remission at the end of an additional 4–6 weeks of acute TMS treatment. Fourth, as with any treatment-resistant patient, failure of benefit from treatment should also raise questions in regard to accuracy of the initial diagnosis.

4.2 Clinical Management During a Treatment Session

TMS is remarkable for its safety and tolerability. Based on the available evidence, it appears to be the most tolerable antidepressant treatment option available for the treatment-resistant patient. Nevertheless, the patient should be advised of the type and pattern of common adverse events experienced in clinical studies. Specifically, cutaneous discomfort at the site of stimulation is common and is expected to be reported in at least one third of all patients. Patients should be advised that, in the majority of patients, these sensations subside rapidly during the course of treatment, usually decreasing in intensity by the second or third session [26]. The other common event that patients should be advised of is the emergence of a post-treatment headache, very often similar in character to a common tension headache. As with cutaneous discomfort, this sensation typically subsides with treatment and may benefit from prophylaxis with over the counter analgesic treatment prior to subsequent treatment sessions [26].

4.3 Considerations in Long-Term Follow-Up

Once a determination has been made that maximal benefit has been obtained, TMS treatment should be tapered and a maintenance regimen should be defined for the patient. In the NeuroStar TMS System clinical studies, patients were discontinued from treatment slowly over a 3-week interval, during which time a maintenance regimen was established with a single antidepressant medication. Follow-up occurred for 6 months [46]. During this time, patients were restricted to continue on this single antidepressant medication only, but were permitted to reaccess TMS upon symptom reemergence. This method of long-term follow-up resulted in a remarkably durable outcome from acute treatment. In this study of nearly 100 patients, about 11 % of patient experienced a recurrence of their illness and left the study. Only about a third of patients required reintroduction of TMS because of symptom recurrence. Of note, of those patients who required TMS reintroduction, most experienced benefit. On average the number of sessions needed to reestablish benefit was about half the number in the original treatment course [46].

At the present time, there are no controlled studies of the use of TMS as a stand-alone maintenance treatment regimen.

5 Conclusions

TMS has now been established as a definitive treatment option and is accepted as a standard of care that is available for patients who have failed to benefit from initial acute phase treatment of major depression [51]. The durability of the acute treatment benefit from TMS is well defined. In this chapter, we have considered several aspects of the clinical use of TMS that remain areas for future study, including the exploration of alternate dosing strategies that may increase the magnitude of clinical benefit and whether TMS can be demonstrated to be effective as a longer term maintenance treatment.

References

1. Speer AM, Kimbrell TA, Wassermann EM et al (2000) Opposite effects of high and low frequency rTMS on regional brain activity in depressed patients. Biol Psychiatry 48: 1133–1141
2. Huang Y-Z, Rothwell JC, Chen R-S, Lu C-S, Chuang W-L (2011) The theoretical model of theta burst form of repetitive transcranial magnetic stimulation. Clin Neurophysiol 122: 1011–1018
3. Drevets WC (1998) Functional neuroimaging studies of depression: the anatomy of melancholia. Annu Rev Med 49:341–361
4. Mayberg HS, Liotti M, Brannan SK, McGinnis S, Mahurin RK, Jerabek PA, Silva JA, Tekell JL, Martin CC, Lancaster JL, Fox PT (1999) Reciprocal limbic-cortical function and negative mood: converging PET findings in depression and normal sadness. Am J Psychiatry 156: 675–682
5. Pizzagalli DA (2011) Frontocingulate dysfunction in depression: toward biomarkers of treatment response. Neuropsychopharmacology 36: 183–206
6. Kito S, Fujita K, Koga Y (2008) Changes in regional cerebral blood flow after repetitive transcranial magnetic stimulation of the left dorsolateral prefrontal cortex in treatment resistant depression. J Neuropsychiatry Clin Neurosci 20:74–80
7. Cohrs S, Tergau F, Riech S, Kastner S, Paulus W, Ziemann U, Ruther E, Hajak G (1998) High-frequency repetitive transcranial magnetic stimulation delays rapid eye movement sleep. Neuroreport 9:3439–3443
8. Pridmore S (1999) Rapid transcranial magnetic stimulation and normalization of the dexamethasone suppression test. Psychiatry Clin Neurosci 53:33–37
9. Sachdev PS, McBride R, Loo C, Mitchell PM, Malhi GS, Croker V (2002) Effects of different frequencies of transcranial magnetic stimulation (TMS) on the forced swim test model of depression in rats. Biol Psychiatry 51: 474–479
10. Post A, Keck ME (2001) Transcranial magnetic stimulation as a therapeutic tool in psychiatry: what do we know about the neurobiological mechanisms? J Psychiatr Res 35: 193–215
11. Holtzheimer PE et al (2001) A meta-analysis of repetitive transcranial magnetic stimulation in the treatment of depression. Psychopharmacol Bull 35:149–169
12. Kozel FA, George MS (2002) Meta-analysis of left prefrontal repetitive transcranial magnetic stimulation to treat depression. J Psychiatr Pract 8:270–275
13. Martin JLR, Barbanoj MJ, Schlaepfer TE, Clos S, Perez V, Kulisevsky J, Gironelli A (2002) Transcranial magnetic stimulation for treating depression. Cochrane Database Syst Rev (2):CD003493
14. Burt T et al (2002) Neuropsychiatric applications of transcranial magnetic stimulation: a meta analysis. Int J Neuropsychopharmacol 5:73–103
15. Aarre TF, Johansen JB et al (2003) Efficacy of repetitive transcranial magnetic stimulation in depression: a review of the evidence. Nord J Psychiatry 57:227–232
16. Couturier J (2005) Efficacy of rapid-rate repetitive transcranial magnetic stimulation in the treatment of depression: a systematic review and meta-analysis. J Psychiatry Neurosci 30(2): 83–90
17. Loo C, Mitchell P (2005) A review of the efficacy of transcranial magnetic stimulation (TMS)

treatment for depression, and current and future strategies to optimize efficacy. J Affect Disord 88:255–267

18. Herrmann L, Ebmeier K (2006) Factors modifying the efficacy of transcranial magnetic stimulation in the treatment of depression: a review. J Clin Psychiatry 67:1870–1876
19. Gross M, Nakamura L et al (2007) Has repetitive transcranial magnetic stimulation (rTMS) treatment for depression improved? A systematic review and meta-analysis comparing the recent vs. the earlier rTMS studies. Acta Psychiatr Scand 116:165–173
20. Lam RW et al (2008) Repetitive transcranial magnetic stimulation for treatment-resistant depression: a systematic review and meta-analysis. Can J Psychiatry 53(9):621–631
21. Schutter DJLG (2009) Antidepressant efficacy of high-frequency transcranial magnetic stimulation over the left dorsolateral prefrontal cortex in double-blind sham-controlled designs: a meta-analysis. Psychol Med 39:65–75
22. Slotema CW, Blom JD, Hoek HW, Sommer IEC (2010) Should we expand the toolbox of psychiatric treatment methods to include repetitive transcranial magnetic stimulation (rTMS)? A meta-analysis of the efficacy of rTMS in psychiatric disorders. J Clin Psychiatry 71(7):873–884
23. Allan C (2011) Transcranial magnetic stimulation in the management of mood disorders. Neuropsychobiology 64:163–169
24. Agency for Healthcare Research and Quality, Effective Health Care Program, Comparative Effectiveness Review Number 33. Nonpharmacologic interventions for treatment-resistant depression in adults. http://www.effectivehealthcare.ahrq.gov/index.cfm/search-for-guides-reviews-and-reports/?pageaction=displayproduct&productid=787
25. Demitrack MA, Thase ME (2009) Clinical significance of transcranial magnetic stimulation (TMS) in the treatment of pharmacoresistant depression: synthesis of recent data. Psychopharmacol Bull 42(2):5–38
26. Janicak PG, O'Reardon JP, Sampson SM, Husain MM, Lisanby SH, Rado JT, Demitrack MA (2008) Transcranial magnetic stimulation (TMS) in the treatment of major depression: a comprehensive summary of safety experience from acute and extended exposure and during reintroduction treatment. J Clin Psychiatry 69(2):222–232
27. O'Reardon JP, Solvason HB, Janicak PG, Sampson S, Isenberg KE, Nahas Z, McDonald WM, Avery D, Fitzgerald PB, Loo C, Demitrack MA, George MS, Sackeim HA (2007) Efficacy and safety of transcranial magnetic stimulation in the acute treatment of major depression: a multi-site randomized controlled trial. Biol Psychiatry 62:1208–1216
28. George MS et al (2010) Daily left prefrontal transcranial magnetic stimulation therapy for major depressive disorder a sham-controlled randomized trial. Arch Gen Psychiatry 67(5): 507–516
29. Avery DH, Isenberg KE, Sampson SM, Janicak PG, Lisanby SH, Maixner MD, Loo C (2008) TMS in the acute treatment of major depression: clinical response in an open-label extension trial. J Clin Psychiatry 69(3):441–451
30. McDonald WM, Durkalski V, Ball ER, Holtzheimer PE, Pavlicova M, Lisanby SH, Avery D, Anderson BS, Nahas Z, Zarkowski P, Sackeim HA, George MS (2011) Improving the antidepressant efficacy of transcranial magnetic stimulation: maximizing the number of stimulations and treatment location in treatment-resistant depression. Depress Anxiety 28(11):973–980
31. Carpenter LL, Janicak PG, Aaronson ST, Boyadjis T, Brock DG, Cook IA, Dunner DL, Lanocha K, Solvason HB, Demitrack MA (2012) Transcranial magnetic stimulation (TMS) for major depression: a multisite, naturalistic, observational study of acute treatment outcomes in clinical practice. Depress Anxiety 29(7):587–596
32. Rossi S, Hallett M, Rossini PM, Pascual-Leone A (2009) Safety, ethical considerations, and application guidelines for the use of transcranial magnetic stimulation in clinical practice and research. Clin Neurophysiol 120:2008–2039
33. Chen R, Gerloff C, Classen J, Wassermann EM, Hallett M, Cohen LG (1997) Safety of different inter-train intervals for repetitive transcranial magnetic stimulation and recommendations for safe ranges of stimulation parameters. Electroencephalogr Clin Neurophysiol 105: 415–421
34. Rush AJ et al (2006) Acute and longer-term outcomes in depressed outpatients requiring one or several treatment steps: a STAR*D report. Am J Psychiatry 163:1905–1917
35. Kennedy SH, Milev R, Giacobbe P, Ramasubbu R, Lam RW, Parikh SV, Patten SB, Ravindran AV (2009) Canadian network for mood and anxiety treatments (CANMAT) clinical guidelines for the management of major depressive disorder in adults. IV. Neurostimulation therapies. J Affect Disord 117:S44–S53
36. Schlaepfer TE, George MS et al (2009) WFSBP guidelines on brain stimulation treatments in psychiatry. World J Biol Psychiatry 1:1–17
37. Sackeim HA (2001) The definition and meaning of treatment-resistant depression. J Clin Psychiatry 16:10–17

38. Hadley D, Anderson BS, Borckardt JJ, Arana A, Li X, Nahas Z, George MS (2011) Safety, tolerability, and effectiveness of high doses of adjunctive daily left prefrontal repetitive transcranial magnetic stimulation for treatment-resistant depression in a clinical setting. J ECT 27(1):18–25
39. Fitzgerald P, Benitez J, de Castella A, Daskalakis ZJ, Brown TL, Kulkarni J (2006) A randomized, controlled trial of sequential bilateral repetitive transcranial magnetic stimulation for treatment-resistant depression. Am J Psychiatry 163:88–94
40. Fitzgerald PB, Hoy K, Gunewardene R, Slack C, Ibrahim S, Bailey M, Daskalakis ZJ (2011) A randomized trial of unilateral and bilateral prefrontal cortex transcranial magnetic stimulation in treatment-resistant major depression. Psychol Med 41(6):1187–1196
41. Fitzgerald PB, Hoy KE, Herring SE, McQueen S, Peachey AV, Segrave RA, Maller J, Hall P, Daskalakis ZJ (2012) A double blind randomized trial of unilateral left and bilateral prefrontal cortex transcranial magnetic stimulation in treatment resistant major depression. J Affect Disord 139:193–198
42. Herwig U, Padberg F, Unger J, Spitzer M, Schoenfeldt-Leucona C (2001) Transcranial magnetic stimulation in therapy studies: examination of the reliability of "standard" coil placement positioning by neuronavigation. Biol Psychiatry 50:58–61
43. Fitzgerald PB, Hoy K, McQueen S, Maller JJ, Herring S, Segrave R, Bailey M, Been G, Kulkarni J, Daskalakis ZJ (2009) A randomized trial of rTMS targeted with MRI based neuronavigation in treatment resistant depression. Neuropsychopharmacology 34(5):1255–1262
44. Herbsman T, Avery D, Ramsey D, Holtzheimer P, Wadjik C, Hardaway F, Haynor D, George MS, Nahas Z (2009) More lateral and anterior prefrontal coil location is associated with better repetitive transcranial magnetic stimulation antidepression response. Biol Psychiatry 66(5): 509–515
45. Pridmore S, Filho JAF, Nahas Z, Liberatos C, George MS (1998) Motor threshold in transcranial magnetic stimulation: a comparison of a neurophysiological method and a visualization of movement method. J ECT 14(1): 25–27
46. Janicak PG et al (2010) Durability of clinical benefit with transcranial magnetic stimulation (TMS) in the treatment of pharmacoresistant major depression: assessment of relapse during a six-month, multisite, open-label study. Brain Stimul 3:187–199
47. Pascual-Leone A, Houser CM et al (1993) Safety of rapid-rate transcranial magnetic stimulation in normal volunteers. Electroencephalogr Clin Neurophysiol 89(2):120–130
48. Loo C, Sachdev P et al (2001) Effects of a 2 to 4-week course of repetitive transcranial magnetic stimulation (rTMS) on neuropsychologic functioning, eletroencephalogram, and auditory threshold in depressed patients. Biol Psychiatry 49:615–623
49. Kroenke K, Spitzer RL, Williams JB (2001) The PHQ-9: validity of a brief depression severity measure. J Gen Intern Med 16(9): 606–613
50. Rush AJ, Trivedi MH, Ibrahim HM, Carmody TJ, Arnow B, Klein DN, Markowitz JC, Ninan PT, Kornstein S, Manber R, Thase ME, Kocsis JH, Keller MB (2003) The 16-item quick inventory of depressive symptomatology (QIDS) clinician rating (QIDS-C) and self-report (QIDS-SR): a psychometric evaluation in patients with chronic major depression. Biol Psychiatry 54:573–583
51. Gelenberg AJ, Freeman MP, Markowitz JC, Rosenbaum JF, Thase ME, Trivedi MH, Van Rhoads RS (eds) (2010) Practice guidelines for the treatment of patients with major depressive disorder, 3rd edn. American Psychiatric Association, Arlington, VA
52. Horvath JC, Mathews J, Demitrack MA, Pascual-Leone A (2010) The NeuroStar TMS device: conducting the FDA approved protocol for treatment of depression. J Vis Exp 45:2345

Chapter 15

Protocol for Depression Treatment Utilizing H-Coil Deep Brain Stimulation

Yiftach Roth and Abraham Zangen

Abstract

The Brainsway Deep TMS System for the treatment of major depressive disorder (MDD) was cleared by the FDA on January 2013 for the treatment of depressive episodes in adult patients suffering from MDD who failed to achieve satisfactory improvement from previous antidepressant medication treatment in the current episode, based on a large, prospective, double blind, randomized, controlled multicenter trial. This chapter presents the detailed protocol of this multicenter trial.

Key words Deep transcranial magnetic stimulation, Major depression disorder, Multicenter trial

1 Multicenter Trial Overview

The safety and effectiveness of the Brainsway Deep TMS System for the treatment of major depressive disorder (MDD) was studied in a prospective, double blind, randomized, controlled, multicenter trial. Institutional review board approval was obtained at all sites. The study was conducted under an Investigational Device Exemption from the U.S. Food and Drug Administration (FDA). An independent Data and Safety Monitoring Board (DSMB) reviewed participant safety and study progress.

Patients were recruited via public media advertisements and physician referrals. Site personnel phone-screened potential participants and those meeting inclusion/exclusion criteria underwent additional on-site screening. All subjects signed an informed consent document before undergoing any study procedures.

A total number of 212 subjects that met inclusion/exclusion criteria were enrolled at 20 study sites in the United States (13 sites), Israel (4 sites), Germany (2 sites), and Canada (1 site), with active enrollment from October 2009 to January 2012.

The study had three phases: (a) An antidepressant taper-off and washout phase (1–2 weeks, no treatment), (b) a 4-week acute

Alexander Rotenberg et al. (eds.), *Transcranial Magnetic Stimulation*, Neuromethods, vol. 89,
DOI 10.1007/978-1-4939-0879-0_15, © Springer Science+Business Media New York 2014

treatment phase (daily treatment with deep TMS or sham TMS), and (c) a 12-week maintenance phase (two treatments per week of deep TMS or sham TMS). Subsequently, subjects were followed for 3 months from the last treatment. Subjects were randomly assigned to either active deep TMS or sham TMS (1:1 ratio) by an Interactive Web Response System (IWRS) system. Subjects were stratified per center by severity of disease as determined by baseline 21-item Hamilton Depression Rating Scale (HDRS-21) scores (<26 vs. ≥26) and the ATHF treatment-resistance levels (ATHF >2 level 1–2 and ATHF 1 level 3 vs. ATHF 2–4, level 3). During the acute treatment phase, TMS sessions were performed daily in a 5-day sequence (5 days per week), for 4 weeks (drop out criteria: less than 18 sessions in 4.5 weeks). During the maintenance phase, subjects were treated twice a week (with at least 48 h between sessions), for 12 weeks. Subjects were discontinued from the study at any point if they were considered by the investigators to be at an elevated risk for suicide. Alternatively, subjects were discontinued if they did not experience a sufficient improvement in depression after 5 weeks of treatment in two consecutive assessments, as defined in the protocol. A sufficient improvement was defined as a decrease of at least one point in the Clinical Global Impression-Severity of Illness (CGI-S) from baseline (i.e., at randomization).

2 Indications for Use of the Device

Based on the multicenter trial results, the FDA cleared the Deep TMS device for the treatment of depressive episodes in adult patients suffering from MDD who failed to achieve satisfactory improvement from previous antidepressant medication treatment in the current episode.

3 Study Population

3.1 General Considerations

Eligible subjects were antidepressant medication-free outpatients, aged 22–68, with a DSM-IV diagnosis of MDD, single or recurrent episode. Duration of current episode was at least 1 month but no more than 7 years. Subjects were required to have a Clinical Global Impressions Severity of Illness (CGI-S) score of at least 4 and a total score of at least 20 on the HDRS-21 at screening visit. Symptom stability was required during the 2-week antidepressant taper-down and washout period. Instability was defined as a change of ±30 % or more from the total HDRS-21 score that was observed at the screening assessment. Antidepressant treatment-resistance during the current episode was assessed using the Antidepressant Treatment History Form (ATHF; 1). Subjects were required to have failed at least one but no more than four adequate antidepressant treatments or to have intolerance to at least two antidepressants in the current episode.

Present suicidal risk, as assessed by the investigator, or significant suicide risk based on HDRS-21 item 3 score of 3 or 4 or suicide risk as assessed using the Scale for Suicide Ideation-Worst, or a history of attempted suicide in the last 3 years were exclusion criteria. Subjects who developed documented suicidal ideation (based on HDRS-21 item 3 score of 3 or 4 or as assessed by the Suicide Ideation Scale) or attempts suicide after receiving active or sham rTMS were discontinued.

Subjects were excluded from the study if they had a lifetime history of psychosis, bipolar disorder; obsessive–compulsive disorder (current or within the past year); posttraumatic stress disorder or eating disorders. Subjects suffering from anxiety or personality disorders were eligible only if this was not the primary diagnosis of the subject. Additional exclusion criteria included any significant neurological disorder or insult; increased risk of seizure for any reason or familial or personal history of epilepsy; lifetime lack of response to an adequate trial of electroconvulsive therapy (ECT); prior treatment with rTMS, deep brain stimulation (DBS), or a vagus nerve stimulator (VNS) implant; pregnancy; presence of intracranial implants or any other metal object within or near the head excluding the mouth, that cannot be safely removed.

3.2 Inclusion Criteria

1. Outpatients.
2. Men and women 22–68 years of age.
3. Primary DSM-IV diagnosis of major depression, single or recurrent episode confirmed by the Structured Clinical Interview for the DSM-IV (SCID-IV), with the additional requirements of a current episode ≥4 weeks and CGI-S ≥4.
4. Current depressive episode is less than 7 years duration (the definition of an episode is demarcated by a period of ≥2 months in which the patient did not meet full criteria for the DSM-IV definition of major depressive episode).
5. Total HDRS-21 ≥ 20 and Item 1 score ≥2 at the screening visit.
6. The patient did not respond to at least one antidepressant treatment, i.e., a minimum of one and a maximum of four antidepressant drug trials, of adequate dose and duration (defined as a minimum level of 3 on the ATHF) in the current episode. Current episode for ATHF level 3 patients should be ≥12 weeks.
7. Patients who have not completed antidepressant trials of adequate dose and duration (defined as a level of 1–2 on the ATHF) due to intolerance to therapy, if they have demonstrated intolerance to two or more antidepressant medications in the current episode.

8. Capable and willing to provide informed consent.
9. Able to adhere to the treatment schedule, and withdrawal of ongoing pharmacotherapy.
10. If currently taking antidepressant pharmacotherapy, the patient must be clinically appropriate to discontinue treatment with those agents. "Clinically appropriate" is defined as removing patients from antidepressant medications without significantly increasing the risk of suicide or aggravating HDRS score by 30 % or more.
11. Able to tolerate psychotropic medication washout and no psychotropics during the H-Coil deep brain rTMS other than benzodiazepine at equivalent daily dose of up to 3 mg lorazepam.
12. Satisfactory safety screening questionnaire for transcranial magnetic stimulation.
13. Patients not suffering from hypo- or hyperthyroidism based on pre-study TSH level or medically stabilized.

3.3 Exclusion Criteria

1. Investigators, site personnel directly affiliated with this study, and their immediate families (immediate family is defined as a spouse, parent, child, or sibling, whether by birth or legal adoption).
2. Individuals diagnosed by the investigator with the following conditions (current unless otherwise stated):
   - Depression secondary to a general medical condition, or substance induced.
   - History of substance abuse or dependence within the past 6 month (except nicotine and caffeine).
   - Schizoaffective disorder.
   - Major depression with psychotic features.
   - Bipolar disorder, except when mania/hypomania was related to administration of antidepressant medications.
   - Eating disorders.
   - Obsessive compulsive disorder.
   - Post-traumatic stress disorder (current or within the past year).
   - Current (within 12 months of baseline) generalized anxiety disorder, panic disorder, or social anxiety disorder as assessed by the investigator to be primary, causing a higher degree of distress or impairment than MDD.
   - Presence (within 12 months of baseline) of a personality disorder (such as antisocial, schizotypal, histrionic, borderline, and narcissistic) as assessed by the investigator to be primary, causing a higher degree of distress or impairment than MDD.

3. Individuals with a significant neurological disorder or insult including, but not limited to:
   - Any condition likely to be associated with increased intracranial pressure.
   - Space occupying brain lesion.
   - Any history of seizure EXCEPT those therapeutically induced by ECT.
   - History of cerebrovascular accident.
   - Transient ischemic attack within 2 years.
   - Cerebral aneurysm.
   - Dementia.
   - Mini Mental State Exam score of less than or equal to 24.
   - Parkinson's disease.
   - Huntington's chorea.
   - Multiple sclerosis.
4. Increased risk of seizure for any reason, including prior diagnosis of increased intracranial pressure (such as after large infarctions or trauma), or history of significant head trauma with loss of consciousness for greater than or equal to 5 min or familial or personal history of epilepsy.
5. Individuals with hearing loss.
6. A positive and unmitigated response to any question on the Transcranial Magnetic Stimulation Adult Safety Screen (TASS) questionnaire.
7. ECT treatment within 3 months prior to the screening visit.
8. Failure to respond to ECT treatment (i.e., consistent with ATHF level 2 or higher) in this or any previous episode.
9. Treatment with rTMS therapy for any disorder within the last 6 months.
10. History of treatment with vagus nerve stimulation (VNS), unless electrodes have been removed.
11. History of treatment with DBS.
12. Use of any investigational drug within 4 weeks of the randomization visit.
13. Use of fluoxetine within 6 weeks of the randomization visit.
14. Use of a Monoamine Oxidase Inhibitor (MAOI) within 2 weeks of the randomization visit.
15. Use of any medication(s) listed on the Excluded Medication List within 1 week of the randomization visit.
16. Present suicidal risk as assessed by the investigator or significant suicide risk based on HDRS-21 item 3 score of 3 or 4 or significant suicide risk as assessed using the Scale for Suicide

Ideation—Worst or a history of attempted suicide in the last 3 years.

17. Any self-inflicted harm in the past 3 months not in the context of suicidical ideation.
18. Cardiac pacemakers, implanted medication pumps, intracardiac lines, or acute, unstable cardiac disease.
19. Intracranial implant (e.g., aneurysm clips, shunts, stimulators, cochlear implants, or electrodes) or any other metal object within or near the head, excluding the mouth, that cannot be safely removed.
20. Implanted neurostimulators.
21. History of abnormal MRI.
22. Known or suspected pregnancy.
23. If participating in psychotherapy, must have been in stable treatment for at least 3 months prior to entry into the study, with no anticipation of change in the frequency of therapeutic sessions, or the therapeutic focus over the duration of the rTMS trial.
24. Clinically significant laboratory abnormality, in the opinion of the investigator based on CBC and biochemistry.
25. Women who are breast-feeding.
26. Women of childbearing potential and not using a medically accepted form of contraception when engaging in sexual intercourse.

4 Randomization and Blinding

The patients were randomized into the study, stratified per center by severity of diseases as determined by the baseline HDRS-21 score and ATHF. After subjects meet the eligibility criteria, they were equally allocated (with a 1:1 ratio) to one of the two treatment groups stratified by HDRS-21 scores (20–25 vs. ≥26), ATHF categories (ATHF ≥2 level 1–2 and ATHF 1 level 3 vs. ATHF 2–4, level 3) and center, based on a stratified randomization scheme using the permuted block method using the SAS (version 9.1) random number generator. The block size was random and study personnel were blinded to the randomization block size.

Central randomization was implemented in conducting this study. The subject number and treatment code were assigned after entering into the IWRS. The users had to use their own user identification (ID) and password, and give the requested subject details (subject's initials, HDRS-21 score on baseline visit, ATHF and center). Based on this information, the IWRS assigned a unique subject number and treatment code, which dictated the treatment assignment for the subject. The unique subject number and treatment code matched one of the pre-programmed treatment cards maintained at the clinical site. The operator inserted the patient's treatment card programmed to activate either real or sham treatment mode according to the treatment group to which the patient has been randomized. The study personnel did not have any knowledge of whether the treatment mode or sham mode is being activated by the patient's treatment card. Thus, all study personnel, including the operator, the independent rater and study subjects were blinded to the treatment being administered. The study patients were asked whether they believe they had received active or sham stimulation after the first treatment session.

5 Device Description

5.1 General Device Description

The H-Coil deep brain rTMS therapy is provided through the Brainsway Deep TMS System that is composed of the following main components:

1. An helmet containing an active TMS Coil (H1-Coil), and a sham coil
2. A TMS Neurostimulator
3. A cooling system
4. A positioning device
5. A cart
6. A controller
7. A cap

A general sketch of the Brainsway H-Coil deep brain rTMS system is shown in Fig. 1.

An outline of system's main components is provided below.

5.2 Device Components Description

The helmet contains an active H1-Coil and a sham coil, which mimics the scalp sensations, facial muscles' activation and acoustic artifact of the real H1-Coil, without any activation of the brain.

5.2.1 H1-Coil Helmet

The H1 Real Coil

The H1-Coil is designed to stimulate neuronal pathways related to the control of motivation, reward, and pleasure, specifically, the prefrontal cortex and fibers connecting the cingulate or the prefrontal cortex with the nucleus accumbens and ventral tegmental area. The H1-Coil is made of insulated copper wires. The total length is about 800 cm, winded into several windings and connected in series. The effective part of the coil, in contact with the patient's head has a shape of half a donut. The frame of the inner rim of the half donut is flexible in order to fit the variability in

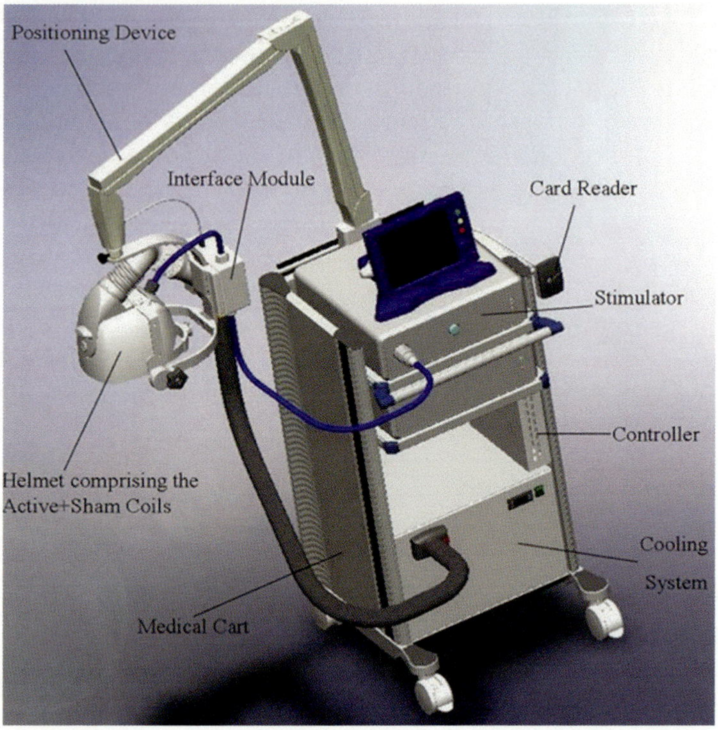

Fig. 1 A general sketch of the Brainsway H-Coil deep brain rTMS system, including a cart, a TMS Neurostimulator, a cooling system, a controller, an arm, a helmet, and an H-Coil

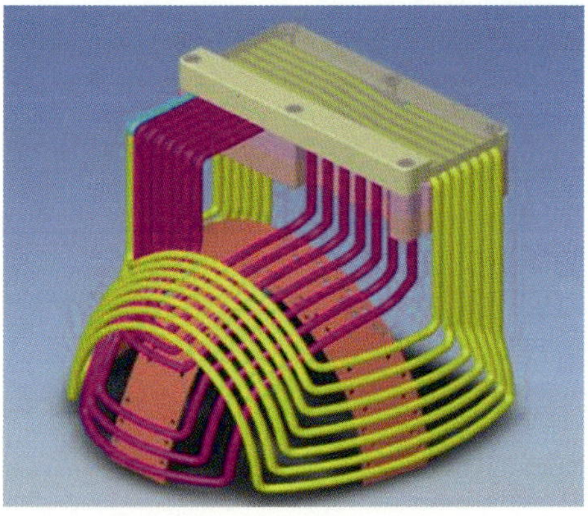

Fig. 2 A sketch of the H1-Coil

human skull shape. The windings are connected to a Magstim cable and a connector. This connector can be connected to the Magstim Rapid or Rapid^2 stimulator. In addition, a temperature sensor is included with an appropriate cable.

A sketch of the H1-Coil is shown in Fig. 2.

The H1 Sham Coil

The H1 Sham Coil is designed to mimic the auditory artifact and the scalp sensations evoked by the real H1-Coil, and to produce activation of facial muscles similar to the effect of a real H1-Coil, without stimulating the brain itself.

The Sham Coil is made of insulated copper wires. The total length is 650 cm, winded into three groups of windings, connected in series. The windings are connected to a Magstim cable and a connector. This connector can be connected to the Magstim Rapid or Rapid^2 stimulator. In addition, a temperature sensor is included with an appropriate cable.

5.2.2 Commercial TMS Neurostimulator

The commercial TMS neurostimulator, the Magstim Rapid, or the Magstim Rapid^2 is used to deliver electrical stimulation to the brain, enabling a controlled output, frequency, pulse duration, and indication of coil temperature. The Magstim Rapid Neurostimulator system contains four boosters, whereas the Magstim Rapid^2 system contains a single power supply unit.

5.2.3 The Positioning System

The positioning system includes a helmet that includes the coils, an adjustable pantograph arm (De-Gotzen, Italy) connected to the helmet, and a device enabling rotation of the helmet around three orthogonal rotation axes. The positioning device enables accurate and comfortable displacement and positioning of the coil over the patient's head.

5.2.4 The Cooling System CS04

The Cooling System CS04 is designed to maintain ambient temperature in the coils during repetitive operation. The cooling system consists of an external unit and an air hose streaming the cooled air into the helmet. The air flow cools the coils during pulse trains and maintains them at ambient temperature (less than 30 °C).

5.2.5 The Cart

The TMS neurostimulator and the cooling system are placed in a movable cart (ITD, Germany). The adjustable arm is connected to the cart.

5.2.6 The Cap

A personal cap is provided for each patient (Fig. 3). The cap is made from a biocompatible material (Fabrifoam). A flexible ruler is attached along the cap midline, with the 0 mark positioned at the patient's nasion. The cap is used for accurate positioning of the coil on the patient's head, by moving the coil from the area above the motor cortex to the treatment location at the prefrontal cortex. In addition the cap guarantees hygienic treatment conditions.

5.2.7 The Controller

The controller is intended for operation of either the sham or an active coil with magnetic cards, while maintaining adequate blinding of the operator as well as the patient.

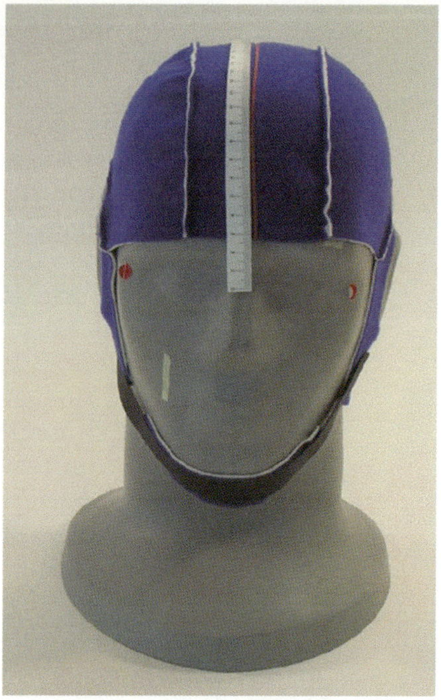

Fig. 3 An image of the Brainsway personal cap on a head model

The operator swipes a magnetic card by the card reader (Fig. 1), and according to the information programmed to the card, the controller connects the correct coil to the stimulator, while the external indications are the same for all coils.

6 Deep TMS Protocol

The treatment administrator determined all motor thresholds (MTs) and administered the treatments. An independent rater evaluated the patient following treatment and completed the subject assessment scales. The principle investigator, treatment administrator, independent rater, and study subjects were all blinded to the treatment being administered. The independent rater could not see the patient until completion of the treatment. All treatments were conducted with minimization of the personal contact with the subject. At the beginning of the trial, individual MTs were measured by placing the assigned coil above the hand area of the motor cortex. All studied groups received stimulations over the prefrontal cortex, 6 cm anterior to the "hot spot" for stimulation of the left hemisphere abductor pollicis brevis (APB) in the hand area.

Each rTMS treatment (for active and sham groups) was conducted as follows: Before starting each treatment, subjects were instructed to insert earplugs to lessen any possible adverse effect on hearing. The resting MT was measured before each treatment by delivering single stimulations to the motor cortex. The MT is measured by gradually increasing the intensity. This is done by using the single pulse mode and applying 1 pulse every 5 s, (0.2 Hz). The electrical activity in APB is then recorded using surface electrodes. Threshold is defined as the lowest intensity of stimulation producing motor-evoked potentials of at least 50 μV in 5 of 10 trials. After defining the motor threshold, the coil was positioned anterior to the hot spot using the ruler on the patient's cap, and an rTMS session was performed at 120 % of the resting motor threshold, similar to previous studies with clinically depressed subjects. The motor threshold determination and positioning of the stimulation coil performed prior to the first treatment session took approximately 5–20 min. In subsequent treatment sessions, this procedure took 2–10 min. During treatment, the operator monitored the patient. In the case of leg, neck, or strong hand motor activation, the session would be stopped, MT re-measured, and, if the MT remained high enough to cause musculature over-activation, the patient would be excluded.

Each treatment session takes about 30 min (of which, the rTMS takes 20 min).

During the first three treatments, sites were allowed to titrate stimulation intensity up, from 100 % MT for the first treatment to 110 % MT for the second treatment and 120 % MT for the third treatment. This was done to improve subjects' tolerability for the treatment. From the third treatment and beyond, patients had to receive treatment at = 120 % MT. During the treatment trial period, the treatment group received the following dose of rTMS: 18 Hz, at 120 % MT, 2 s pulse train, 20 s inter-train interval, 55 trains, (a total of 1,980 pulses per session) for a total of 44 sessions (a total number of 87,120 pulses in 16 weeks). The control group received inactive/sham treatment with identical parameters. Patients were told that face and hand twitching may occur due to either sham or active treatment and they were instructed to sit with their arms crossed and folded.

During the treatment procedure, the treatment administrator was instructed to observe the subject closely for any sign of imminent seizure activity or muscle twitching. The presence of a physician or nurse trained in seizure management, emergency equipment (oxygen, suction, blood pressure monitor, and CPR equipment), and antiepileptic medications was readily available in the immediate vicinity. Subjects were informed of the risk of permanent hearing damage if an earplug should become detached or fall out. Subjects were asked to immediately report any loosening or detachment of an earplug during rTMS and treatment was stopped if a subject reported or an investigator observed that an earplug has loosened or has fallen out. Auditory threshold testing was also conducted at screening and during the course of the study.

In the case of coil overheating, a warning was displayed on the stimulator UI screen and the treatment administrator was instructed to remove the coil from the patient. The coil typically cools down within several minutes at which point treatment could be resumed. Any events of coil overheating, removal, and treatment resumption were recorded. Patients that participated in the study were not allowed to meet with each other before, during, or after assessment or treatments in order to maintain study blinding.

7 Compliance

TMS treatment was administered in a controlled environment of a clinical research center under direct observation of the administrator and the study staff to ensure compliance with the study requirements. The date and time of each TMS session were recorded in the CRF.

8 Concomitant Therapy

The following medications were not allowed at least 2 weeks prior to the first rTMS treatment and throughout the study: antidepressants, mood stabilizers, and antipsychotics.

Insomnia medications prescribed prior to commencement of treatment could be continued during the study (up to 3 mg Lorivan, or equivalent).

Only the following insomnia medications could be prescribed during the course of the study:

- Zolpidem (p.r.n. up to 10 mg/day orally or Zolpidem CR (6.25 or 12.5 mg/day orally).
- Zaleplon (p.r.n. up to 20 mg/day orally).
- Zopiclone (p.r.n up to 15 mg/day orally or eszopiclone (p.r. up to 3 mg/day orally).
- Benzodiazepines up to a dose equivalent to 3.0 mg of lorazepam or equivalent.

Acetaminophen or other medications for the treatment of local pain, dental pain, or headaches were allowed, as necessary.

All medications taken by a subject (prescription or nonprescription) were documented in the concomitant therapy section of the CRF. This includes medications taken within 40 days before treatment initiation and during the course of the study.

For any medication given as a treatment for a new or worsening condition, the condition had to be documented on the Adverse Event Form of the CRF.

9 Study Procedures and Evaluations

9.1 Study Procedures

The trial consisted of two periods.

9.1.1 Overview

1. Pre-study screening and antidepressant medication taper-down/washout period (up to 2 weeks or at the discretion of the investigator).
2. Treatment trial period (16 weeks).

During the 18 weeks of the trial, patient visits may occur within a ±3-day deviation from the scheduled time.

Efficacy rating scales to be used in the study include:

- HDRS-21
- Hamilton Anxiety Rating Scale (HARS)
- Quick Inventory of Depressive Symptomatic Self-Report (QIDS-SR)
- CGI-S
- Clinical Global Impression Improvement Scale (CGI-I)
- Patient Global Impression Improvement Scale (PGI-I)
- Quality of Life Enjoyment and Satisfaction Questionnaire (Q-LES-Q)
- SF-36
- GAF

Safety evaluations include adverse event monitoring (including auditory changes), vital signs, physical and neurological examination, urine pregnancy test, Young Manic Rating Scale (YMRS), Mini-Mental State Exam (MMSE), Buschke Selective Reminding Test (BSRT), Autobiographical Memory Interview—Short Form (AMI-S), cognitive tests, and SSI.

9.1.2 Screening and Medication Taper-Down/Washout Period

Potential subjects were given complete information describing the study treatment and their role in the trial, and they were encouraged to ask any questions regarding the trial. The risks and requirements of this clinical research trial were explained to each potential subject. Those volunteering to take part read and signed the Informed Consent Form for participation in the clinical research trial before any trial-related procedures are performed.

Upon obtaining the signed informed consent, inclusion and exclusion criteria were reviewed to verify the subject's eligibility. Subjects can only be outpatients at the time of screening. All screening procedures are shown in the Time and Events Schedule.

Once the investigator determined that the subject was eligible for participation in the trial and after all assessments had been done, previous medication dosages were tapered down to zero during a 2-week taper-down period (or at the discretion of the investigator), with 5–8 days of medication washout prior to treatment. To ensure patient safety during the washout period, patients were closely monitored by assessing CGI-S, HDRS, and SSI at a second screening visit after 1 week (S2), and patients unable to tolerate medication taper-down were excluded. Patients who could tolerate a quicker taper-down period could skip the second screening visit (S2), provided there was at least 5 days without antidepressant medications (i.e., washout) between the first screening visit (S1) and the baseline visit (B1). Patients who did not need to be tapered down from antidepressant medication could have the first screening visit (S1) and the baseline visit (B1) performed within 1 week of each other and the second screening visit (S2) excluded. Enrolled patients were dropped from the study if they show ≥±30 % changes in the HDRS-21 score from the time of screening (Visit S1) to the baseline assessment (Visit B1). Baseline data are defined as the data generated during Visit B1 before the first rTMS/sham treatment. Subjects who tolerate the taper-down/washout medication, and who have completed all assessments on baseline visit, were randomized to either the deep brain rTMS treatment group or the sham control group.

9.1.3 Treatment Trial Period

The treatment regimen in both study groups consisted of a total of 16 weeks of treatment, consisting of five daily treatment sessions per week for 4 weeks, followed by 2 weekly treatment sessions from weeks 5 to 16. The daily treatments (weeks 1–4) were delivered over a period of 4.5 weeks to enable make-up treatments in place of missed treatments, although no more than two missed days of treatment are allowed. The two weekly treatment sessions (weeks 5–16) had to be administered with at least 48 h between treatments. Treatments were administered on holidays/vacations, as well. Most evaluation visits were performed once a week during the treatment trial period. Some of the evaluations were performed less frequently. All procedures performed during the different visits are detailed in Sect. 8.2. Patients were discontinued if they did not experience a sufficient improvement in depression after 5 weeks of treatment, where a sufficient improvement is defined as a decrease in the CGI-S from baseline (i.e., at randomization) of at least one point, in two consecutive assessments or if there is presentation of suicidal ideation (based on HDRS-21 item 3 score of 3 or 4 or as assessed by the Suicide Ideation Scale or mental status examination) or attempted suicide after receiving active or sham rTMS.

Procedures performed at Visit 44 would also be performed in case of early discontinuation of the study, except if a patient dropped out of the study before visit 11.

9.2 Efficacy and Safety Evaluations

All study efficacy and safety evaluation assessments were performed on the morning of the next treatment day.

9.2.1 Hamilton Depression Rating Scale

Subjects were interviewed at the screening visit (Visit S1 and S2), baseline visit (Visit B1), every other week (and visit 22) during the treatment period and at 12 weeks follow-up. The HDRS was published more than 40 years ago for the purpose of "quantifying the results of an interview" [1]. Although not designed for use in treatment studies, Hamilton anticipated that the scale would have value in evaluating the impact of treatment. During the past 40 years, the HDRS has been the most widely used outcome measure in antidepressant efficacy trials [2].

9.2.2 Hamilton Anxiety Rating Scale

Subjects were interviewed at the screening visit (Visit S1), baseline visit (Visit B1), every other week (and visit 22) during the treatment period and at 12 weeks follow-up.

HARS is a rating scale developed to quantify the severity of anxiety symptomatology. It consists of 14 items, each defined by a series of symptoms. Each item is rated on a 5-point scale, ranging from 0 (not present) to 4 (severe). Seven of the items specifically address psychic anxiety and the remaining seven items address somatic anxiety. The HARS was developed by Max Hamilton in 1959. It provides measures of overall anxiety, psychic anxiety (mental agitation and psychological distress), and somatic anxiety (physical complaints related to anxiety). The HARS is administered by an interviewer who asks a semi-structured series of questions related to symptoms of anxiety.

9.2.3 Quick Inventory of Depressive Symptomatic Self-Report

Subjects completed the self-rated, QIDS-SR questionnaire at the screening visit (Visit S1), baseline visit (Visit B1), every other week (and visit 22) during the treatment period and at 12 weeks follow-up.

The 16-item QIDS-SR [3] is designed to assess the severity of depressive symptoms. The QIDS-SR assesses the entire criterion symptom domains designated by the American Psychiatric Association Diagnostic and Statistical Manual of Mental Disorders, fourth edition, to diagnose a major depressive episode. The assessments can be used to screen for depression, although they have been used predominantly as measures of symptom severity. The QIDS-SR is easy to administer and requires minimal training. QIDS-SR is sensitive to change with medications, psychotherapy, or somatic treatments, making it useful for both research and clinical purposes.

9.2.4 Clinical Global Impression–Severity Score

Subjects were assessed for the severity of illnesses at the screening visit (Visit S1 and S2), baseline visit (Visit B1), once a week during the treatment trial period and at 12 weeks follow-up.

CGI-S is a commonly used tool for assessing the severity of the patient's depression. It uses seven levels, ranging from normal to severely ill. The CGI-S rating scale is used to rate the severity of a subject's depressive condition on a 7-point scale ranging from 1 (not ill) to 7 (extremely ill). This scale permits a global evaluation of the subject's condition at a given time.

9.2.5 Clinical Global Impression Improvement Scale

The clinician had to assess how much the patient's illness has improved or worsened relative to a baseline state, at the baseline visit (Visit B1), every other week (and visit 22) during the treatment trial period and at 12 weeks follow-up.

The Global Improvement item requires the clinician to rate how much the patient's illness has improved or worsened relative to a baseline state. Compared to condition at baseline, a patient's illness is compared to change over time, and rated as: very much improved; much improved; moderately improved; minimally improved; no change; minimally worse; moderately worse; much worse; or very much worse.

9.2.6 Patient Global Impression Improvement Scale

Subjects completed the self-rated, PGI-I questionnaire. The PGI-I appraises how much the patient's illness has improved or worsened relative to a baseline state, at the baseline visit (Visit B1), every other week (and visit 22) during the treatment trial period and at 12 weeks follow-up.

9.2.7 Self-Rated Health Status Survey (SF-36)

Subjects completed the self-rated health status survey, SF-36, at the baseline visit (Visit B1), and during the treatment period at visits 22 and 44.

This is a short-form general health status survey with 36 questions. All questions are scored on a scale from 0 to 100, with 100 representing the highest level of functioning possible, and then averaged together, for a final score within eight subscales. The eight subscale scores are subsequently aggregated into two Summary scales, the PCS and the MCS (Physical and MENTAL Component Summary).

9.2.8 Quality of Life Enjoyment and Satisfaction Questionnaire

Subjects completed the self-rated questionnaire, Q-LES-Q, at the baseline visit (Visit B1), every 2 or 4 weeks during the treatment period (Visits 20, 22, 24, 28, 36, and 44) and at 12 weeks follow-up.

The Q-LES-Q is a self-report measure designed to enable investigators to easily obtain sensitive measures of the degree of enjoyment and satisfaction experienced by subjects in various areas of daily functioning. Higher scores indicate better quality of life.

9.2.9 Personal and Social Functioning (GAF)

Subjects were assessed for personal and social functioning using the GAF, at the baseline visit (Visit B1), every 2 or 4 weeks during the treatment period (Visits 20, 22, 24, 28, 36, and 44) and at 12 weeks follow-up.

GAF is a commonly used tool for assessing the daily functioning of the patient and to evaluate to what extent the depressive symptoms effects his daily functioning. These global interviewer-rated measures have scores ranging from 0 to 100, with lower scores denoting poorer functioning. The GAF rates symptoms and functioning during the past month.

9.2.10 Transcranial Magnetic Stimulation Adult Safety Screen

Subjects were assessed for potential TMS risk factors using the TASS questionnaire, at the screening visit (Visit S1) to ensure patient safety.

The safety questionnaire is a safety tool meant to screen potential subjects for risks of adverse events during TMS. A positive screen is any "yes" answer, indicates further investigation by the clinician (but not indicating exclusion from TMS).

9.2.11 Young Manic Rating Scale

Subjects were assessed for symptoms of mania using the YMRS questionnaire at the baseline visit (Visit B1), every other week (and visit 22) during the treatment period and at 12 weeks follow-up.

The YMRS is an 11-item instrument used to assess the severity of mania in patients with a diagnosis of bipolar disorder. The 11 items are: Elevated Mood, Increased Motor Activity Energy, Sexual Interest, Sleep, Irritability, Speech (Rate and Amount), Language—Thought Disorder, Content, and Disruptive—Aggressive Behavior, Appearance, and Insight. Ratings are based on patient self-reporting, combined with clinician observation. According to Suppes T. et al. [4] a YMRS score of 12 or higher was considered reflective of at least mild hypomania (symptoms adequate to meet DSM-IV-TR hypomania criteria); a YMRS score of 15 or higher corresponds with moderate hypomania; and a YMRS score of 21 or higher corresponds with severe hypomania or mild mania.

9.2.12 Mini-Mental State Exam

Subjects were assessed for cognitive changes using the MMSE, at the screening (Visit S1), at the baseline visit (Visit B1), once during the daily treatment period at week 4 (Visits 20), four times during weeks 5–16 (Visits 22, 24, 28, and 44) and at 12 weeks follow-up.

The MMSE or Folstein test is a brief 30-point questionnaire test that is used to assess cognition. It is commonly used in medicine to screen for dementia. In the time span of about 10 min, it samples various functions, including arithmetic, memory, and orientation. The MMSE test includes simple questions and problems in a number of areas: the time and place of the test, repeating lists of words, arithmetic, language use and comprehension, and copying a drawing. Any score over 24 (out of 30) is effectively normal. The normal value is also corrected for degree of schooling and age. It was introduced by Folstein et al. in 1975 [5], and is widely used with small modifications.

9.2.13 Scale for Suicide Ideation

The Scale for Suicide Ideation (SSI; 4) is a 21-item interviewer—administered rating scale that measures the current intensity of patients' specific attitude, behaviors, and plans to commit suicide on the day of the interview.

Subjects were assessed for suicide risk using the SSI questionnaire at the screening visits (Visit S1 and S2), Baseline visit (Visit B1), and every week during the daily treatment period (Visits 5, 10, 15, and 20), every second week during the twice-weekly treatment period (Visits 22, 24, 24, 28, 32, 36, 40, and 44) and at 12 weeks follow-up. Each item consists of three options graded according to suicidal intensity on a 3-point scale ranging from 0 to 38. The SSI consists of five screening items. Three items assess the wish to live or the wish to die and two items assess the desire to attempt suicide. If the respondent reports any active or passive desire to commit suicide, then 14 additional items are administered. Individual items assess suicidal risk factors such as the duration and frequency of ideation, sense of control over making an attempt, number of deterrents, and amount of actual preparation for a contemplated attempt. Two additional items record incidence and frequency of previous suicide attempts. The SSI takes approximately 10 min to administer. The predictive validity of the SSI for completed suicide has been established for patients seeking outpatient psychiatric treatment. Specifically, patients who scored in the higher risk category (i.e., SSI total score greater than two) were approximately seven times more likely to commit suicide than those who scored in the lower risk category [6].

9.2.14 Scale for Suicide Ideation-Worst

The 19-item Scale for Suicide Ideation-Worst (SSI-W: 6) is an interviewer-administered rating scale that measures the intensity of patients' specific attitudes, behaviors, and plans to commit suicide during the time period that they were the most suicidal. Subjects were interviewed at the Screening Visit (S1).

The instrument was developed to obtain a more accurate estimate of suicide risk. Specifically, interviewers instructed patients to recall the approximate date and circumstances when they were experiencing the most intense desire to commit suicide. Patients are then asked to keep this experience in mind while the interviewer rates patients' responses to the 19 items regarding how suicidal they were at that time. As with SSI, each SSI-W item consists of three options graded according to the suicidal intensity on a 3-point scale ranging from 0 to 2. The ratings are summed to yield a total score, which ranges from 0 to 38. Individual items assess characteristics such as wish to die, desire to make an active or passive suicide attempt, duration and frequency of ideation, sense of control over making an attempt, number of deterrents, and amount of actual preparation for a contemplated attempt. The SSI-W takes approximately 10 min to administer. In a prospective study, psychiatric patients who scored in the higher risk category, i.e., SSI-W total score greater than 14 were 14 times more likely to commit suicide than patients who scored in the lower risk category [7].

9.2.15 Auditory Threshold Testing

The absolute threshold of hearing (ATH) is the minimum sound level of a pure tone that an average ear with normal hearing can hear in a noiseless environment. The absolute threshold is classed as the point at which a response is elicited a specified percentage of the time. The patients' auditory threshold was determined using an audiometer that meets the requirements of one of the following standards; IEC 60645, ISO 8253, or ANSI S3.6 or is FDA cleared. Auditory threshold testing was performed at the screening visit (Visit S1), at week 6 (Visit 24) and at the end of the study, at week 16 (Visit 44), or at the time of discontinuation.

9.2.16 Buschke Selective Reminding Test

The BSRT [8] is used to provide a traditional measure of verbal learning and memory using 12-word lists, ten trials, and 2-h delayed free recall, followed by reacquisition of the list (ten trials) (four alternate forms). Since subjects are only reminded of words not recalled on the previous trial, this task provides more information on encoding and retention than other list learning tasks. This task is especially sensitive to anterograde amnestic effects of ECT and other interventions [9]. The primary dependent measure was total words correctly reported at the second administration (reacquisition). The BSRT test was performed at the baseline visit (Visit B1), at week 5 (Visit 22) and at the end of the study, at week 16 (Visit 44), or at the time of discontinuation.

9.2.17 Autobiographical Memory Interview—Short Form

The AMI-S assesses memory for events in the past and the development of retrograde amnesia. This common side effect of ECT is not expected to be manifest with DTMS and a single task was used for this purpose. The AMI-S is a brief version of the instrument that has been repeatedly found especially sensitive to the short- and long-term effects of ECT on retrograde memory [10–12]. The original AMI distinguished depressed patients from controls in baseline recall productivity, and methods of ECT administration in short- and long-term retrograde amnesia. Validity as a measure of retrograde amnesia has been established both by studies using family member corroboration of autobiographical events, and by demonstrating temporal gradients in amnesia and recover of recent memories with time from ECT. The short form involves inquiries about the details of six events (e.g., last New Year's Eve). Five questions are asked about each event, yielding 30 items. Following each phase patients are only asked about items for which they produced a response at baseline. Inconsistency in responses relative to baseline (including "don't know" responses) is the measure of retrograde amnesia. The specific events and items were selected from the original form of the AMI to maximize (a) depressed patient productivity at baseline and (b) overall sensitivity to ECT amnestic effects. Total amnesia scores on the long- and short form of the AMI are highly correlated. This measure is included largely for comparative purposes and was analyzed separately. The AMI-S test was performed at the baseline visit (Visit B1), at week 5 (Visit 22) and at the end of the study, at week 16 (Visit 44), or at the time of discontinuation.

9.3 Efficacy Measures

All efficacy outcome measures were assessed by a blinded study rater who was not permitted access to the treatment sessions. Raters were required to pass the study Rater Certification Program, which was developed to ensure adequate scoring reliability and rating skills. Subjects were instructed not to disclose any details of the treatment session to the study raters during rating sessions. Furthermore, subjects were instructed to report all adverse events only to the device operator. Efficacy ratings were administered at baseline and once weekly until the end of the study (week 16).

9.3.1 Study Measures

- *HDRS* measured by change from baseline was evaluated. As the main efficacy endpoint, this measure refers to the mean change in HDRS-21 score from baseline to end of week 5. A response is defined as ≥50 % improvement from baseline HDRS-21 score.
- *Level of anxiety* was based on the change from baseline in HARS.
- *Subjective improvement* as measured by changes in QIDS-SR.
- *Global improvement,* which was based on the changes from baseline in Clinical and Patients Global Impression-Severity of Illness (CGI-S, CGI-I, and PGI-I).
- *Quality of life* as measured with SF-36 and Q-LES-Q.
- *Personal and social functioning* as measured by change in GAF.

Study measures and the statistical analysis are further discussed in the statistics section.

9.4 Safety Evaluations

A physical and neurological examination, including body weight was completed at the screening visit (S1) at 4, 5, 6, 8, and 12 weeks, at the study end or at the time of discontinuation, and at 12-week follow-up. A urine pregnancy test was performed in all females of childbearing age at screening and at study end or at early discontinuation. If a subject develops documented suicidal ideation (based on HDRS-21 item 3 score of 3 or 4 or as assessed by the Suicide Ideation Scale) or attempts suicide after receiving active or sham rTMS, the investigator should discontinue his or her participation in the study. Subjects were questioned regarding auditory changes and any changes were reported. Subjects were informed of the risk of permanent hearing loss if an earplug should become detached or fall out. Subjects were asked to immediately report any loosening or detachment of an earplug during rTMS, and rTMS had to immediately be stopped if a subject reports or if an investigator observes that a subject's earplug has loosened or has fallen out. Auditory threshold testing was performed at the screening visit, at week 6 and at week 16 or at the time of discontinuation. A screening questionnaire for transcranial magnetic stimulation was completed at the screening visit. YMRS that looks for manic/hypomanic symptoms was assessed at baseline and then weekly during the treatment period. Evaluation using the MMSE, BSRT, and AMI-S cognitive tests for detection of cognitive changes were completed at baseline, at week 22 and at the study end or at the time of discontinuation (MMSE was also conducted at weeks 22, 24 and 28). Subjects were instructed to report AEs as they occur. AEs were assessed at each study visit.

10 Subject Completion/Withdrawal

10.1 Completion

A subject was considered to have completed the study if he or she completed all required assessments at week 16 of the treatment trial period. Subjects who prematurely discontinue study treatment for any reason (other than withdrawal as defined below) before completion of the 16 weeks were not considered to have completed the study.

10.2 Withdrawal from the Study

A subject was withdrawn from the study for any of the following reasons:

1. Inability to locate and quantify a motor threshold (MT) as defined in the protocol.
2. Lost to follow-up.
3. Withdrawal of consent.
4. Subject is not compliant with requirements of the study, including inclusion criteria and exclusion criteria.
5. The study is prematurely stopped or halted (e.g., clinical halt).
6. Discontinuation of the study treatment (final assessments will be obtained). A subject was discontinued from study treatment if:
   - If a subject develops documented suicidal ideation (based on HDRS-21 item 3 score is 3 or 4 or as assessed by the Suicide Ideation Scale) or attempts suicide after receiving active or sham rTMS.
   - A subject that does not experience a sufficient improvement in depression after 5 weeks of treatment was discontinued, where a sufficient improvement is defined as a decrease in the CGI-S from baseline (i.e., at randomization) of at least one point, in two consecutive assessments. That is, if the patient's CGI-S score is equal to or greater than the CGI-S at baseline, after 5 weeks of treatment in two consecutive assessments, he/she was discontinued.

- Subject has significant tremor in any limb.
- Subject experiences a seizure.
- The investigator believes that for safety reasons (e.g., an adverse event) it is in the best interest of the subject to stop treatment.
- The subject becomes pregnant.
- The subject missed more than 2 days of treatment during weeks 1–4 of the treatment trial period.
- The subject missed more than 3 days of treatments during weeks 5–16 of the treatment trial period.

The above safety discontinuation criterion, i.e., suicide ideation and worsening of depression after 5 weeks, were meant to ensure similar discontinuation at the primary efficacy endpoint measured at 5 weeks.

If a subject is lost to follow-up, every possible effort must be made by the study center personnel to contact the subject and determine the reason for discontinuation. The measures taken to follow-up must be documented.

When a subject withdraws before completing the study, the reason for withdrawal is to be documented on the CRF and in the source document.

11 Study Endpoints

The primary endpoint was the change in the total score on the HDRS-21 from baseline to week 5. The secondary efficacy endpoints were the response and remission rates according to the following HDRS-21 criteria: Response was defined as a reduction of at least 50 % in the total HDRS-21 score at week 5 compared to baseline, and remission was defined by a total HDRS-21 score <10 at week 5. Other efficacy measurements including the Clinical Global Impression (Severity and improvement), patient-reported outcomes (the Quick Inventory of Depressive Symptoms—Self-Report version (QIDS-SR)), and the Patient Global Impressions Improvement Scale (PGI-I) were obtained weekly. Daily functioning score and quality-of-life measures were also collected.

12 Statistical Analysis

It was assumed that 85 subjects per group (170 in total) would provide 90 % power at a significance level of 5 % (two-tailed) in detecting a difference of 3.75 points in the mean change from baseline HDRS-21 scores between treatment and sham groups, considering a standard deviation of 7.5 points (data from pilot study), and assuming an effect size of 0.5. Allowing for a 15 % dropout from the study irrespective of study arm and success of treatment, 200 randomized subjects were required. Approximately 20 % of the subjects were expected to leave the study between screening and randomization, thus it was necessary to screen approximately 250 subjects in order to arrive at the point of randomization with at least 200 subjects. The patients were randomized into the study, stratified by center and severity of disease as determined by the baseline HDRS-21 score and ATHF category. After a subject met the eligibility criteria, he/she was equally allocated (with a 1:1 ratio) to one of the two treatment groups stratified by HDRS-21 scores (<26 vs. ≥26), ATHF category (ATHF ≥2 level 1–2 and ATHF 1 level 3 vs. ATHF 2–4, level 3), and clinical center, based on a stratified randomization scheme using the permuted block method (with a random block size) programmed with the SAS (version 9.2.) random number generator.

Comparisons of baseline demographic and clinical characteristics between the study groups were performed to ensure that the study groups were balanced at baseline and that the randomization was successful. For comparison of means (continuous variables), the two-sample *t*-test was used or a nonparametric equivalent. For comparison of proportions (categorical variables), the Chi-square test or Fisher's exact test were used, as appropriate. Kaplan–Meier curves of the time to dropout up to week 5 and up to week 16 are presented, using a log-rank test to compare study groups.

The change in HDRS-21 from baseline to week 5 (primary endpoint) was compared between the treatment groups using a repeated measures analysis (RMA) of covariance (SAS® MIXED procedure). The analysis, which aims to compare the slopes of the changes from baseline in HDRS-21 scores between study arms, included the following fixed effects: time from randomization (in weeks), treatment group, time–treatment group interaction, center (site of study), baseline HDRS-21 score, and ATHF category at baseline. Baseline HDRS-21 score was entered as a continuous variable so as to minimize the potential for colinearity problems. Individual subjects' intercepts and time effects were also included in the model as random effects (random intercept and slope model). The principal statistical analysis was a comparison between the slopes of the treatment groups, derived from the time–treatment interaction term from the RMA model described above. The adjusted mean slope of change from baseline in HDRS-21 scores at week 5 post-randomization is estimated from the model (LS-means) for each study group as well as for the difference between the groups' adjusted mean slopes, and these are presented together with 95 % confidence intervals.

Secondary outcome measures were the response and remission rates at week 5 post-randomization. Tertiary endpoints were the change from baseline to week 16, as well as response and remission rates at week 16. The change in HDRS-21 from baseline to week 16 was compared between the treatment groups using analysis of covariance of the change from baseline to the last observed value (LOV). Baseline HDRS-21, ATHF category at baseline, and site were entered as covariates. The LOV is defined as the last available post-baseline visit data up to and including the last treatment visit or termination visit. The adjusted mean change (LS-means) from baseline in HDRS-21 scores at week 16 (LOV) is estimated from the model (LS-means) for each study group, as well as the difference between the adjusted means. These are presented together with 95 % confidence intervals.

The overall significance level for this study is 5 % using two-tailed tests, except for treatment–site interaction that was tested at a significance level of 10 %. Nominal *P*-values are presented. Statistical analyses were performed using SAS® V9.2 (SAS Institute, Cary, NC, USA). Effect sizes were calculated using methods described in Feingold [13] where ES = difference between slopes/Pooled SD of baseline HDRS-21 score.

References

1. Hamilton M (1960) A rating scale for depression. J Neurol Neurosurg Psychiatry 23:56–62
2. Prien RF, Carpenter LL, Kupfer DJ (1991) The definition and operational criteria for treatment outcome of major depressive disorder: a review of the current research literature. Arch Gen Psychiatry 48(9):796–800
3. Rush AJ, Trivedi MH, Ibrahim HM, Carmody TJ, Arnow B, Klein DN, Markowitz JC, Ninan PT, Kornstein S, Manber R, Thase ME, Kocsis JH, Keller MB (2003) The 16-Item Quick Inventory of Depressive Symptomatology (QIDS), clinician rating (QIDS-C), and self-report (QIDS-SR): a psychometric evaluation in patients with chronic major depression. Biol Psychiatry 54(5):573–583
4. Suppes T, Mintz J, McElroy SL, Altshuler LL, Kupka RW, Frye MA, Keck PE Jr, Nolen WA, Leverich GS, Grunze H, Rush AJ, Post RM (2005) Mixed hypomania in 908 patients with bipolar disorder evaluated prospectively in the Stanley Foundation Bipolar Treatment Network: a sex-specific phenomenon. Arch Gen Psychiatry 62(10):1089–1096
5. Folstein MF, Folstein SE, McHugh PR (1975) "Mini-mental state". A practical method for grading the cognitive state of patients for the clinician. J Psychiatr Res 12(3):189–198
6. Brown GK, Beck AT, Steer RA, Grisham JR (2000) Risk factors for suicide in psychiatry outpatients: a 20—year prospective study. J Consult Clin Psychol 68:371–377
7. Beck AT, Brown GK, Steer RA, Dahlsgaard KK, Grisham JR (1999) Suicide ideation at its worst point: a predictor of eventual suicide in psychiatric outpatients. Suicide Life Threat Behav 29:1–9
8. Buschke H (1973) Selective reminding for analysis of memory and learning. J Verbal Learning Verbal Behav 12:543–550
9. Datto CJ, Levy S, Miller DS, Katz IR (2001) Impact of maintenance ECT on concentration and memory. J ECT 17:170–174
10. Kopelman MD (1994) The autobiographical memory interview (AMI) in organic and psychogenic amnesia. Memory 2:211–235
11. McElhiney MC, Moody BJ, Steif BL, Prudic J, Devanand DP, Nobler MS, Sackeim HA (1995) Autobiographical memory and mood: effects of electroconvulsive therapy. Neuropyschology 9:501–517
12. McElhiney M, Moody B, Sackeim H (1997) The autogiographical memory interview—short form. New York State Psychiatric Institute, New York
13. Feingold A (2009) Effect sizes for growth-modeling analysis for controlled clinical trials in the same metric as for classical analysis. Psychol Methods 14(1):43–53

Chapter 16

Navigated Transcranial Magnetic Stimulation: Principles and Protocol for Mapping the Motor Cortex

Jari Karhu, Henri Hannula, Jarmo Laine, and Jarmo Ruohonen

Abstract

Transcranial magnetic stimulation (TMS) is a unique method for studying the human brain. Whereas the majority of imaging tools detect and map all the brain areas that participate during a given task (both primary and secondary network activations), TMS, when used to evoke a measurable physiological response, maps only those areas that are *mandatory* for the observed reaction. As such, TMS is particularly suitable for mapping cortical motor areas and for assessing the functional status of the motor tracts, both in normal subjects and in patients. In this chapter, we explore the physical and mechanistic background of using TMS to map the motor cortex. In addition, we outline a detailed protocol for mapping the cortical representation of various muscles—a protocol which can be used in basic research or as part of a clinical diagnostic or treatment procedure.

Key words Transcranial magnetic stimulation (TMS), Motor mapping, MEP, Hand muscles, Leg muscles, Presurgical mapping

1 Introduction

1.1 TMS is a Unique Tool for Cortical Mapping

Transcranial magnetic stimulation (TMS) is a unique method for studying the human brain. The fundamental difference between TMS and all other functional brain imaging tools concerns functionality Whereas the majority of imaging tools detect and map all the brain areas that participate during a given task (both primary and secondary network activations), TMS, when used to evoke a measurable physiological response, maps only those areas that are *mandatory* for the observed reaction. As such, TMS is particularly suitable for mapping cortical motor areas and for assessing the functional status of the motor tracts, both in normal subjects and in patients.

1.2 Why Do We Need Individual Navigation of TMS?

In clinical settings, the need for accuracy and repeatability is critical. The size and shape of the head and brain, the distance between the stimulating coil and responding neuronal tissue, and the location

Alexander Rotenberg et al. (eds.), *Transcranial Magnetic Stimulation*, Neuromethods, vol. 89,
DOI 10.1007/978-1-4939-0879-0_16,

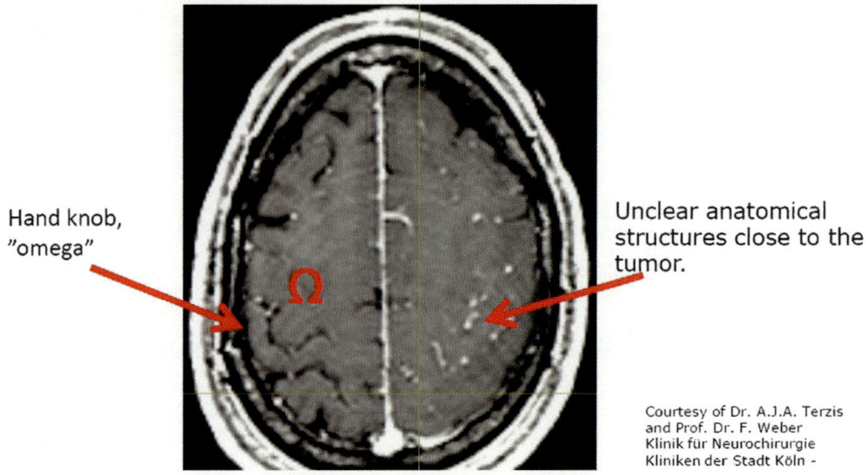

Fig. 1 Due to tumor and peritumoral oedema, structure of the brain on the right it unclear

and orientation of anatomical structures must be defined individually—a standardized location with respect to external landmarks of the skull is not sufficient (e.g., for preoperative mapping of eloquent cortical areas or for delivery of therapeutic stimulation). Anatomical measurements of in vivo brain macroscopic anatomy have shown that the anterior–posterior variation of central sulcus location is on the order of ±1.5–2 cm with respect to the Talairach coordinate system [1]. This disparity is likely to be much larger with respect to the external landmarks of the skull. Individual patterns of sulci and gyri cannot be determined by any proportional coordinates, and certainly not by any morphometric landmarks. Moreover, in clinical diagnostics the anatomical structures are often smeared by brain pathology (Fig. 1), tumors, oedema, bleeding, and/or vascular alterations.

1.3 Navigating the Electric Field (E-Field) in the Brain, Not the Coil on the Scalp

Individualized navigation and targeting solves many of the critical issues associated with TMS reproducibility and reliability. Navigation combines data of anatomical structures with controlled delivery of stimulation thereby forming the basis for dose determination and targeting. The core concept of navigation is similar to GPS-like navigators, where structural brain maps are acquired from individual MRIs (Fig. 2). This is crucial for clinical applications, because each person has his/her own individual brain topography with unique shapes, structures, and cortical convolutions.

Despite the name "transcranial magnetic stimulation," the magnetic field that originates from the TMS coil has no effect on human cells. Instead, TMS directly affects the membrane potential of neurons. A principal law of nature is that any time-varying

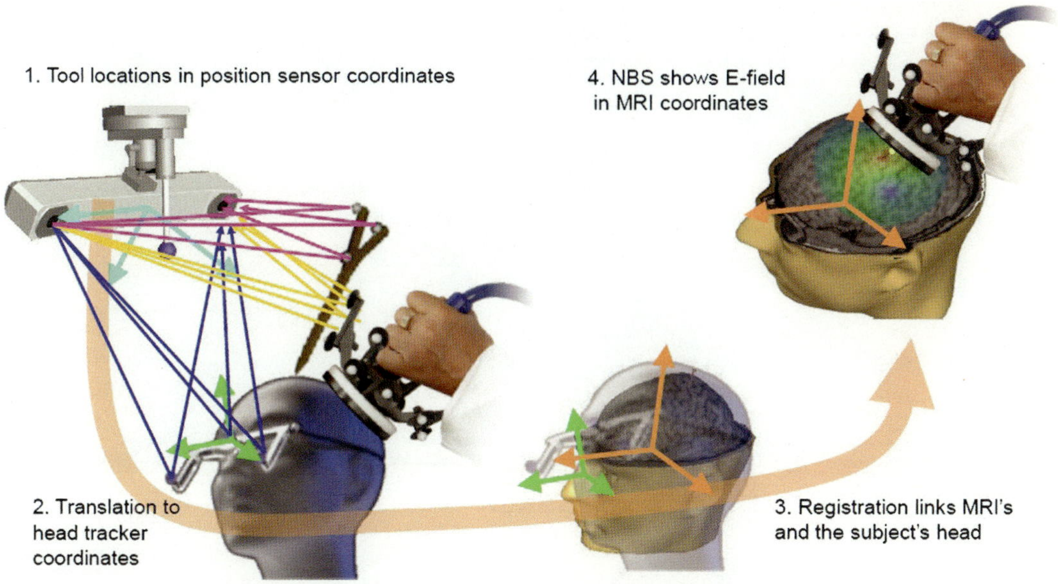

Fig. 2 The coil and stimulation-induced intracranial electric field are visualized online

magnetic field is always accompanied by a secondary electric field. It is this electric field, or E-field, that moves the same charged particles and molecules that are the principle elements of traveling depolarization fronts, or action potentials, in neuronal cells.

Although the electric field is generated by the magnetic field of the coil, a multitude of variables will affect its shape and strength. Accordingly, intuitive thinking concerning where stimulation takes place in the brain is very likely to fail. Yet, without knowing the fields, there is no simple way of knowing which part of the brain the TMS is stimulating. As such, online navigation of the TMS-evoked electric field is required to reliably link macroscopic biological responses to the individual anatomical structures and neuronal regions reached by the E-field.

A navigated TMS examination, which utilizes online calculation and targeting of stimulus-evoked E-fields in the brain starts with registration—a process that links the structural MRIs to the actual head of the patient (cf. Fig. 2). Once successfully completed, the TMS coil can be guided in relation to the brain structures using optical tracking (Fig. 2). The E-field is displayed online over the MRIs to precisely guide the stimuli to the target structures. The physiological response to a stimulus is recorded, often as an EMG measurement, and temporally linked to the stimulus. Finally, diagnostic and other considerations can be determined from the relation between the characteristics of evoked responses as related to the location and spread of the E-field.

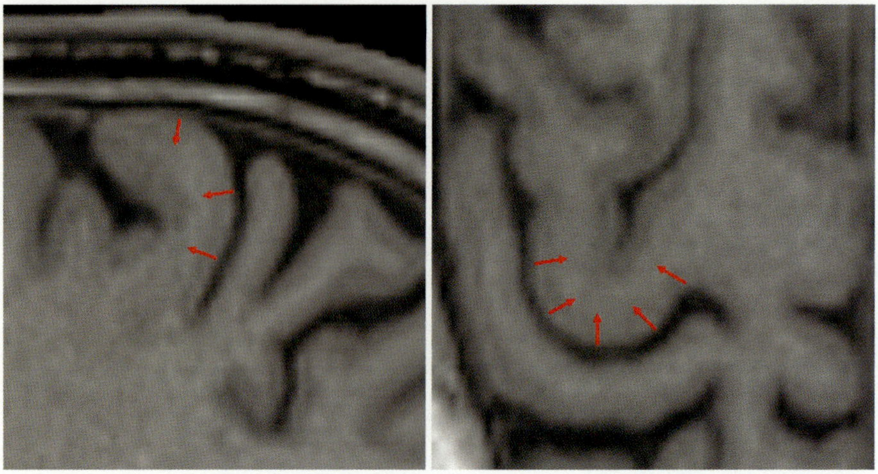

Fig. 3 E-field alignment perpendicular to underlying sulcus

1.4 Physics

At this point, let us elaborate on some important questions: How does one know the exact location of the neurons that are the source of the observed EMG response? In other words, how can one relate the extracranial coil location to the resulting intracranial physiological processes? How strong and widespread is the stimulation? These questions of physiology and physics were partly addressed during the early days of TMS (e.g., ref. [2]) and the explanations have since contributed to current navigation solutions.

TMS has many similarities to intraoperative stimulation during open craniotomy and cortical electrodes (DCS, Direct intraoperative Cortical Stimulation). In practice, very close agreement is found between localization results using these two techniques [3, 4]. At a conceptual level, TMS creates "virtual electrodes" in the brain which serve to stimulate neurons. Indeed, the mechanisms of action are the same for TMS and DCS: a potential difference (physicists will use the term E-field) acts as a force in the tissue that serves to move electrical charges (particles, ions, molecules). Because the neuronal cell membrane is intrinsically sensitive to local changes in the electric potential along the path of the axon, wherever the electric field is of adequate strength and suitable direction, it will trigger an action potential.

While much of the literature uses the terms "electrical current" and "electric field" interchangeably, they are not the same: the electrical current is a by-product of the electric field forcing the movement of electrically charged ions in the tissue. An electric field can exist without an electrical current, for instance where a cellular membrane prevents a flow of ions. As such, it is not primarily the electric current that needs to be determined when utilizing TMS, but rather the E-field.

In the case of electrodes placed directly in the brain, it is correct to assume that the field is the strongest in the immediate vicinity of the electrodes. However, the spread of electric fields from the electrodes is very complex as the E- field and current will follow paths of least electrical impedance in the tissue and are greatly influenced by macroscopic (e.g., sulci, cerebrospinal fluid) and microscopic (e.g., cortical layers, preferred orientation of cells) factors. In TMS, the interaction is relatively simple. The magnetic field from the coil is undisturbed by any tissue variation and this magnetic field will generate (induce) a stimulating electric field in each and every intracranial location reached. A significant part of the electric field in the brain is characterized by the magnetic field and, accordingly, by the location and orientation of the TMS coil. Macroscopic (e.g., skull shape) and microscopic (e.g., changes in resistivity along the path of the E-field) factors will affect the electrical field also in TMS [5], but less-so than in DCS. This is the main reason why TMS can be modeled precisely, perhaps better than direct intraoperative electrical stimulation.

1.5 E-Field Calculation

The E-field can be closely estimated, when the shape and size of the copper windings in the coil, the shape and size of the head, the electrical characteristics of the stimulator, and the exact location and orientation of the coil with respect to the head are characterized [5]. Note that the electrical conductivity of the tissue is *not* listed here. In fact, the nearly spherical shape of the head over motor areas reduces the absolute value of electrical conductivity to a secondary role, at least when spherical head models are used that account for local variations in head curvature [6]. Brain lesions do have an effect on E-fields, but even then the prominent features of the E-field are determined primarily by coil-to-head distance, coil orientation, and local skull shape.

1.6 Physiology

For electrical stimulation, it has been established that neurons are excited at lower thresholds when applied voltages induce currents oriented longitudinally along the axon rather than transversely across the axon [7–9]. In fact, electrical stimulation has been found most effective when the applied current has the same orientation and timing as the normal flow of postsynaptic current during depolarization: from dendrites, through the soma, and on to the axon. Early TMS studies suggested that the threshold for excitation is sensitive to orientation [10, 11], with optimal responses when the induced current is oriented 45° medial to the anterior–posterior plane. Although neither of these early TMS studies used any navigation method to confirm the underlying surface anatomy, both interpreted this orientation as indicating that optimal stimulation of the primary motor cortex was achieved when the induced current was perpendicular to the central sulcus. This assumption was subsequently confirmed in many later studies.

TMS coils are flat and, when placed tangential to the scalp, the induced E-field is tangential to the scalp. Hence, at the crown of gyri, the E-field is in the plane horizontal to the radial alignment of the cortical columns. Accordingly, Day et al. [7] hypothesized that TMS stimulation must be exciting the tangentially oriented neural elements at the gyral crown, such as horizontal interneurons or horizontal collaterals of pyramidal track axons. This claim was based on the assumption that the induced currents are stronger when proximal to the coil. Despite the importance of coil-to-target distance to the induced E-field, this assumption clearly disagrees with the clear orientation selectivity observed by Brasil-Neto et al. [11] and Mills et al. [10] Horizontal fibers extend uniformly in all directions within a plane parallel to the cortical surface. The isotropism of the horizontal fibers should translate into a lack of a preferred orientation for TMS, as the induced E-field should excite an equivalent fraction of the fiber population in any orientation.

The clear sensitivity of single pulse TMS responses to the orientation of the coil suggests that the predominant activation mechanisms are related to the trajectories of pyramidal tract axons and their direction along the cortical columns. Suprathreshold stimulation may lead to a combined activation of transsynaptic pathways and direct stimulation of axonal pathways deeper in the gray matter or in the bending/tapering of white matter structures [12]. When the TMS-induced E-field causes sufficient membrane depolarization, action potentials are generated in the entire stimulated neuronal volume thereby leading to synaptic transmission and excitatory or inhibitory postsynaptic potentials (EPSPs or IPSPs, respectively).

Physiological principles imply that navigated TMS is most effectively applied by orienting the E-field longitudinal and orthodromic to the greatest possible number of neurons at the site of interest. Due to rather uniform columnar structure everywhere in human cortex, in practice this means targeting the E-field perpendicular to the closest underlying sulcus (Fig. 3).

1.7 Accounting for the Local Anatomy

Even with easily recognizable individual anatomical landmarks, our ability to provide quantitative guidelines for functional representation is limited. One well-known landmark, the "pli de passage fronto-pariétal moyen" [13]. Manifests as an elevation in the floor of the sulcus at its midpoint and is currently thought to localize the somatotopic hand area which is often easily seen on the cortical surface as "omega" or "hand knob" [14–16]; (Fig. 1). However, no quantitative guidelines exist for the recognition of the PPFM relative to functional cortical representations in the central sulcus. Indeed, the variation of functional representations of other body parts or Broca's motor speech area may be on the order of ±1.5 cm in Talairach coordinates [1].

In addition to spatially accurate information regarding cortical representations, information pertaining to the "dose" of stimulation is needed for clinical diagnostics and follow-up visits. It is evident that no simple measure (such as the strength of the delivered magnetic pulse) can account for the observed large individual anatomical variation. The most common, and so far only, solution to control the stimulus strength is to derive it from the individually measured motor threshold (MT), i.e., to the lowest stimulus intensity sufficient to activate a peripheral muscle when the stimulus is delivered to the presumably optimal cortical motor representation area of said muscle.

There are several theoretical and practical caveats in this approach. First, the optimal location for the stimulating coil is difficult to define and reproduce while measuring the threshold. Second, the thresholds of even neighboring muscles may differ greatly. Finally, the real strength of the stimulus actually reaching the cortex is completely unknown when we move away from the motor cortical areas. For example, in aging, the frontal cortical areas tend to atrophy more than the motor areas. This may lead to practically no effective stimulation of frontal areas as the stimulus intensity is determined via measurements made over non-atrophied parietal areas.

1.8 Accuracy of Navigation

The key performance criterion for useful navigation is spatial accuracy. Current literature on navigated (as well as non-navigated) TMS lacks a discussion of accuracy. The determination of accuracy needs to be approached from an accepted definition: “Accuracy is the closeness of the predicted and true locations” or “The measure of an instrument’s capability to approach a true or absolute value” (FDA).

With TMS, accuracy can be defined as the distance (in millimeters) between the visualized “hotspot” location to the in vivo location of the stimulated neurons. The hotspot, in turn, can be defined as the location of the calculated strongest E-field that the navigator visualizes on the individual MRI of the cortex. By necessity, the accuracy of a navigation unit is an aggregate sum of multiple factors. Below, we divide the components of accuracy into four groups, each group independently contributing to the visualized location of the hotspot (Table 1):

1. *Coil localization*:
   - 3D technique to locate the coil and reference trackers (normally optical or electromagnetic).
   - Manufacturing tolerances of the coil and trackers.
2. *Movement of the reference tracker during examinations*:
   - Movement affects the detected coil position and, accordingly, the E-fields.
3. *E-field computation model*:
   - Errors in the coil’s output field, unknown and/or varying coil characteristics: shape, size, details of the copper windings.

Table 1
Error groups affecting the accuracy of navigated transcranial magnetic stimulation

Error group	Error source
Group 1: Localization of the coil, optical tracking	• Optical tracking of the coil • Manufacturing tolerances for the coil and trackers • Head tracker movement
Group 2: Registration	• MR image quality and resolution • Registration algorithm • Pinpointing of landmarks using the digitizer pen
Group 3: Electromagnetic field modeling	• Fitting of the conductor model to the head shape • Modeling the coil • Head conductor model

For an integrated system, total accuracy can be derived from each error group
For a combination of partial systems, total accuracy can be established only by testing each individual combination separately

- Computational model of the E-field (e.g., numerical uncertainties, simplified tissue characteristics).
- Fitting of the computational model to the individual head.

4. *Errors in alignment of anatomical images to live situation*:
   - Registration algorithm that aligns the individual head MRIs and the live head.
   - Human errors in alignment procedures.

Total accuracy is obtained from the root sum of the squares of the above uncertainties from all sources, quantified for use under true or simulated clinical (rather than laboratory) conditions. In the published literature, only the accuracy of optical tracking is usually reported—not the total accuracy. Optical tracking error may be as good as 0.2 mm (RMS), but this is clearly misleading as this error source represents but a negligible component of the total error. Precision, defined as the degree to which a previous location can be redetermined, is not the same as accuracy. With regard to TMS navigation, precision is significantly better than accuracy, as many of the errors remain stable across repeat stimulations; including uncertainties inherent in E-field calculation and MR-image distortions [17]. It is important to note that the accuracy of a TMS navigation system without E-field-based navigation cannot be determined: if the predicted location of stimulation is not visualized, a given measure for the distance to the true location is meaningless.

Table 2 lists the factors contributing to the accuracy of a specified navigated TMS system, the eXimia NBS System (Nexstim Oy, Helsinki, Finland; Table 2). The authors were not able to find other manufacturers' equivalent specifications, but hopefully they are or will become available to users and researchers.

Table 2
Error sources contributing to accuracy

Error source	Mean error in E-field hotspot (mm)
Coil localization – Optical tracking – Manufacturing tolerances of the coil and trackers	1.6
Movement of the head tracker during an examination	3.1
E-field computation – The coil's output and characteristics – Model of the intracranial E-field – Fitting of the model to individual head	3.8
Registration to anatomical MRIs – MRI image imperfections – Registration algorithms	2.5
System accuracy (Root square sum of all error sources) $\sqrt{(1.6)^2+(3.1)^2+(3.8)^2+(2.5)^2}$	= 5.7

These values are for the Nexstim eXimia NBS system
Listed values are the effect of each error or error group on the exact location of the maximal E-field in the cerebral cortex

2 Practical Protocol for Motor Mapping

2.1 Selecting Target Muscles

The target muscles for recording motor-evoked potentials (MEPs) naturally depend on the individual study design and/or clinical question. The muscles listed below have proven to be easily identifiable and often provide distinct, artifact-free responses to navigated TMS. The relative placement of surface EMG electrodes can be found in most textbooks of clinical neurophysiology.

Hand:

1. Abductor Pollicis Brevis (APB).
2. Abductor Digiti Minimi (ADM) or First Dorsal Interosseus (FDI).
3. Extensor Carpi Radialis (ECR) and Flexor Carpi Radialis (FCR), in practice often Flexor Digitorum Superficialis (FDS).
4. Biceps Brachii (BB).
5. Deltoideus (D).

Foot:

1. Tibialis Anterior (TA) and Soleus (S).
2. Abductor Hallucis Brevis (AHB).
3. Extensor Digitorum Brevis (EDB).

Face:

1. Mentalis (M) or Nasalis (N).
2. Orbicularis Oris (OO).

2.2 MRI to Head Co-Registration

Each navigated TMS imaging session begins with aligning the MRI images to the subject's head via the "MRI to head" co-registration. This is followed by determination of subject's individual motor threshold (MT).

Bridge of the nose and the tragus of the ear (crux helicis) are easily identifiable landmarks which are visible on both MRI and subject's head (Fig. 4). 3D-locations are first marked on the MRIs and then located on subject's head using a digitizing pen with an optical tracking system.

This type of registration allows for real-time monitoring of the coil position without restraining the subject's head during the TMS experiment. The optical tracking system uses infrared sensors to measure the 3D-location of infrared light reflecting passive markers attached to the coil and to the head tracker ("glasses") worn by the subject (see Fig. 2).

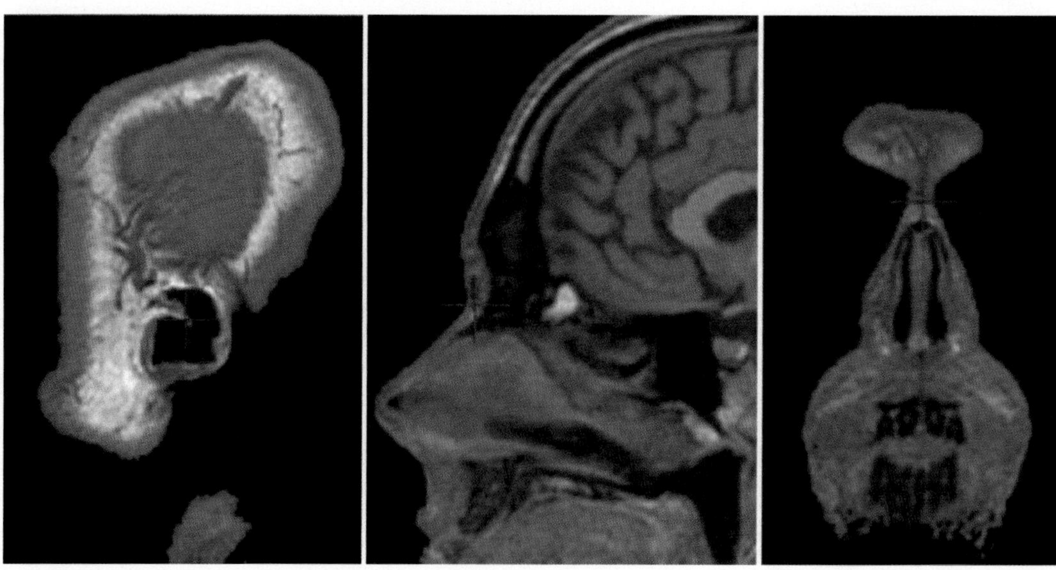

Fig. 4 The tragus and the bridge of the nose targeted on an MRI

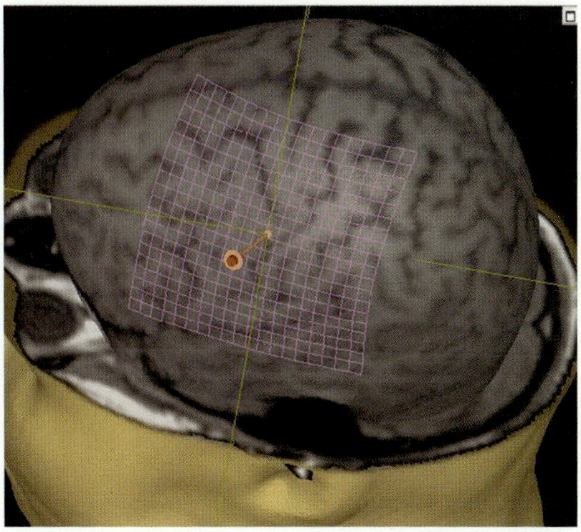

Fig. 5 Marker placed on the "motor knob" of the cortex with surrounding area grid

2.3 Finding and Using the Anatomical "Hand Knob" as Starting Point

Arguably the best known landmark on the cortical surface, "omega" or "hand knob" is currently thought to contain the functional hand area. If in doubt when observing the 3D rendering of an individual hemisphere, macroanatomically the "pli de passage fronto-pariétal moyen" (PPFM) manifests as an elevation in the floor of the sulcus at its midpoint and can usually be seen as a "passage" through the sulcus when searching systematically for the depth of suspected anatomical sulcus from cortical surface.

A marker can be positioned on the motor cortex at 20–25 mm depth from skin to aid in the visualization and planning the delivery of stimuli (Fig. 5). The exact position of this marker is best set to the temporal end of the anatomical "motor knob" on the central sulcus. Stimulus intensity in terms of E-field delivered to anatomical target is set initially to 80–100 V/m.

2.4 Finding the Optimal Location of Small Hand Muscles by Rough Mapping

A preparing stimulus pulse is given over the marker and the target APB/ADM/FDI response is noted. Two stimuli are then given over the target and the resultant muscle response latencies (normal 15–25 ms) and amplitudes are checked. During this procedure, the hand muscles should be relaxed (as assessed by continuous EMG tracking; peak-to-peak noise level must be under 50 μV).

IF EMG RESPONSE >500 μV: The stimulation intensity is decreased by 1–2 % of stimulator output and stimulation is repeated until the APB peak-to-peak response is in the range of 100–500 μV.

IF EMG RESPONSE <100 μV: The stimulation intensity is increased until the APB amplitude is 100–500 μV.

With the final determined stimulation intensity, stimulation is continued by following the central sulcus towards the interhemispheric central fissure so that the stimulus orientation is always perpendicular to the central sulcus and the stimulus direction is always frontal (Fig. 6). Stimulation is ended when there is no response.

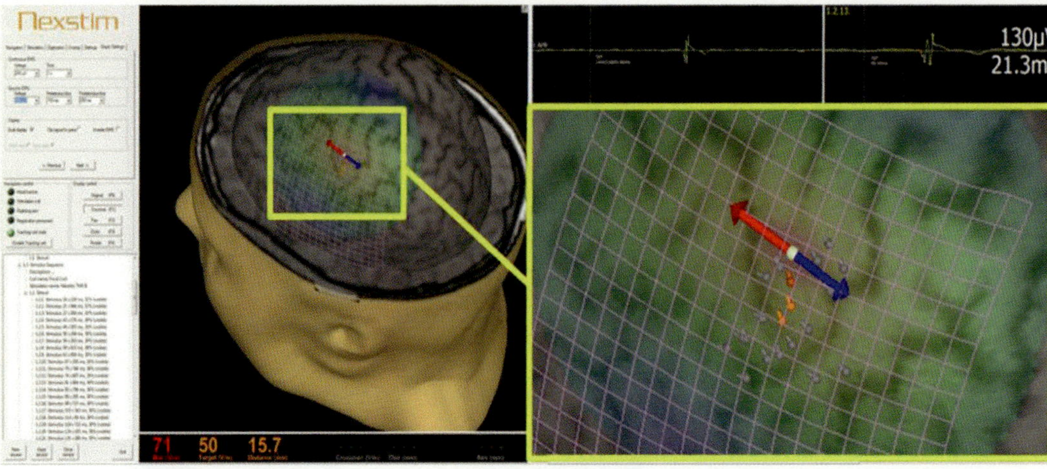

Fig. 6 Stimulation orientation should maintain perpendicularity to the central sulcus

Note: Make sure that you stimulate also directly over the motor gyrus.

The optimal location for motor threshold determination is defined by finding the largest target muscle response and marking it as a repeated stimulation target.

2.5 Defining the Optimal Orientation of the E-Field

The first stimulus is given on the zero-degree position, i.e., in the same orientation as that which elicited the best MEP response during initial, rough mapping. The recorded EMG response amplitude becomes the new reference.

Using the available aiming/orienting tools, stimulation on the marker is continued so that after every stimulus the coil is turned 20° counterclockwise. The target muscle response amplitude (μV) is checked after each pulse. The stimulation is ended when the response is less than 100 μV.

After returning to the original orientation on the marker, stimulation is continued so that after every pulse the coil is turned 20° clockwise. After every pulse the intensity (μV) is checked. Stimulation is ended when the response is less than 100 μV.

The optimal coil orientation gives the largest EMG response and is selected for defining the motor threshold.

Note: Optimization of the rotation means that once the hot spot has been located, stimuli will be delivered in the same spot, but in other rotations. The aim of different rotation is to find whether a better response can be obtained from the same location (bearing in mind that the tilting should remain optimal).

2.6 Defining the Resting Motor Threshold (rMT) at Optimal Location and Orientation

Start the motor threshold software (if not using any automated system, consult Rossini et al.). Stimulate over the repeated stimulus target. Be sure to leave 5–10 s between each pulse to avoid summation. After each stimulus, note the peak-to-peak amplitude of muscle response in motor threshold software and reject responses with non-physiologcal latency or amplitude as well as those with preceding muscle tension software and proceed with automatically adjusted intensity. Eventually, the MT software will determine and display the patient's motor threshold as a percentage of the maximum stimulator output. Verify the determined MT by stimulating at the suggested value ten times—5 out of 10 stimuli should evoke responses larger than 50 μV.

Note: If you generate fewer than four responses out of 10 stimuli, increase the stimulation intensity by 1% and start the ten-pulse verification procedure over again. If you get ten responses out of 10 stimuli, decrease the stimulation intensity by 1% and start the ten-pulse verification procedure over again.

It is of utmost importance that MT is defined from a relaxed muscle. The spontaneous EMG must be observed throughout the procedure to exclude responses elicited while the target muscle is tensed or patient is moving to avoid lowered active MT values.

Subject's talking lowers hand MTs and should be avoided during the procedure.

2.7 Mapping of M1

2.7.1 Hand and Arm

For hand area mapping, the starting point is the hand knob ("omega") of central sulcus (see the functional location in Fig. 7). First stimulation intensity is set at 110 % of hand rMT (stimulation intensity may also be chosen to be at lower level, depending on the exact purpose of the mapping and the compliance of the subject). Stimulation is continued by following the central sulcus towards the interhemispheric central fissure. The stimulus orientation is always perpendicular to the central sulcus and the stimulus direction is always frontal. Stimulation is ended when there is no response. *Note: Make sure that you stimulate also directly over the motor gyrus.* After central sulcus mapping both the sulcus posterior to and the sulcus anterior to central sulcus should be mapped, preferrably spacing each stimulus manually in 2–5 mm distance from each other or with the aid of 5 mm spacing grid (at least 1 stimulus to each software grid square). In unclear/critical areas, for example close to brain lesion, it is essential to deliver at least 3 stimuli with differing orientations. Premotor areas are mapped as far frontally as visible MEPs emerge.

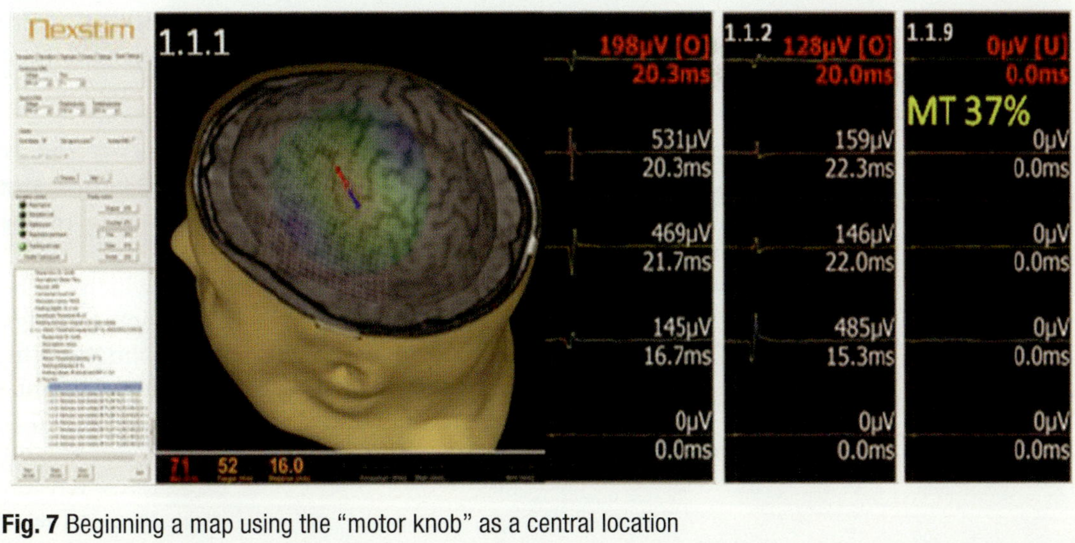

Fig. 7 Beginning a map using the "motor knob" as a central location

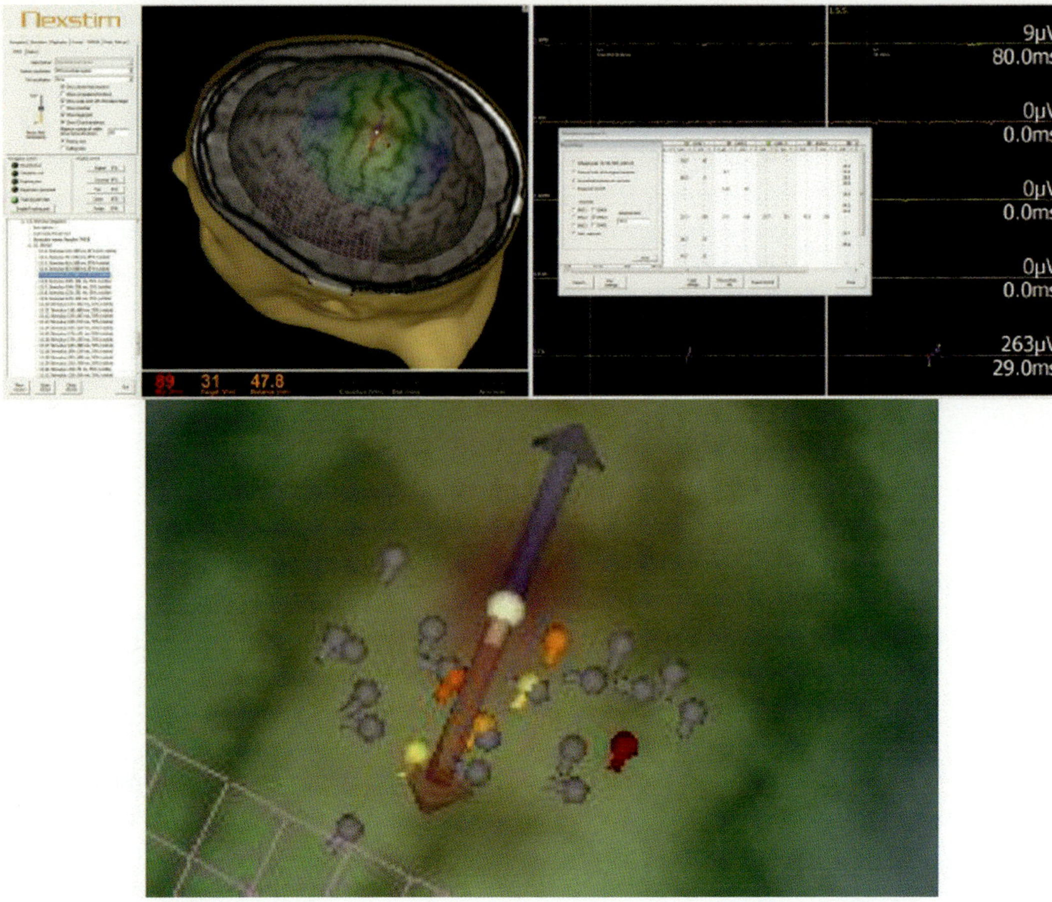

Fig. 8 Leg area map (colors correspond to motor response: *gray* = smallest response/*red* = largest response)

During the mapping, small markers (cf. Fig. 12 for hand and Fig. 8 for mapping of leg area) can be utilized to show each delivered stimulus location and stimulus direction overlayed to the grid and underlying cortical anatomy. The markers colored white (>1,000 µV), yellow (500–1,000 µV), and red (50–500 µV) are stimulus locations evoking EMG responses over threshold limit 50 µV (Nexstim eXimia software).

2.7.2 Face

Target muscle (Mentalis (M)/Nasalis (N)/Orbicularis Oris (OO)/ any specified muscle) representation area between hand knob and Sylvian fissure is stimulated with 2–5 mm steps according to the procedure described for hand and arm.

First stimulation intensity is set at 110 % hand MT which can be raised immediately to +20 V/m to hand MT stimulation intensity at rest. The face area of central sulcus is of a different configuration, "shallower" than more central areas often demanding higher stimulation intensities for adequate mapping. Of course, time allowing, defining facial MT for a specific muscle is an optimal solution.

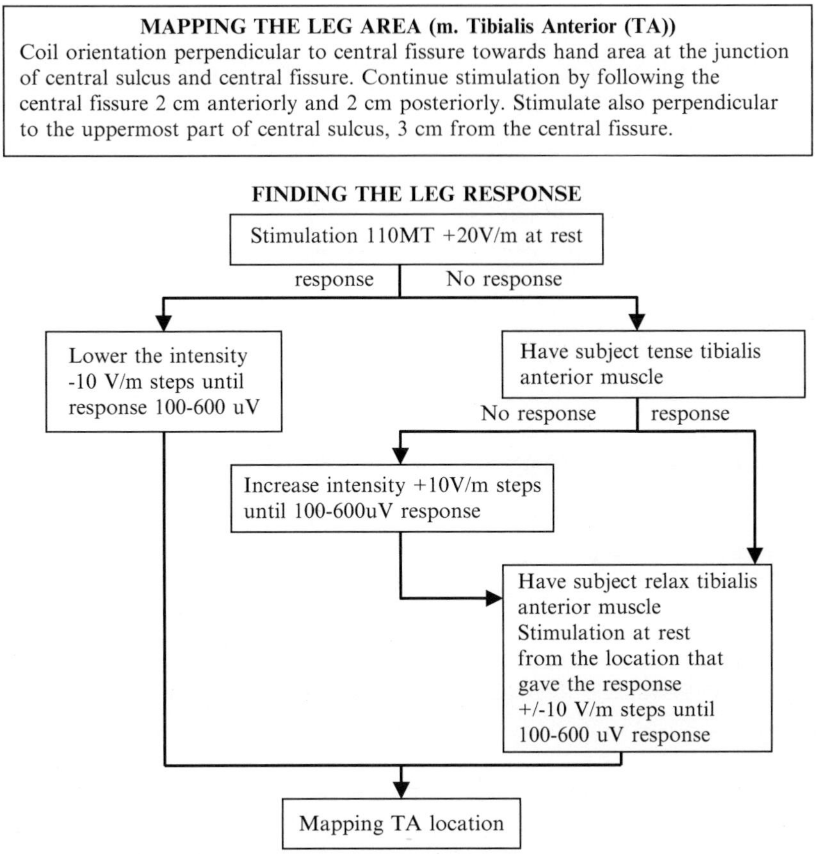

Fig. 9 Flow-chart for leg mapping

2.7.3 Leg

Target muscle (TA/Soleus/ADB/EDB/AHB) representation area near of the central fissure is stimulated first with 110 % hand MT +20 V/m stimulation intensity at rest. If no responses are found, the mapping can be continued according to Fig. 9.

2.8 Mapping Outside M1: The Extent of Map?

A common mistake when mapping motor areas is to limit the map only to the cortical strip containing motor cortex proper, M1. In practice, space-occupying lesions, which are a common reason to perform mapping alter the cortical excitation–inhibition balance. These alterations, which may be further complicated by plastic changes and simple "pushing" of functional cortex by the lesion, may lead to greatly altered motor maps. The most common change is an enlargement of the reacting cortex in the direction of premotor areas. Moreover, recent findings in NBS studies indicate that non-primary prefrontal motor areas contain a cortical patch with direct connections to hand muscles and conduction latencies comparable to those elicited by the stimulation of M1 [18].

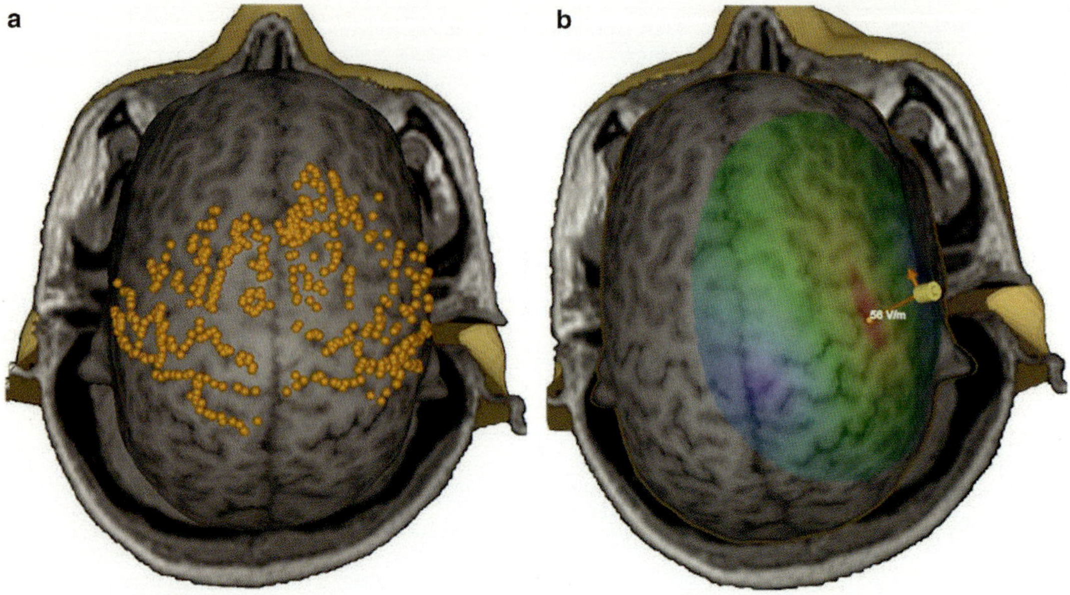

Fig. 10 (**a**) In addition to M1, an optimal motor map will cover the postcentral gyrus and precentral motor areas reaching premotor and SMA cortices. (**b**) Strong response will typically extend for several centimeters around optimal representation of the muscle. (Courtesy of Selja Vaalto, NBS Laboratory, Kuopio University Hospital)

Optimal motor mapping will cover, in addition to M1, postcentral gyrus and precentral motor areas reaching premotor and SMA cortices (often several centimeters from the optimal representation of hand in M1; Fig. 10).

2.9 Off-Line Analysis

The primary purpose of analysis is to verify that only those cortical locations which elicit real MEPs are accepted to the map and that maps are visualized according to the needs of the specific clinical situation/study. The following steps are usually sufficient.

1. Verify the latencies and amplitudes of the MEP responses of the mapping session. These must be physiologically meaningful and represent "real" motor-evoked responses.
2. Establish optimal APB (ADM, FDI, TA...) location (the stimulus location evoking the largest EMG responses). Coordinates are defined on a chosen stimulation surface (3D-MR image) for individual subjects, but it is useful to store coordinates at fixed depth (in mm from the scalp surface) for further comparisons.
3. Establish rMT for follow-up or comparisons, defining E-field (V/m) at target location of APB (FDI) at fixed depth (25 mm).
4. Capture all stimulus locations eliciting muscle responses with normal latencies and greater than or equal to 50 μV in relaxed muscles.

5. Make separate analysis files for visualization of the motor maps, e.g., containing all hand muscle locations and face muscle locations.

2.10 What Are "Real" Motor-Evoked Potentials (MEPs)?

Latencies and amplitudes of responses to stimulation are the main determinants when deciding whether muscle responses observed in surface EMG are "real" MEPs or artifacts. In normal subjects, MEP latency and amplitude are determined by two stimulation variables and two physiological variables. The stimulation variables, strength and location, are by definition optimized (or at least comparable from subject to subject) by standard motor mapping procedure. This leaves the physiological factors, age and height, to consider when verifying MEPs for motor maps.

In practice, all elicited MEPs must be verified visually before accepting the corresponding stimulus locations to a cortical motor map. The variation of MEP latencies is not overly large, and normative materials for NBS have recently been published (Fig. 11) [19]. Naturally, the clinical condition may alter MEPs significantly, but the purely artifactual deflections are in most cases clearly outside the physiological range of real responses (<10 ms or >50 ms).

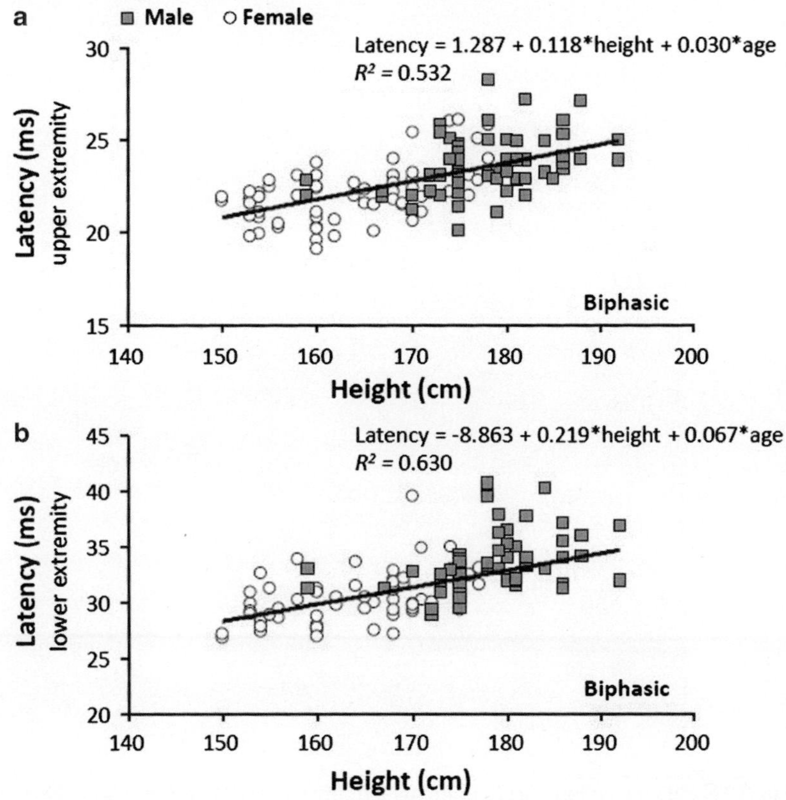

Fig. 11 Scatter plots showing the upper (**a**) and lower (**b**) normative MEP latency values for men and women correcting for height. These values were obtained using the standard motor mapping protocol

However, sometimes the only clue for whether to accept an MEP is the reproducibility of the response latency and waveform. In a poorly functioning motor system or when studying areas critically close to lesions, it is recommended to repeat stimuli over the same cortical target to reproduce MEPs several times.

2.11 Visualizing Motor Maps

The visualization of motor maps can be done by "thresholding" the map on the basis of the MEP voltage, where each of the stimuli/response relationships is represented by a colored marker. For example, in Fig. 12, the color-coding is: Gray—no response (<50 μV), Red—response between 50 and 500 μV, Yellow—response

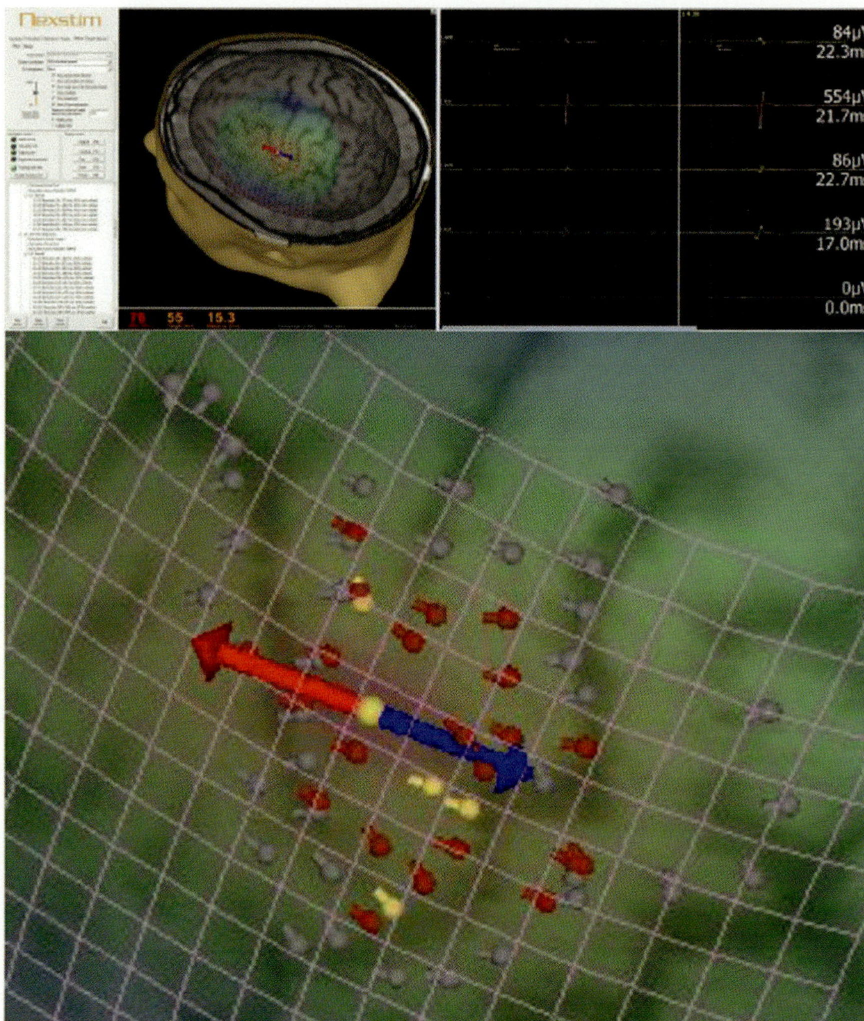

Fig. 12 Hand area Voltage Scale map, each of the stimuli/response relationships are shown as a colored markers (*Gray*: No response, meaning <50 μV, *Red*: Response between 50 and 500 μV, *Yellow*: Response between 500 and 1,000 μV, *White*: Response >1,000 μV). The "tail" of each stimulus marker on the cortical surface shows the direction of E-field

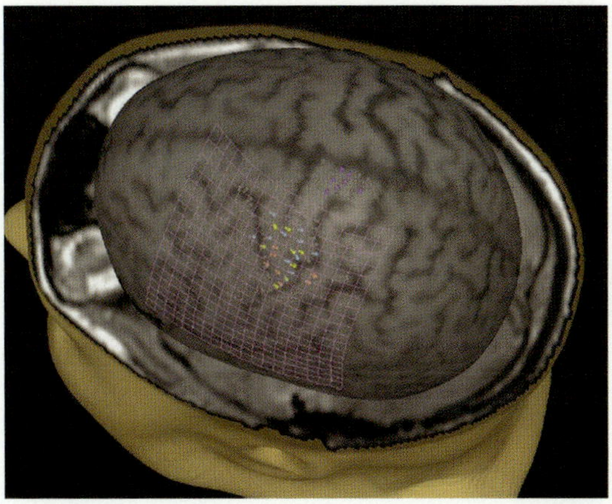

Fig. 13 Map according the channel color of the strongest response. This map assigns the EMG channel-specific color to each response exceeding the selected response limit ("threshold") value (50 μV). Values below the response are shown as *gray*. This feature intends to display an overview of the cortical muscle representations. The stimuli are colored by comparing the amplitudes of motor-evoked potentials recorded simultaneously from different EMG channels and then assigning each stimulus the color of the EMG channel giving the strongest normalized response (*Green*: APB, *Orange*: FDI, *Yellow*: ADM, *Blue*: ECR, *Purple*: TA)

between 500 and 1,000 μV, White—response >1,000 μV. The map changes with respect to selected MEP thresholds and, obviously, the higher the voltage threshold the smaller the cortical area providing large responses. The value of 50 μv is the standard starting point for thresholding the maps, as well as for determining rMT.

Another example of visualization is shown in Fig. 13. This map is constructed according to the channel color of the strongest response. This map assigns the EMG channel-specific color to each response exceeding the specified response limit, or "threshold" value (50 μv). Values below the response limit will be shown as gray. This feature displays an overview of the cortical muscle representations.

Finally, for clinical purposes and for determining eloquent motor cortices, a simple on–off map is sufficient (Fig. 14). This is the simplest form of map that can be transferred to neuronavigators or third-party visualization softwares.

3 Mapping Process as Part of Clinical Workflow and Requirements for Accurate Results

The use of navigated TMS utilizing the motor mapping protocol described above as part of a clinical process is shown in Table 3. The protocol has been demonstrated to work well when Nexstim NBS system is used but may not be applicable with other TMS systems. General software and hardware features

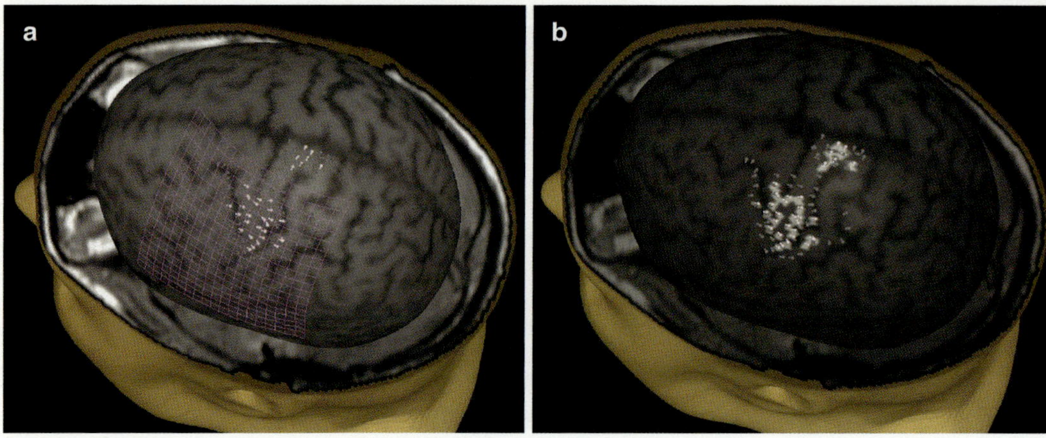

Fig. 14 (**a**) On–Off map (On = *white* markers, stimuli above 50 μV peak-to-peak MEP; Off = *Gray* markers, stimuli below 50 μV peak-to-peak MEP). (**b**) DICOM export of On–Off map on superficial cortex to be used in pre-procedural planning

that facilitate the use of navigation system for motor mapping are also indicated in Table 3.

The accuracy of Nexstim NBS in clinical use has been evaluated by comparing the localization of cortical representation areas of individual muscles obtained noninvasively by the NBS System in patients with brain tumors to the corresponding locations obtained invasively during surgery by direct cortical stimulation (DCS) [3]. The nTMS and DCS hotspots were located on the same gyrus in all cases. The mean ± SEM distance between the nTMS and DCS hotspots was 7.83 ± 1.18 mm for the APB muscle ($n = 15$) and 7.07 ± 0.88 mm for the tibialis anterior (TA) muscle ($n = 8$). Interestingly, when a low number of DCS stimulations were performed due to demand of the time pressure or clinical situation, the distance between the nTMS and DCS hotspots increased substantially ($r = -0.86$ for APB). After excluding the cases with <15 DCS APB responses, the mean ± SEM distance between the hotspots was only 4.70 ± 1.09 mm for APB ($n = 8$).

In another pilot study of nTMS in neurosurgical mapping, Forster et al. [4] compared the results of noninvasive, navigated TMS, and fMRI with the results of perioperative DCS in patients with brain tumors located in or adjacent to the motor cortex. By comparing (pair-wise Euclidian) distances between the DCS coordinates, corresponding nTMS coordinates and centroid of fMRI activation coordinates for each specific muscle MEP, the clinicians were able to measure average distances from nTMS to DCS and fMRI to DCS. Results showed that distances from nTMS to DCS (10.5 ± 5.67 mm) were significantly smaller than those from fMRI to DCS (15.0 ± 7.6 mm). In their conclusion, the authors note that “nTMS is more precise than fMRI when correlated with intraoperative DCS” and, in addition, “nTMS anticipates information usually only enabled by direct cortical stimulation.”

Table 3
Example of use of motor mapping as part of clinical process to obtain results that can be used for pre-procedural planning purposes

	Task	Time (min) to perform task (estimate)	Software (SW) features facilitating task	Hardware (HW) features facilitating task
1	Preparing the patient for examination (interview, attaching EMG-leads, explaining the procedure)	10 min		
2	Uploading the patient's MRI to the system	3 min	SW generates a 3D-model of the patients head based on the individual MRI	
3	Co-registering the patient's head with the 3D-model	40 s	To ensure accurate co-registration a guided registration process of a 3-point basic registration and a 9-point advanced registration is used Typical alignment accuracy is 2 mm	Headtrackers worn by patient act as fiducials and allow free head movement by patient Tracking system (infrared camera) follows fiducials Typical tracking system accuracy is <1.5 mm (mean)
4	Determination of Motor Threshold (MT)			
4a	Determining optimal location for MT assessment	5 min	SW calculates, based on a validated head model utilizing 40,000 locally fitted spheres the electric field produced by the TMS at the chosen target location EMG responses to TMS portrayed in real time provide instant feedback SW shows user coil orientation in relation to brain anatomy and allows coil positioning perpendicular to the sulci	Coil design and manufacturing specifications allow focal delivery of magnetic stimulation which generates an electric field that corresponds to the one calculated by SW

(continued)

Table 3
(continued)

	Task	Time (min) to perform task (estimate)	Software (SW) features facilitating task	Hardware (HW) features facilitating task
4b	Finding the optimal coil orientation for MT determination	2 min	SW tool allowing stimulation of the same anatomic location with the coil in different orientations ensures that MT determination is performed with the coil in an orientation producing largest responses	
4c	MT determination at optimal location and orientation	3 min	To ensure that motor threshold determination is performed with the coil the predetermined location and orientation, the SW provides the user with a targeting tool that allows stimulation only if the location is within 2 mm and coil orientation within 2° of those of the target position	
5	Motor mapping	10–20 min	See 4a–c above	An integrated diagnostic system comprising the navigation system, TMS and EMG allow stimulation with a known accuracy
6	Analysis and reporting of results		Can be performed either online or after the patient has left SW allows the user to verify that results are correct and disregard false responses	
7	Export of results to third-party software		Results can be exported in DICOM format to third-party software such as neuronavigators	

References

1. Keller SS, Highley JR, Garcia-Finana M, Sluming V, Rezaie R, Roberts N (2007) Sulcal variability, stereological measurement and asymmetry of Broca's area on MR images. J Anat 211(4):534–555
2. Ravazzani P, Ruohonen J, Grandori F, Tognola G (1996) Magnetic stimulation of the nervous system: induced electric field in unbounded, semi-infinite, spherical, and cylindrical media. Ann Biomed Eng 24:606–616
3. Picht T, Schmidt S, Brandt S, Frey D, Hannula H, Neuvonen T, Karhu J, Vajkoczy P, Suess O (2011) Preoperative functional mapping for rolandic brain tumor surgery: comparison of navigated transcranial magnetic stimulation to direct cortical stimulation. Neurosurgery 69(3):581–588, discussion 588
4. Forster MT, Hattingen E, Senft C, Gasser T, Seifert V, Szelényi A (2011) Navigated transcranial magnetic stimulation and functional magnetic resonance imaging—advanced adjuncts in preoperative planning for central region tumors. Neurosurgery 68(5):1317–1324, discussion 1324–5
5. Ruohonen J, Ilmoniemi RJ (2005) Basic physics and design of TMS devices and coils. In: Hallett M, Chokroverty S (eds) Magnetic stimulation in clinical neurophysiology. Butterworth, Philadelphia
6. Hämäläinen MS, Sarvas J (1987) Feasibility of the homogeneous head model in the interpretation of neuromagnetic fields. Phys Med Biol 32(1):91–97
7. Day BL, Dressler D, Maertens de Noordhout A, Marsden CD, Nakashima K, Rothwell JC, Thompson PD (1989) Electric and magnetic stimulation of human motor cortex: surface EMG and single motor unit responses. J Physiol 412:449–473
8. Ranck JB Jr (1975) Which elements are excited in electrical stimulations of mammalian central nervous system: a review. Brain Res 98:417–440
9. Rushton WA (1927) Effect upon the threshold for nervous excitation of the length of nerve exposed and the angle between current and nerve. J Physiol 63:357–377
10. Mills KR, Boniface SJ, Schubert M (1991) Origin of the secondary increase in firing probability of human motor neurons following transcranial magnetic stimulation. Studies in healthy subjects, type I hereditary motor and sensory neuropathy and multiple sclerosis. Brain 114(Pt 6):2451–2463
11. Brasil-Neto JP, McShane LM, Fuhr P, Hallett M, Cohen LG (1992) Topographic mapping of the human motor cortex with magnetic stimulation: factors affecting accuracy and reproducibility. Electroencephalogr Clin Neurophysiol 85:9–16
12. Amassian VE, Eberle L, Maccabee PJ, Cracco RQ (1992) Modelling magnetic coil excitation of human cerebral cortex with a peripheral nerve submersed in a brain shaped volume conductor: the significance of fiber-bending excitation. Electroencephalogr Clin Neurophysiol 77:390–401
13. Broca P (1888) Mémoires d'anthropologie. Reinwald, Paris
14. Yousry TA, Schmid UD, Alkadhi H, Schmidt D, Peraud A, Buettner A, Winkler P (1997) Localization of the motor hand area to a knob on the precentral gyrus. A new landmark. Brain 120(Pt 1):141–157
15. White LE, Andrews TJ, Hulette C, Richards A, Groelle M, Paydarfar J, Purves D (1997) Structure of the human sensorimotor system. I: morphology and cytoarchitecture of the central sulcus. Cereb Cortex 7:18–30
16. Boling W, Olivier A, Bittar RG, Reutens D (1999) Localization of hand motor activation in Broca's pli de passage moyen. J Neurosurg 91:903–910
17. Danner N, Julkunen P, Könönen M, Säisänen L, Nurkkala J, Karhu J (2008) Navigated transcranial magnetic stimulation and computed electric field strength reduce stimulator-dependent differences in the motor threshold. J Neurosci Methods 174(1):116–122
18. Teitti S, Määttä S, Säisänen L, Könönen M, Vanninen R, Hannula H, Mervaala E, Karhu J (2008) Non-primary motor areas in the human frontal lobe are connected directly to hand muscles. Neuroimage 40:1243–1250
19. Säisänen L, Julkunen P, Niskanen E, Danner N, Hukkanen T, Lohioja T, Nurkkala J, Mervaala E, Karhu J, Könönen M (2008) Motor potentials evoked by navigated transcranial magnetic stimulation in healthy subjects. J Clin Neurophysiol 25(6):367–372 doi: 10.1097/WNP.0b013e31818e7944

Chapter 17

Speech Mapping with Transcranial Magnetic Stimulation

Phiroz E. Tarapore

Abstract

The identification of eloquent cortical regions associated with language function is of great relevance to basic scientists as well as clinicians. Repetitive transcranial magnetic stimulation (rTMS) is a relatively novel technique that allows for noninvasive, lesion-based identification of critical language sites. The integration of frameless stereotactic navigation systems with rTMS allows for precise targeting and identification of the cortical region that is being stimulated. Integration of this navigational data also allows for accurate, real-time analysis of magnetic field strength in the stimulated region of interest. Thus, navigated rTMS allows the investigator to identify precisely the threshold of stimulation necessary to interrupt the language function of a given cortical locus. The most common rTMS language mapping protocols are safe, with a nominal risk of seizure being the predominant serious adverse effect. They are also well-tolerated by the patient. Validation studies have found that navigated rTMS language maps correlate well with language maps obtained using direct cortical stimulation (DCS), the current gold-standard. Future studies will focus on improving the protocols used for rTMS language mapping, as well as exploring and validating the language tasks which yield the most relevant and accurate language maps.

Key words Language mapping, Speech mapping, Transcranial magnetic stimulation, TMS, Repetitive transcranial magnetic stimulation, rTMS, Navigated transcranial magnetic stimulation, Direct cortical stimulation, DCS, rTMS safety, rTMS tolerability

1 Theoretical Background

1.1 Speech Mapping, and Why it is Done

The localization of essential cortical language regions is of great relevance across a wide spectrum of basic, translational, and clinical neuroscience. Retrospective localization studies relied largely on lesion-symptom mapping, in which existing injuries were correlated with clinical symptoms [1–6]. The advent of intraoperative direct cortical stimulation (DCS) mapping revolutionized this field, allowing clinicians to interrogate a region with a temporary electrical lesion. While this technique is effective, it is also highly invasive, requiring a craniotomy for cortical exposure, and is thus performed only when necessary for the surgical management of an existing lesion. A technique for noninvasive, lesion-based identification of critical language sites would therefore be of great

Alexander Rotenberg et al. (eds.), *Transcranial Magnetic Stimulation*, Neuromethods, vol. 89,
DOI 10.1007/978-1-4939-0879-0_17, © Springer Science+Business Media New York 2014

benefit to clinicians. Such a technique would also be important for neurophysiologists interested in testing cognitive models of language, including language streams [7] and connectome-based models [8].

Transcranial magnetic stimulation (TMS) is a technique that has demonstrated promise in this regard. Based on Faraday's principle of electromagnetic induction, TMS applies a brief pulse of high-strength magnetic field over the scalp which passes through the skull and induces an electrical current in the underlying brain region [9]. These pulses of current, if applied appropriately, are sufficient to depolarize a population of neurons, inducing an action potential [10]. Repetitive trains of these pulses, so-called repetitive TMS (rTMS), can have either an inhibitory or a stimulatory effect on cortical excitability, depending on the frequency of the rTMS trains [11].

Furthermore, the development of *navigated* TMS (nTMS) has allowed for this technology to arrive rapidly at the forefront of non-invasive language mapping modalities. What sets apart the nTMS system from a non-nTMS system is its demonstration, in real-time, of the precise location and strength of the magnetic pulse. By integrating a frameless stereotactic navigational system (such as those used commonly utilized in neurosurgical and other procedures) with a TMS coil, one can co-register a structural MRI or CT brain scan to a subject's anatomy using fiducial markers or anatomical landmarks. This advancement allows the investigator to deliver TMS pulses with unprecedented precision under image guidance [12–15]. Furthermore, in some nTMS systems, the strength and directionality of each TMS pulse is calculated on-the-fly according to a dynamic spherical model which takes into account the preset parameters of stimulation as well as the subject's scalp/skull thickness [16, 17]. As a result, when the stimulation coil is positioned over the subject's scalp, the investigator can visualize the targeted cortical region, the strength and orientation of the magnetic dipole, and the cone of activation generated by the magnetic pulse.

nTMS therefore allows the investigator, for the first time, to pinpoint precisely the cortical region that is being targeted. This capacity for accurate localization suggests a number of novel applications for TMS. In particular, it offers the possibility of mapping essential cortical regions associated with language function. Prior studies [18–24] have examined the use of rTMS to cause speech arrest and lateralize language (see below for details). However, these efforts were not stereotactically guided, and showed rTMS to be an unreliable technique for determining language laterality, largely because of a high false-positive rate for speech arrest sites on the supposedly non-dominant hemisphere. It should be noted that most of these studies did report finding dominant-hemisphere positive language disruption sites in almost all their subjects (see Table 1, "# Pts with + SA").

Table 1
Studies exploring the speech arrest capabilities of rTMS

	Date	No. of Pts	No. of Pts with +SA	rTMS frequency (Hz)	Train length (s)	Task	Primary pathology	Coil	Design
Pascual-Leone	1991	6	6	25	10	Counting	Epilepsy	CRRMS	15 positions bilat (10/20), c/w IAT
Michelucci	1994	14	7	16–20	6–10	Counting	Epilepsy	CRRMS	9 positions bilat (10/20), c/w IAT
Jennum	1994	21	14	30	1	Counting, reading	Epilepsy	MagPro biphasic	4 positions bilat (10/20), c/w IAT
Wassermann	1999	14	13	15, 5–18	2–3	Naming, reading	Epilepsy	CHSMS	1 cm grid, c/w IAT in 10 pts
Epstein	1996	5	4	4	1–5	Counting	None	CHSMS	Continuous over lat frontal region
Epstein	1999	10	10	4	2–5	Counting	None	Custom circular	Continuous over lat frontal region
Epstein	2000	17	16	4	N/A	Counting	Epilepsy	Custom circular	Continuous over lat frontal region

Note: none of these studies utilized stereotactic neuronavigation

1.2 Prior Studies of Speech Using TMS

Although the use of nTMS for language mapping is in its infancy, in the last three decades, several eminent research groups have examined TMS as a technique for studying the lateralization of language function [18–24]. In the first reported study of rTMS and the interruption of language function, Pascual-Leone et al. showed that speech arrest could be obtained in all six of six test subjects. Subsequent studies by Michelucci & Valzania [19] and Jennum & Friberg [20] had less convincing results, finding speech arrest sites in 50 % and 67 % of their study subjects, respectively. Wassermann & Blaxton [21] showed that rTMS of left-sided targets would preferentially interfere with object naming. Finally, Epstein & Woodard [24] showed that rTMS was inferior to the Wada test in determining language laterality. One other report of particular significance was that of Epstein and colleagues, which showed that low-frequency rTMS (4–5 Hz) can be equally effective as high frequency rTMS (15–30 Hz) in causing language disruption, because most subjects are unable to tolerate the discomfort of high-frequency rTMS [22].

1.2.1 Strengths and Weaknesses of These Prior Studies

Navigated vs. Non Navigated

Although rTMS technology has been around for more than five decades, the integration of rTMS with a neuronavigation system is a relatively recent development [13, 14]. By co-registering identifiable anatomical points on the patient's MRI with those same points on the patient's head, one can "teach" the system the exact location of the patient's head in 3-dimensional space. In this way, the MRI scan is precisely related to the actual brain, and one can target specific points on the cortical surface accurately and reproducibly [12]. Only in the last few years have reliable, commercially-available navigated rTMS systems become available outside the laboratory.

The advantages of integrated navigation are many. First and foremost, it allows for accurate targeting of a particular cortical region: cortical anatomy can vary from patient to patient and, without navigation, the cortical region being stimulated is at best an educated guess. Second, navigated rTMS delivers more consistent magnetic field strength: with new algorithms, the magnetic field strength incident on a cortical region can be calculated in real-time [16, 17]. Third, navigation allows for very high inter-session reliability: The co-registration process is repeated at the beginning of every session and, unless there is significant change in the patient's anatomy, every session is set up and targeted in the same way. nTMS has been verified in numerous studies as safe, reliable, and more accurate than non-navigated techniques [15, 25]. Successful speech mapping requires that a specific cortical region be targeted. Therefore, the ability to navigate accurately to that target is of paramount importance.

Image guidance is a critical component of any modern TMS-based language mapping. Precise targeting of the stimulus pulse is clearly important, but we have found that the orientation of the dipole is also critical. In fact we have noted on several occasions that small rotations of just 10–15° can alter the amplitude of a motor evoked potential (MEP) or the positivity of a language site. In general, to maximize patient comfort, it is ideal to maintain the dipole in perpendicularity with fibers of the temporalis muscle and the adjacent sulcus.

Coil Orientation

Two other improvements in the current technique are the use of the figure-of-8 coil and the ability to optimize the electrical field strength in real-time. A figure-of-8 coil offers improved ability to focus the magnetic field on a small cortical region. Relatedly, the navigated system allows the operator to visualize in real-time the theoretical field strength for a proposed stimulation pulse given the angle of the coil and the targeted cortical region. It is thus possible to maximize the effective cortical depolarization for a given stimulator output intensity, thus ensuring improved consistency between pulse trains (Fig. 1).

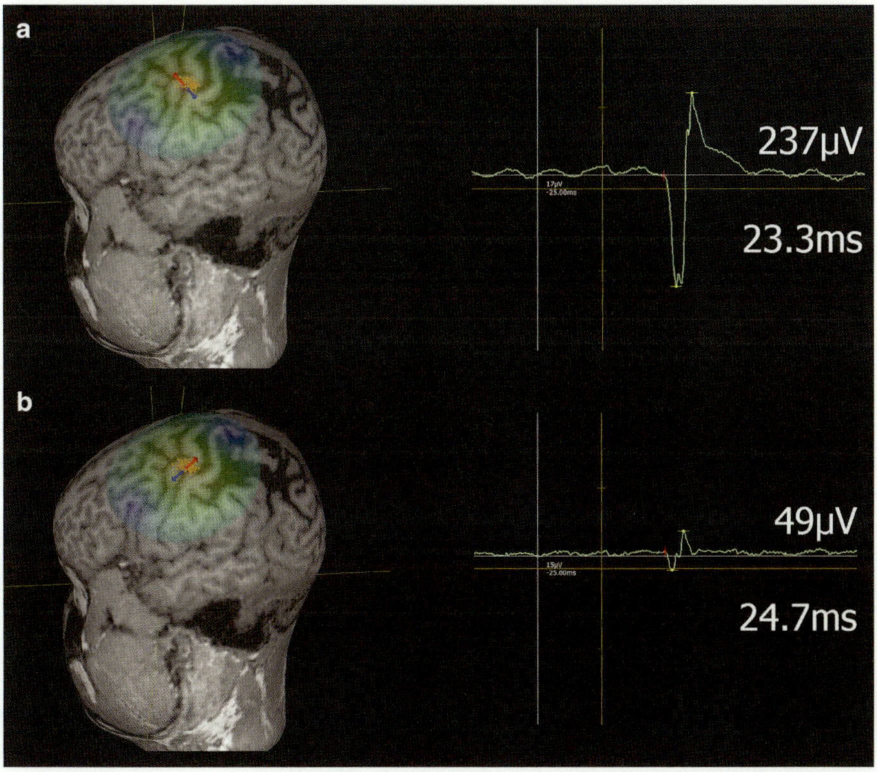

Fig. 1 An example of the effect of magnetic dipole orientation on motor-evoked potential (MEP). In (**a**), the MEP for extensor pollicis brevis is 237 μV, while in (**b**), with 90° rotation, the MEP for the same muscle is 49 μV. Both pulses were of the same intensity

Stimulator Settings

Another improvement in the modern TMS protocol is the change in stimulator parameters. The initial TMS language studies utilized high frequencies of 15–30 Hz with long trains of up to 10 s in their stimulation protocols. These parameters, in addition to being intensely uncomfortable to subjects, would not be approved by most modern ethical review committees because of the risk of provoking seizure. Indeed, Wassermann and colleagues chose to change the stimulation protocol in the midst of their study when it became known that low frequency stimulation was equally effective [21].

1.3 Current Protocol for nTMS Speech Mapping

The most widely used protocol for TMS speech mapping consists of a train of 10 pulses at 5 Hz lasting a total of 2 s. The inter-stimulus interval is maintained at 3–5 s, with the cue (if visual) being visible for 1 s less than the total inter-stimulus interval. The rTMS pulse train should be stimulated by an electronic trigger, with the pulse train starting at the onset of the task stimulus.

Many tasks may be used during speech mapping. The most common is object naming, perhaps because it is also the most commonly used task in the operating room during DCS mapping of speech function. Early studies of TMS speech mapping, however, suggest that the usage of a battery of "region-specific" tasks may yield better results than a single task (see below: "Region Specific Tasks"). Other tasks that may be useful in this regard are counting, verb generation (with auditory or visual cue presentation), word reading, auditory word repetition, word/non-word categorization, and free description.

1.4 Analysis of Data

The analysis of TMS speech mapping data is often more complicated than its acquisition. Language errors are often subtle, and involving a speech and language pathologist is often the best way to achieve the most accurate categorization of speech errors. Additionally, the use of audio-visual recording equipment to capture every speech mapping session is essential to getting the most accurate and detailed speech maps. In many cases, speech errors are so subtle that it is only after multiple viewings of a particular trial with direct comparison to baseline that an error can be picked up. Louimis and colleagues found, in a series of four patients, that results were more accurate and reproducible using this system [26].

Equally important to identifying the speech errors is tabulation of their location. Once the TMS sites with associated errors have been identified, those sites must be recorded and shared, both for the patient's own surgery and for future publication of results. The easiest method for transferring TMS speech maps to the operating room is by exporting the patient's map as a DICOM stack. Most nTMS systems allow for native exporting of the results to a DICOM stack, and this facility should be utilized where possible.

The usage of TMS speech maps for research purposes poses more of a challenge. To be of use in most research applications, each patient's speech map must be standardized, and the results tabulated into an aggregated database. In this fashion, the results of many patients, often with multiple modalities of imaging, may be compared against each other. There are many methods for achieving anatomical standardization, and a detailed discussion of these is beyond the scope of this chapter. One of the most common methods uses the toolbox Statistical Parametric Mapping 8 (SPM8) for MATLAB. SPM8 can standardize the brain to the Montreal Neurological Institute atlas template. In cases where tumor growth or peritumoral edema, distort the brain anatomy, it may be necessary to place reference points manually on identifiable anatomical structures such as the Sylvian fissure, inferior frontal sulcus, superior temporal sulcus, and central sulcus to assist the normalization procedure.

Once normalized, the speech maps may be superimposed on any standardized cortical atlas; a freely available and highly featured

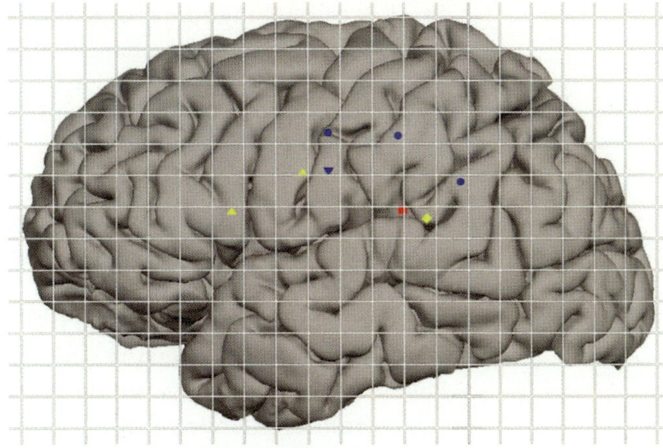

Fig. 2 A completed speech map superimposed on a template brain ("Bert," from Freesurfer). Different types of speech errors may be represented with unique markers to allow for visual representation of a complete speech mapping session

such atlas is Freesurfer (Fig. 2). The Freesurfer standardized brain, nicknamed "Bert," can be modified to display standard cortical regions, thus easily representing the anatomical location of points on the speech map. Another reference system, commonly used by neurosurgeons, is the standard 1 cm^2 grid superimposed on the standardized lateral view of the cortex. While this method of representation lacks any anatomical reference, it offers a categorization system of higher granularity. For example, while several points might fall on the MNI superior temporal gyrus and be lumped together, the grid system divides the same area into more than 12 individual boxes. Each of these representation systems has its advantages and disadvantages. Once standardized, it is a trivial step to represent each speech map using either one of these systems.

1.5 Safety

1.5.1 Repetitive TMS

In the report of the NIH Consensus Conference [27], rTMS was recognized to be associated with potentially more frequent (although still rare) adverse effects than single-pulse TMS. The frequency of adverse effects was noted to increase in parallel with the frequency (Hz) and intensity (% of motor threshold) of the rTMS trains used. The workshop issued safety guidelines where 10 Hz rTMS was considered safe for use in the non-medical setting. These guidelines have been reassessed and updated with no change to the safety recommendations [28]. Protocols using 10 Hz stimulation are commonly in use for treatment of depression in the non-medical environment, and rTMS devices are FDA-approved for rTMS-based mapping of speech and language. Furthermore, there is good evidence that no deleterious cognitive effects result from the standard rTMS speech mapping protocol [29]. There are also potential risks of microvascular changes, seizures, and hearing loss, as described below.

1.5.2 Microvascular Changes

Studies of prolonged exposure to TMS were performed in animal models. Microvascular changes in the cortex were found in rats exposed to over 100 pulses at 2.8 T, but these findings were not evident in rabbits exposed to 100–200 pulses per day at 2.4 T over 30–42 days, nor were they seen in rabbits exposed to 1,000 pulses at 2.0 T over a period of months [30]. Furthermore, rats exposed to 50–60 1.9 T pulses delivered over 15 min had no significant changes in blood–brain barrier permeability [31].

1.5.3 Seizure Activity

Seizure activity is a theoretical risk of TMS. Although single pulse TMS has not caused seizure activity in healthy volunteers, seizures have been produced by single-pulse TMS in several patients with large cerebral infarcts or other structural lesions [32–34]. Epilepsy patients without gross lesions may also be at a minor degree of risk from single pulse TMS. For example, in a series of studies by Hufnagel and colleagues, single-pulse TMS was occasionally able to induce seizures or activate epileptic electroencephalography (EEG) foci [35–37]. In this study slow frequency trains of stimulation were used. However, these results may not apply in patients in whom epilepsy is medically controlled. To wit, Tassinari et al. [38] found that TMS did not incite seizure activity in a cohort of patients receiving anti-epileptic therapy.

rTMS has been reported to cause seizures in at least seven individuals in whom frequencies of 3 Hz or higher have been utilized [27]. However, these patients were mostly patients with underlying seizure disorders. Only one patient reported had a seizure without any prior history of epilepsy. Thus, given the 1,000 of patients treated with rTMS protocols for stroke, depression, and other investigational purposes, only a handful of reported seizures have taken place as long as the safe stimulation parameters are observed.

1.5.4 Auditory Changes

During TMS there is a loud clicking sound from the coil (up to 120–130 dB). Single-pulse TMS induces permanent increases of the auditory threshold [39] in animals and transient increases in humans [40], but these changes can be prevented with the use of foam earplugs [41].

1.5.5 Psychiatric Changes

Treatment-emergent mania has been reported for low and high frequency rTMS in patients with uni- and bipolar depression after stimulation of the left prefrontal cortex [28, 42]. However, in these studies, there was no significant difference between rates of treatment-emergent mania in active vs. sham groups, suggesting that rTMS was not a factor in its emergence.

1.5.6 EMG

There may be a small amount of irritation of the skin from the electrode paste (less than 5 % chance). There is no known risk involved in recording EMG activity from these sensors and electrodes.

1.5.7 Headaches

Another risk that has been cited is a higher frequency of experiencing mild headaches, which resolve soon after the procedure. It is speculated that the headaches may be a result of maintaining a fixed position of the head and neck for the duration of the procedure. Administration of a standard over-the-counter dose of oral Tylenol may be useful for treating post-procedure headache.

1.6 Tolerability

Optimizing subject comfort is of great importance to make nTMS a viable modality for language mapping. The inherent issue is that the majority of language sites fall in the inferior frontal and temporal regions, underneath the temporalis muscle and facial nerve. In targeting these regions, the magnetic pulse must travel through the muscle fibers; in so doing it triggers a brief but powerful spasm of the temporalis muscle. It can also depolarize the facial nerve, resulting in similar contractions of the orbicularis oculus muscle. Thus, in a rTMS train of 10 pulses at 5 Hz, the subject can experience a jaw-rattling sensation, the description of which ranges from "uncomfortable" to "painful." One may mitigate this effect by maintaining the dipole of our induced field in perpendicularity to the temporalis muscle fibers. Nevertheless, half of subjects required some reduction in the stimulation intensity.

Not only is the jaw contraction unpleasant for the subject, but it can also confound the examiner's ability to distinguish true speech errors from subject withdrawal. Indeed, this side effect is the reason why we chose not to use dysarthria as a specific type of speech disturbance; it was often impossible to differentiate between true dysarthria from cortical disruption and "secondary" dysarthria from temporalis interference. Thus, when a questionable site was encountered, it is important to repeat stimulation; eventually, with a few repetitions, the vast majority of sites will resolve themselves as speech arrest or as temporalis interference.

1.7 Limitations

While nTMS mapping is a promising new modality, there are several limitations that bear mentioning. The first of these relate to the precision of the navigation itself. The tolerance of registration on most modern nTMS systems is estimated at 2–3 mm; given that the guidance system is not frame-based it is possible that the actual error is higher. Care must therefore be taken to ensure that preoperative maps match up with anatomical landmarks in the operating room: gyri and sulci, for example, and blood vessels can be used to ensure an accurate co-registration between the preoperative and the intraoperative findings.

Another inherent limitation in the specificity of nTMS is transsynaptic excitation of downstream (and, possibly, upstream) neuronal units. Neurophysiological and neuroimaging studies have shown that rTMS of a given brain region induces distributed activation of neural circuitry via transsynaptic spread, which follows established functional networks [43–45]. Thus, behavioral effects may be a result of activation not in the target region but in a distant, functionally connected region. The "over-calling" seen in many nTMS validation studies is likely a result of this limitation. Nevertheless, it should be pointed out that the overall nTMS language map does reflect the distribution of DCS language sites in both of the published series to date [46–48]. Additionally, it seems to correlate with prior studies of un-navigated rTMS mapping of language sites.

Another inherent limitation to TMS is the spread of the magnetic field itself. The figure-of-8 coil used in most modern systems generates a conical magnetic field. The field is therefore roughly circular at the cortical surface with a diameter of about 2 cm (and greatest intensity at the center with a sharp fall-off at the edges) and tapers toward its apex which occurs approximately 4.5 cm from the coil surface. As a result, nTMS might be disrupting subcortical white matter tracts, while the overlying cortex is inappropriately identified as a site of language disruption.

Furthermore, a given neuron's orientation, volume, axonal and dendritic organization, and innate threshold affect the likelihood of a magnetic pulse generating an action potential [49]. Thus, nTMS positive sites are not "points"—that is a misnomer; it is more accurate to say that they are regions, and to be aware that closely approximated regions of positivity on the map may all be associated with a single eloquent cortical site.

Finally, the basic parameters of nTMS stimulation should be examined more systematically and thoroughly. It is possible that relatively small adjustments in the frequency or number of pulses could improve results. Similarly, the timing of the onset of the pulse train is potentially important. In one published series, the pulse train was initiated just before the presentation of the stimulus in an object naming task, largely because that is the protocol used for DCS [48]. It is possible, however, that initiating the pulse train with, or just after the stimulus might improve the specificity of nTMS mapping. These variations in task and parameter must be methodically explored in future studies of nTMS protocols.

1.8 Future Directions

1.8.1 Improvement on Protocol

As mentioned in the previous section, although there are inherent limitations to nTMS speech mapping, many "limitations" are merely a result of new technology and early experiences. Foremost amongst those is the protocol of stimulation itself. The most widely used protocol is a 5 Hz rTMS train of 2 s duration. This protocol seems to strike a balance between effective temporary lesioning of the underlying cortex and patient discomfort. It does stand to reason, however, that other protocols may achieve more accuracy with different frequencies, train durations, or stimulator intensities. Furthermore, the rTMS protocol itself may be open for discussion: preliminary results of language mapping with theta-burst protocols have yielded promising results. The next few years of TMS speech mapping research will shed much light on these issues, and the reader is encouraged to review current literature because the pace of discovery is rapid.

1.8.2 Region-Specific Tasks

Recent studies have shown that the convergence between nTMS and DCS language maps is relatively high for the opercular region, but lower for the posterior region [47, 48]. This discrepancy likely represents an opportunity for refining the nTMS mapping protocol, perhaps by selecting tasks that are particularly suited for testing a particular region. Object naming, for example, while typically associated with temporal activity, does seem to involve inferior frontal sites as well in a subset of patients [50]. Thus, one might hypothesize that a different, purely temporal task may work better for posterior temporal sites—a semantic task, for example, such as identification of famous faces [51], or categorization of objects [52].

It is also possible that a task with auditory stimuli might be needed, such as the verb-generation task that is often used for fMRI and MEG studies. Further experimentation with more specialized tasks will likely improve the accuracy of nTMS mapping, and should be carefully explored and validated with patient outcomes. Refinement of the nTMS tasks serves a dual purpose: not only does it improve the accuracy of language maps, it also keeps the actual mapping process as short as possible. As many of the subjects will be undergoing mapping as part of a full day of pre-surgical tests, appointments, and evaluations, efficiency in the mapping process will be appreciated, and will yield the best possible results.

1.9 Conclusion

nTMS is a useful modality for generating language maps noninvasively. It is thus of immediate interest in the clinical management of subjects with eloquent brain tumors; it also has wide-ranging implications in basic science and translational studies of cortical language representation and physiology.

2 Methods

2.1 Patient Selection

2.1.1 Inclusion and Exclusion Criteria

As with any procedure, a successful outcome relies first on appropriate subject selection. Speech mapping should be considered in any case of a lesion involving eloquent or perieloquent speech regions including dominant temporal, inferior and middle frontal, supramarginal and angular gyri. For practical purposes, with speech mapping, the "lesion" is defined broadly: it may be neoplastic, as with primary glial neoplasms, electrophysiologic, as with epileptic foci determined by EEG or MEG, ischemic or hemorrhagic cerebrovascular accidents, or any other intracranial abnormality which is perceived to threaten the aforementioned areas. Additionally, nTMS-based speech mapping should be considered in any patient being considered for intraoperative awake language mapping with DCS.

Although the rTMS protocol is well-tolerated, there are contraindications to its application. These include the following:

- Implanted metallic parts of implanted electronic devices, including pacemakers, defibrillators, or implant medication pump.
- Pregnancy.
- History of uncontrolled epilepsy with >1 seizure per day.
- Implanted brain stimulator.
- Aneurysm clip or other metal in body.
- Scalp wounds or infection.

2.2 Patient Preparation

2.2.1 What to Expect

Prior to starting the mapping, all subjects must be prepared for the experience with a short verbal description of the expected sensations associated with single-pulse and repetitive nTMS. These sensations should be explicitly defined as muscle contraction, tingling, and pain. Subjects should be encouraged to give feedback about the level of discomfort associated with the nTMS mapping process so that the stimulation intensity could be decreased if necessary. They should also be encouraged at multiple points to ask for a break if needed, and to opt out of the mapping altogether if they feel any significant discomfort. Finally, subjects should be prepared for a possible speech arrest with an explanation of the purpose of mapping, a basic description of the underlying neurophysiological processes, and an assurance that any language disturbance is a result of the stimulation and would spontaneously resolve.

2.2.2 Time of Procedure

The procedure should take approximately 45–90 min, depending on the size of the mapped region and the experience of the operator. It is recommended to keep study duration under 1 h, as this seems to represent a limit in the ability of the patient to cooperate with sometimes demanding cognitive tasks associated with language mapping.

2.2.3 Imaging Requirements

The patient must undergo a high resolution, thin-slice MRI scan prior to nTMS speech mapping. The navigation will be based off this scan. Optimal slice thickness is 1.5 mm, with matrix size $256 \times 256 \times (108–140)$, and field of view $= 260 \times 260$ mm with skin-to-skin coverage including the nasion and preauricular points.

2.3 Mapping Procedure

2.3.1 MRI to Head Registration

Each TMS imaging session will begin with aligning the MRI images to the subject's head via the MRI to Head Registration process.

Using the bridge of the nose and the tragus of the ear, landmarks that are visible on both the subject's MRI and the head, the 3D-locations of the landmark are measured using a digitizing pen with an optical tracking system (Fig. 3). This registration allows for real-time monitoring of the coil position without restraining the subject's head during the TMS procedure [53]. The optical tracking system uses two cameras to triangulate the location in 3D space

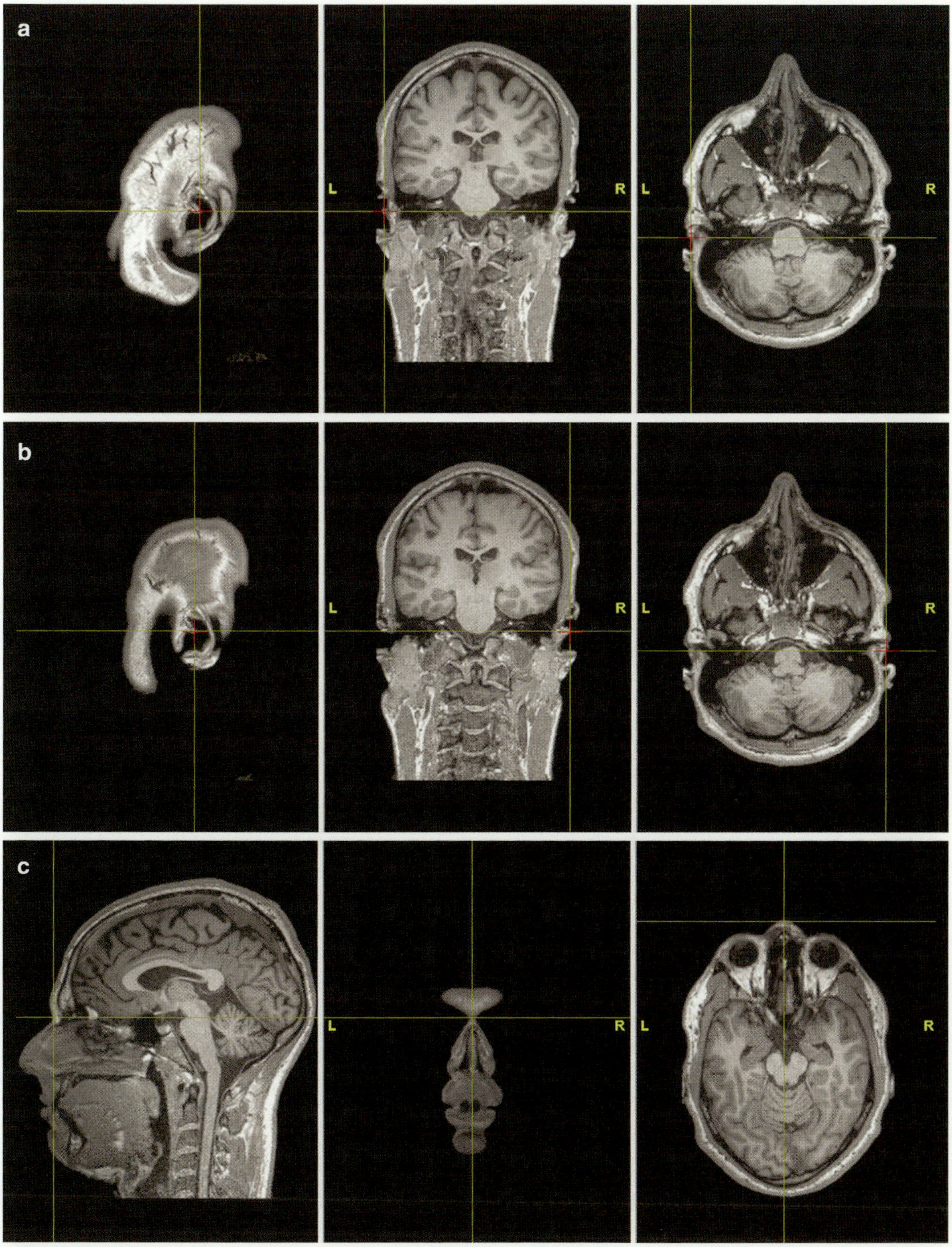

Fig. 3 The sagittal, coronal, and axial cuts of commonly used registration points (marked with a *red cross*). In (**a**) is the left crus of helix, in (**b**) the right crus of helix, and (**c**) the nasion

of infrared reflectors attached to the coil and subject's head. The mapping surface should be located between 23 and 28 mm deep to the scalp. This "peeling depth" will vary amongst subjects and must be set individually.

2.3.2 Determination of Resting Motor Threshold

Once the co-registration has been completed, a resting motor threshold (RMT) is calculated. RMT is defined as the minimum stimulation intensity capable of generating a MEP in 50 % of cases [54]. Because there is variation between subjects in the intensity of stimuli required to elicit a response, the RMT standardizes the stimulator intensity to each patient's physiological requirement. Surface electrodes (Neuroline 720; Ambu, Ballerup, Denmark) are attached to the patient's muscles for EMG recording. Typically, EMG electrodes are affixed to abductor digiti quinti (ADQ) and abductor pollicis brevis (APB). The most likely location of the hand knob is identified. This area is then stimulated in a grid pattern with 5 mm raster while systematically varying the rotation, tilt, and yaw of the magnetic stimulator. Once the location of maximal threshold is identified (which may correspond to either ADQ or APB), the stimulator output is adjusted until an MEP is evoked 50 % of the time. The speech mapping procedure is then conducted starting at 110 % of this RMT value.

2.3.3 Baseline Run

After RMT has been calculated, the subject will first perform each task *without* TMS mapping to establish a baseline. All subject responses must be captured on audio-visual recording equipment for offline analysis after the mapping procedure. This baseline run must include all the cues that will be used within a given task. If multiple tasks are being employed, the baseline and mapping should be concluded for one task before proceeding on to the next. For example, if a subject is undergoing mapping with object naming and word reading, he should do the object naming baseline, then object naming mapping, before proceeding on to word reading baseline and word reading mapping. This baseline serves several purposes. First, it familiarizes the subject with the task. For some tasks such as verb generation, it can take a few repetitions before the subject is familiarized with the task. Second, it establishes a baseline for comparison under mapping conditions. For example, in object naming, if a subject fails to recognize an object during baseline testing, it should be omitted during mapping. Often, speech errors are subtle and it is only through close comparison of the baseline with the test condition that an error can be detected.

2.3.4 Mapping Runs

Having completed the baseline, the mapping process begins. The subject is presented with a sequence of stimuli, and instructed to perform some task based on those stimuli. For example, in object naming, the subject is shown images at 3–5 s intervals, depending on the task and cognition. The subject should be able to complete the task comfortably within the given time but not have excess time

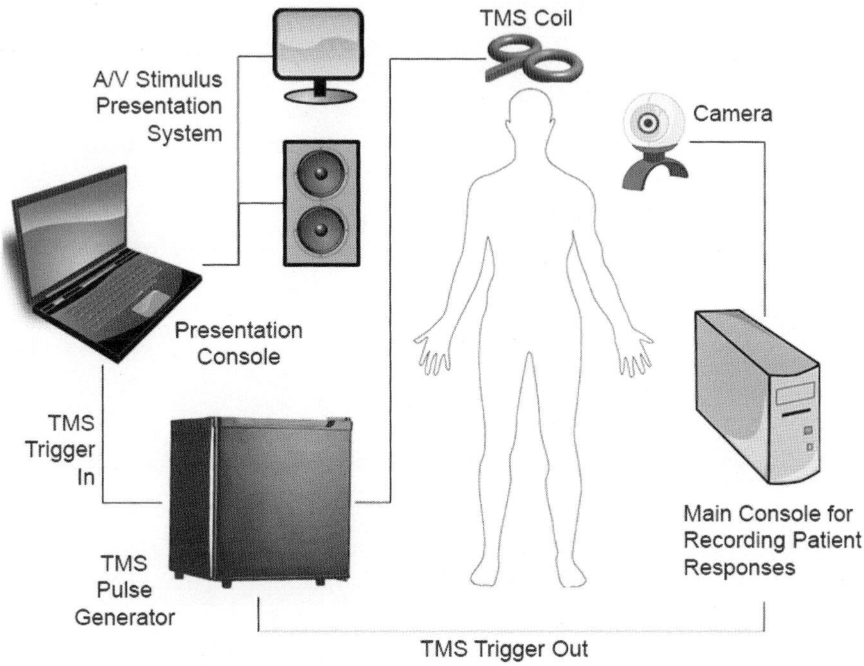

Fig. 4 A schematic of the key components in a TMS-based speech mapping setup. Note the presence of two unique computers: one controls the presentation software and triggers the TMS pulses timelocked to the stimuli; the other records patient responses and correlates those responses with the cortical location of the associated TMS pulse train

between trials, so that the mapping run proceeds expediently. The TMS coil should be automatically triggered by the presentation software, such that TMS trains are stimulation-locked. Stimulator intensity is set to 100 % of RMT, and onset delay of TMS stimulation trials may be set anywhere from 0 to 300 ms (Fig. 4).

As the sequence of cues is presented, the investigator moves the coil systematically around the region to be mapped, delivering a single train at each relevant point, and following a grid with 1 cm raster (Fig. 5). Each session should continue until every grid point has been covered once. If the subject is demonstrating discomfort during mapping, the investigator can alternate between more uncomfortable and less uncomfortable sites within the region of interest so that the subject is not confronted with multiple uncomfortable trials in a row. If the subject is still unable to continue due to discomfort, stimulator intensity may be reduced by 10 % at a time until the subject is able to continue. Note must be made of the intensity used during mapping in case the results become questionable due to very low intensity (i.e., <70 % RMT).

In most cases, three mapping runs are adequate to achieve a reliable language map. Each run must cover the entire region. Having completed mapping for a given task, the subject should be offered a break, a glass of water, and the patient's comfort level

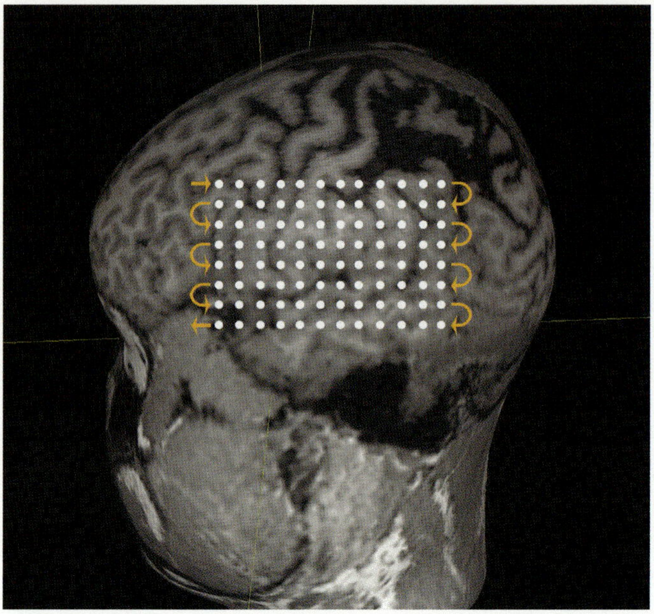

Fig. 5 A sample pattern covering the left fronto-temporal region during a typical speech mapping session. The grid utilizes a 1 cm raster

must be assessed. Usually a break of 3–5 min is appreciated. Following the break, the study should continue with the next task: baseline first, and mapping to follow.

2.4 Analysis

Once the mapping process is complete, the patient is excused and analysis takes place. The baseline and mapping trials, all of which are recorded, must be studied and analyzed for language disruptions. In many cases, disruptions are subtle and multiple viewings of the baseline and mapping conditions may be necessary to categorize the error appropriately. Offline analysis of speech errors has been shown to improve the accuracy of speech error classification [26]. It is optimal for the error analysis to be conducted by two speech pathologists working independently; any disagreements between these two observers may then be decided by a third, independent expert. It is also imperative that the analysis of language disruptions takes place with the analysts blinded to the cortical region being stimulated.

Once the trials have been analyzed and language disruptions have been categorized, the investigator must then return to the cortical language map and identify those points which had associated errors. In most cases, this can be done using the native software of the nTMS system. These points may then be saved for future reference, or downloaded and exported to a DICOM stack which may be fused with an intraoperative frameless stereotactic navigation system and used in the operating room (Fig. 6).

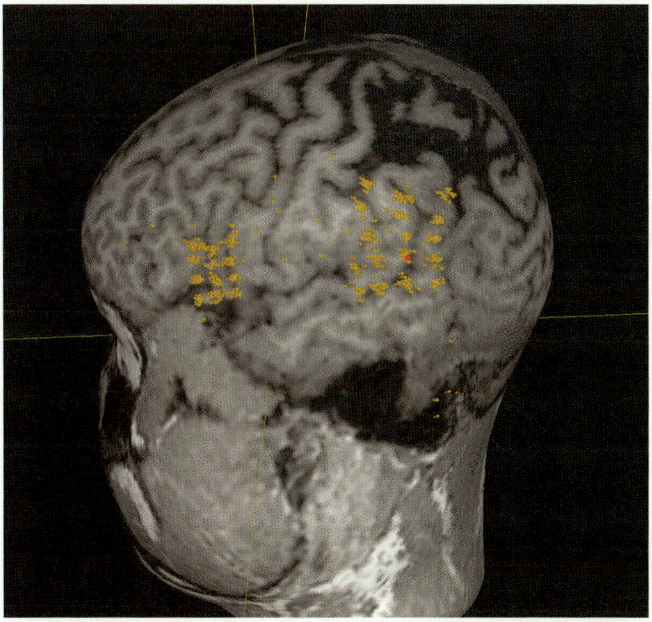

Fig. 6 A completed speech map prior to standardization. The *red circle* represents a site of anomia whilst the *blue circle* represents a site of complete speech arrest

References

1. Bates E, Wilson SM, Saygin AP et al (2003) Voxel-based lesion-symptom mapping. Nat Neurosci 6(5):448–450, PubMed PMID: 12704393. Epub 2003/04/22. eng
2. Adolphs R, Damasio H, Tranel D, Cooper G, Damasio AR (2000) A role for somatosensory cortices in the visual recognition of emotion as revealed by three-dimensional lesion mapping. J Neurosci 20(7):2683–2690, PubMed PMID: 10729349. Epub 2000/03/24. eng
3. Friedrich FJ, Egly R, Rafal RD, Beck D (1998) Spatial attention deficits in humans: a comparison of superior parietal and temporal-parietal junction lesions. Neuropsychology 12(2): 193–207, PubMed PMID: 9556766. Epub 1998/04/29. eng
4. Chao LL, Knight RT (1998) Contribution of human prefrontal cortex to delay performance. J Cogn Neurosci 10(2):167–177, PubMed PMID: 9555105. Epub 1998/06/20. eng
5. Dronkers NF (1996) A new brain region for coordinating speech articulation. Nature 384(6605):159–161, PubMed PMID: 8906789. Epub 1996/11/14. eng
6. Naeser MA, Hayward RW (1978) Lesion localization in aphasia with cranial computed tomography and the Boston Diagnostic Aphasia Exam. Neurology 28(6):545–551, PubMed PMID: 565884. Epub 1978/06/01. eng
7. Hickok G, Poeppel D (2004) Dorsal and ventral streams: a framework for understanding aspects of the functional anatomy of language. Cognition 92(1–2):67–99, PubMed PMID: 15037127. Epub 2004/03/24. eng
8. Lemaire JJ, Golby A, Wells WM III et al (2013) Extended Broca's area in the functional connectome of language in adults: combined cortical and subcortical single-subject analysis using fMRI and DTI tractography. Brain Topogr 26(3):428–441, PubMed PMID: 23001727. Epub 2012/09/25. Eng
9. Wagner T, Valero-Cabre A, Pascual-Leone A (2007) Noninvasive human brain stimulation. Annu Rev Biomed Eng 9:527–565, PubMed PMID: 17444810. Epub 2007/04/21. eng
10. Mills KR, Murray NM, Hess CW (1987) Magnetic and electrical transcranial brain stimulation: physiological mechanisms and clinical applications. Neurosurgery 20(1):164–168, PubMed PMID: 3543723. Epub 1987/01/01. eng
11. Kobayashi M, Pascual-Leone A (2003) Transcranial magnetic stimulation in neurology. Lancet Neurol 2(3):145–156, PubMed PMID: 12849236. Epub 2003/07/10. eng
12. Julkunen P, Saisanen L, Danner N et al (2009) Comparison of navigated and non-navigated transcranial magnetic stimulation for motor cortex mapping, motor threshold and motor evoked

potentials. Neuroimage 44(3):790–795, PubMed PMID: 18976714. Epub 2008/11/04. eng

13. Krings T, Chiappa KH, Foltys H, Reinges MH, Cosgrove GR, Thron A (2001) Introducing navigated transcranial magnetic stimulation as a refined brain mapping methodology. Neurosurg Rev 24(4):171–179, PubMed PMID: 11778822. Epub 2002/01/10. eng
14. Krings T, Foltys H, Reinges MH et al (2001) Navigated transcranial magnetic stimulation for presurgical planning–correlation with functional MRI. Minim Invasive Neurosurg 44(4):234–239, PubMed PMID: 11830785. Epub 2002/02/07. eng
15. Picht T, Mularski S, Kuehn B, Vajkoczy P, Kombos T, Suess O (2009) Navigated transcranial magnetic stimulation for preoperative functional diagnostics in brain tumor surgery. Neurosurgery 65(6 Suppl):93–98, PubMed PMID: 19935007; discussion 8–9. Epub 2009/12/16. eng
16. Sarvas J (1987) Basic mathematical and electromagnetic concepts of the biomagnetic inverse problem. Phys Med Biol 32(1):11–22, PubMed PMID: 3823129. Epub 1987/01/01. eng
17. Tarkiainen A, Liljestrom M, Seppa M, Salmelin R (2003) The 3D topography of MEG source localization accuracy: effects of conductor model and noise. Clin Neurophysiol 114(10):1977–1992, PubMed PMID: 14499760. Epub 2003/09/23. eng
18. Pascual-Leone A, Gates JR, Dhuna A (1991) Induction of speech arrest and counting errors with rapid-rate transcranial magnetic stimulation. Neurology 41(5):697–702, PubMed PMID: 2027485. Epub 1991/05/01. eng
19. Michelucci R, Valzania F, Passarelli D et al (1994) Rapid-rate transcranial magnetic stimulation and hemispheric language dominance: usefulness and safety in epilepsy. Neurology 44(9):1697–1700, PubMed PMID: 7936299. Epub 1994/09/01. eng
20. Jennum P, Friberg L, Fuglsang-Frederiksen A, Dam M (1994) Speech localization using repetitive transcranial magnetic stimulation. Neurology 44(2):269–273, PubMed PMID: 8309572. Epub 1994/02/01. eng
21. Wassermann EM, Blaxton TA, Hoffman EA et al (1999) Repetitive transcranial magnetic stimulation of the dominant hemisphere can disrupt visual naming in temporal lobe epilepsy patients. Neuropsychologia 37(5):537–544, PubMed PMID: 10340313. Epub 1999/05/26. eng
22. Epstein CM, Lah JJ, Meador K, Weissman JD, Gaitan LE, Dihenia B (1996) Optimum stimulus parameters for lateralized suppression of speech with magnetic brain stimulation. Neurology 47(6):1590–1593, PubMed PMID: 8960755. Epub 1996/12/01. eng
23. Epstein CM, Meador KJ, Loring DW et al (1999) Localization and characterization of speech arrest during transcranial magnetic stimulation. Clin Neurophysiol 110(6):1073–1079, PubMed PMID: 10402094. Epub 1999/07/13. eng
24. Epstein CM, Woodard JL, Stringer AY et al (2000) Repetitive transcranial magnetic stimulation does not replicate the Wada test. Neurology 55(7):1025–1027, PubMed PMID: 11061262. Epub 2000/11/04. eng
25. Tarapore PE, Tate MC, Findlay AM et al (2012) Preoperative multimodal motor mapping: a comparison of magnetoencephalography imaging, navigated transcranial magnetic stimulation, and direct cortical stimulation. J Neurosurg 117(2):354–362, PubMed PMID: 22702484. Epub 2012/06/19. Eng
26. Lioumis P, Zhdanov A, Makela N et al (2012) A novel approach for documenting naming errors induced by navigated transcranial magnetic stimulation. J Neurosci Methods 204(2):349–354, PubMed PMID: 22108143
27. Wassermann EM, Cohen LG, Flitman SS, Chen R, Hallett M (1996) Seizures in healthy people with repeated "safe" trains of transcranial magnetic stimuli. Lancet 347(9004):825–826, PubMed PMID: 8622349. Epub 1996/03/23. eng
28. Rossi S, Hallett M, Rossini PM, Pascual-Leone A (2009) Safety, ethical considerations, and application guidelines for the use of transcranial magnetic stimulation in clinical practice and research. Clin Neurophysiol 120(12):2008–2039, PubMed PMID: 19833552. Pubmed Central PMCID: 3260536. Epub 2009/10/17. eng
29. Guse B, Falkai P, Wobrock T (2010) Cognitive effects of high-frequency repetitive transcranial magnetic stimulation: a systematic review. J Neural Transm 117(1):105–122, PubMed PMID: 19859782. Pubmed Central PMCID: 3085788. Epub 2009/10/28. eng
30. Counter SA (1993) Neurobiological effects of extensive transcranial electromagnetic stimulation in an animal model. Electroencephalogr Clin Neurophysiol 89(5):341–348, PubMed PMID: 7691574. Epub 1993/10/01. eng
31. Ravnborg M, Knudsen GM, Blinkenberg M (1990) No effect of pulsed magnetic stimulation on the blood-brain barrier in rats. Neuroscience 38(1):277–280, PubMed PMID: 2255398. Epub 1990/01/01. eng
32. Fauth C, Meyer BU, Prosiegel M, Zihl J, Conrad B (1992) Seizure induction and magnetic brain stimulation after stroke. Lancet 339(8789):362, PubMed PMID: 1346431. Epub 1992/02/08. eng
33. Homberg V, Netz J (1989) Generalised seizures induced by transcranial magnetic stimulation of

motor cortex. Lancet 2(8673):1223, PubMed PMID: 2572937. Epub 1989/11/18. eng

34. Kandler R (1990) Safety of transcranial magnetic stimulation. Lancet 335(8687): 469–470, PubMed PMID: 1968184. Epub 1990/02/24. eng
35. Hufnagel A, Elger CE, Klingmuller D, Zierz S, Kramer R (1990) Activation of epileptic foci by transcranial magnetic stimulation: effects on secretion of prolactin and luteinizing hormone. J Neurol 237(4):242–246, PubMed PMID: 2391546. Epub 1990/07/01. eng
36. Hufnagel A, Elger CE, Marx W, Ising A (1990) Magnetic motor-evoked potentials in epilepsy: effects of the disease and of anticonvulsant medication. Ann Neurol 28(5):680–686, PubMed PMID: 2175574. Epub 1990/11/01. eng
37. Hufnagel A, Jaeger M, Elger CE (1990) Transcranial magnetic stimulation: specific and non-specific facilitation of magnetic motor evoked potentials. J Neurol 237(7):416–419, PubMed PMID: 2273410. Epub 1990/11/01. eng
38. Tassinari CA, Michelucci R, Forti A et al (1990) Transcranial magnetic stimulation in epileptic patients: usefulness and safety. Neurology 40(7):1132–1133, PubMed PMID: 2113205. Epub 1990/07/01. eng
39. Counter SA, Borg E, Lofqvist L, Brismar T (1990) Hearing loss from the acoustic artifact of the coil used in extracranial magnetic stimulation. Neurology 40(8):1159–1162, PubMed PMID: 2381522. Epub 1990/08/01. eng
40. Pascual-Leone A, Cohen LG, Shotland LI et al (1992) No evidence of hearing loss in humans due to transcranial magnetic stimulation. Neurology 42(3 pt 1):647–651, PubMed PMID: 1549231. Epub 1992/03/01. eng
41. Pascual-Leone A, Houser CM, Reese K et al (1993) Safety of rapid-rate transcranial magnetic stimulation in normal volunteers. Electroencephalogr Clin Neurophysiol 89(2):120–130, PubMed PMID: 7683602. Epub 1993/04/01. eng
42. Xia G, Gajwani P, Muzina DJ et al (2008) Treatment-emergent mania in unipolar and bipolar depression: focus on repetitive transcranial magnetic stimulation. Int J Neuropsychopharmacol 11(1):119–130, PubMed PMID: 17335643. Epub 2007/03/06. eng
43. Paus T, Jech R, Thompson CJ, Comeau R, Peters T, Evans AC (1997) Transcranial magnetic stimulation during positron emission tomography: a new method for studying connectivity of the human cerebral cortex. J Neurosci 17(9):3178–3184, PubMed PMID: 9096152. Epub 1997/05/01. eng
44. Valero-Cabre A, Pascual-Leone A (2005) Impact of TMS on the primary motor cortex and associated spinal systems. IEEE Eng Med Biol Mag 24(1):29–35, PubMed PMID: 15709533. Epub 2005/02/16. eng
45. Bestmann S (2008) The physiological basis of transcranial magnetic stimulation. Trends Cogn Sci 12(3):81–83, PubMed PMID: 18243042. Epub 2008/02/05. eng
46. Sanai N, Mirzadeh Z, Berger MS (2008) Functional outcome after language mapping for glioma resection. N Engl J Med 358(1):18–27, PubMed PMID: 18172171. Epub 2008/01/04. eng
47. Picht T, Krieg SM, Sollmann N et al (2013) A comparison of language mapping by preoperative navigated transcranial magnetic stimulation and direct cortical stimulation during awake surgery. Neurosurgery 72(5):808–819, PubMed PMID: 23385773. Epub 2013/02/07. Eng
48. Tarapore PE, Findlay AM, Honma SM et al (2013) Language mapping with navigated repetitive TMS: proof of technique and validation. Neuroimage 82:260–272, PubMed PMID: 23702420
49. Pashut T, Wolfus S, Friedman A et al (2011) Mechanisms of magnetic stimulation of central nervous system neurons. PLoS Comput Biol 7(3):e1002022, PubMed PMID: 21455288. Pubmed Central PMCID: 3063755. Epub 2011/04/02. eng
50. Lubrano V, Filleron T, Demonet JF, Roux FE (2014) Anatomical correlates for category-specific naming of objects and actions: a brain stimulation mapping study. Hum Brain Mapp 35(2):429–443, PubMed PMID: 23015527. Epub 2012/09/28. Eng
51. Gorno-Tempini ML, Price CJ (2001) Identification of famous faces and buildings: a functional neuroimaging study of semantically unique items. Brain 124(pt 10):2087–2097, PubMed PMID: 11571224. Epub 2001/09/26. eng
52. Peelen MV, Caramazza A (2012) Conceptual object representations in human anterior temporal cortex. J Neurosci 32(45):15728–15736, PubMed PMID: 23136412. Epub 2012/11/09. eng
53. Hannula H, Ylioja S, Pertovaara A et al (2005) Somatotopic blocking of sensation with navigated transcranial magnetic stimulation of the primary somatosensory cortex. Hum Brain Mapp 26(2):100–109, PubMed PMID: 15864816. Epub 2005/05/03. eng
54. Rossini PM, Barker AT, Berardelli A et al (1994) Non-invasive electrical and magnetic stimulation of the brain, spinal cord and roots: basic principles and procedures for routine clinical application. Report of an IFCN committee. Electroencephalogr Clin Neurophysiol 91(2):79–92, PubMed PMID: 7519144. Epub 1994/08/01. eng

Index

Alexander Rotenberg et al. (eds.), *Transcranial Magnetic Stimulation*, Neuromethods, vol. 89,
DOI 10.1007/978-1-4939-0879-0, © Springer Science+Business Media New York 2014

D

E

F

N

O

P

T

U

V

W

Printed by Printforce, the Netherlands